Spatial Statistics for Remote Sensing

Remote Sensing and Digital Image Processing

VOLUME 1

Series Editor:

Freek van der Meer, *International Institute for Aerospace Survey and
   Earth Sciences, ITC, Division of Geological Survey, Enschede, The Netherlands*

Editorial Advisory Board:

Michael Abrams, *NASA Jet Propulsion Laboratory, Pasadena, CA, U.S.A.*
Paul Curran, *University of Southampton, Department of Geography,
   Southampton, U.K.*
Arnold Dekker, *CSIRO, Land and Water Division, Canberra, Australia*
Steven de Jong, *Wageningen University and Research Center,
   Center for Geoinformation, Wageningen, The Netherlands*
Michael Schaepman, *University of Arizona, Optical Sciences Center,
   Tucson, AZ, U.S.A.*

SPATIAL STATISTICS FOR REMOTE SENSING

edited by

ALFRED STEIN

FREEK VAN DER MEER

and

BEN GORTE

*International Institute for Aerospace Survey and Earth Sciences,
ITC, Enschede, The Netherlands*

KLUWER ACADEMIC PUBLISHERS
DORDRECHT / BOSTON / LONDON

A C.I.P. Catalogue record for this book is available from the Library of Congress.

ISBN 0-7923-5978-X

Published by Kluwer Academic Publishers,
P.O. Box 17, 3300 AA Dordrecht, The Netherlands.

Sold and distributed in North, Central and South America
by Kluwer Academic Publishers,
101 Philip Drive, Norwell, MA 02061, U.S.A.

In all other countries, sold and distributed
by Kluwer Academic Publishers,
P.O. Box 322, 3300 AH Dordrecht, The Netherlands.

Printed on acid-free paper

All Rights Reserved
© 1999 Kluwer Academic Publishers
No part of the material protected by this copyright notice may be reproduced or
utilized in any form or by any means, electronic or mechanical,
including photocopying, recording or by any information storage and
retrieval system, without written permission from
the copyright owner

Printed in the Netherlands.

Preface

This book is a collection of papers on spatial statistics for remote sensing. The book emerges from a study day that was organized in 1996 at the International Institute for Aerospace Survey and Earth Sciences, ITC, in Enschede, The Netherlands. It was by several means a memorable event. The beautiful new building, according to a design by the famous modern Dutch architect Max van Huet was just opened, and this workshop was the first to take place there. Of course, much went wrong during the workshop, in particular as the newest electronic equipment regularly failed. But the workshop attrackted more than hundred attendants, and was generally well received. The results of the workshop have been published in Stein et al. (1998).

The aim of the workshop was to address issues of spatial statistics for remote sensing. The ITC has a long history on collecting and analyzing satellite and other remote sensing data, but its involvement into spatial statistics is of a more recent date. Uncertainties in remote sensing images and the large amounts of data in many spectral bands are now considered to be of such an impact that it requires a separate approach from a statistical point of view. To quote from the justification of the study day, we read:

> Modern communication means such as remote sensing require an advanced use of collected data. Satellites collect data with different resolution on different spectral bands. These data all have a spatial extension and are often related to each other. In addition, field data are collected to interpret and validate the satellite data, and both are stored and matched, using geographical information systems. Often, statistical inference is necessary, ranging from simple descriptive statistics to multivariate geostatistics. Classification, statistical sampling schemes and spatial interpolation are important issues to deal with. Maximum likelihood and fuzzy classification are now used intermixedly. Careful attention must be given as to where and how to sample efficiently. Interpolation from points to areas of land deserves thorough attention to take into account the spatial variability and to match different resolutions at different scales. Finally, the data have to be interpreted for making important environmental decisions.

This book reflects the set-up of the study-day. It addresses issues on remote sensing, interpolation, modeling spatial variation, sampling, classification and decision support systems, to name just a few. Several of the authors were also speakers during the workshop, but some topics were at that day not addressed, and hence the set of authors has been enlarged to guarantee a broad coverage of aspects of spatial statistics for remote sensing purposes.

The book would not have been possible without contributions from various people. First, we wish to thank the authors, who have all been working very hard to give the book the level as it has just now. Second, we like to thank the ITC management for making book and workshop possible. We thank Dr. Elisabeth Kosters, Mr. Bert Riekerk and Mrs. Ceciel Wolters for their contributions at various stages. Finally, we like to thank Mrs. Petra van Steenbergen at Wolters Kluwer Academic Publishers, for her patience and her continuing assistance of giving the book the appearance it has by now.

Alfred Stein
Freek van der Meer
Ben Gorte
Enschede, April 1999

Contributors and editors

Peter Atkinson
 Dept of Geography, Univ of Southampton, Highfield, Southampton
 SO17 1BJ, United Kingdom
 `p.m.atkinson@soton.ac.uk`

Sytze de Bruin
 Wageningen University, Dept of GIRS, PO Box 33, 6700 AH Wageningen, The Netherlands
 `sytze.debruin@staff.girs.wau.nl`

Paul Curran
 Dept of Geography, Univ of Southampton, Highfield, Southampton
 SO17 1BJ, United Kingdom
 `swc295@soton.ac.uk`

Jennifer Dungan
 Johnson Controls World Services Inc, NASA Ames Research Center, MS242-2, Moffet Field, CA 94035-1000. USA
 `dungan@gaia.arc.nasa.gov`

Ben Gorte
 ITC, Division of Geoinformatics, PO Box 6, 7500 AA Enschede, The Netherlands
 `ben@itc.nl`

Jaap de Gruijter
 SC-DLO, PO Box 125, 6700 AC Wageningen, The Netherlands
 `j.j.de.gruijter@sc.dlo.nl`

Cees van Kemenade
 CWI, PO Box 04079, 1090 GB Amsterdam, The Netherlands
 `cees.van.kemenade@cwi.nl`

Freek van der Meer
 ITC, Division of Geological Survey, PO Box 6, 7500 AA Enschede, The Netherlands
 `vdmeer@itc.nl`

Robert J Mokken
 University of Amsterdam, Sarphatistraat 143, 1018 GD Amsterdam, The Netherlands
 `mokken@ccsom.uva.nl`

Martien Molenaar
> ITC, Division of Geoinformatics, PO Box 6, 7500 AA Enschede, The Netherlands
> `molenaar@itc.nl`

Andreas Papritz
> ETH Zurich, Inst. Terrestrial Ecology, Soil Physics, Grabenstraße 11a, CH-8952 Schlieren, Switzerland `papritz@ito.umnw.ethz.ch`

Han la Poutré
> CWI, PO Box 04079, 1090 GB Amsterdam, The Netherlands
> `han.la.poutre@cwi.nl`

Ali Sharifi
> ITC, Division of Social Science, PO Box 6, 7500 AA Enschede, The Netherlands
> `sharifi@itc.nl`

Andrew Skidmore
> ITC, Division of Agriculture, Conservation and Environment, PO Box 6, 7500 AA Enschede, The Netherlands
> `skidmore@itc.nl`

Alfred Stein
> ITC, Division of Geoinformatics, PO Box 6, 7500 AA Enschede, The Netherlands
> `alfred.stein@bodlan.beng.wau.nl`

Contents

Preface	v
Contributors and editors	vii
Introduction	xv

I   1

1 Description of the data used in this book — Ben Gorte   **3**
    1.1   The Landsat Program 4
    1.2   Image radiometry 4
    1.3   Image geometry 6
    1.4   Study area 7

2 Some basic elements of statistics — Alfred Stein   **9**
    2.1   Population vs. sample 10
    2.2   Covariance and correlation 15
            2.2.1   Significance of correlation 17
            2.2.2   Example 18
    2.3   Likelihood 18
    2.4   Regression and prediction 21
    2.5   Estimation and Prediction 22

3 Physical principles of optical remote sensing — Freek van der Meer   **27**
    3.1   Electromagnetic Radiation 27
    3.2   Physics of Radiation and Interaction with Materials 28
    3.3   Surface Scattering 29
    3.4   Reflectance properties of Earth surface materials 30
            3.4.1   Minerals and Rocks 30
            3.4.2   Vegetation 34
            3.4.3   Soils 35
            3.4.4   Man-made and other materials 36
            3.4.5   Spectral re-sampling 36

	3.5	Atmospheric attenuation	37
	3.6	Calibration of LANDSAT TM to at-sensor spectral radiance	38

4 Remote sensing and geographical information systems — Sytze de Bruin and Martien Molenaar — 41

4.1	Data modeling		42
	4.1.1	Geographic data models of real world phenomena	42
	4.1.2	GIS data structures	42
4.2	Integrating GIS and remote sensing data		45
	4.2.1	Geometric transformation	46
	4.2.2	GIS data used in image processing and interpretation	47
	4.2.3	Geographic information extraction from RS imagery	49
4.3	Propagation of uncertainty through GIS operations		53
4.4	Software implementation		54

II — 55

5 Spatial Statistics — Peter M. Atkinson — 57

5.1	Spatial variation		57
	5.1.1	Remote sensing	57
	5.1.2	Spatial data and sampling	58
	5.1.3	Remotely sensed data	59
	5.1.4	Spatial variation and error	60
	5.1.5	Classical statistics and geostatistics	61
5.2	Geostatistical background		61
	5.2.1	Structure functions	62
	5.2.2	The Random Function model	64
	5.2.3	The sample variogram	66
	5.2.4	Variogram models	68
5.3	Examples		71
	5.3.1	Remotely sensed imagery	71
	5.3.2	Sample variograms	74
	5.3.3	Indicator variograms	77
	5.3.4	Cross-variograms	79
5.4	Acknowledgments		81

6 Spatial prediction by linear kriging — Andreas Papritz and Alfred Stein — 83

6.1	Linear Model		87
6.2	Single Variable Linear Prediction		90
	6.2.1	Precision criteria	90
	6.2.2	Simple kriging	90
	6.2.3	Universal kriging	95
	6.2.4	Ordinary Kriging	99

		6.2.5	The practice of kriging	102
	6.3	Multi-Variable Linear Prediction		106
		6.3.1	Co-kriging weakly stationary processes	106
		6.3.2	Co-kriging intrinsic processes	107
		6.3.3	Benefits of co-kriging	108
	6.4	Mapping the Variable *BAND1* of the Twente Images		109

7 Issues of scale and optimal pixel size — Paul J. Curran and Peter M. Atkinson   115

	7.1	Introduction		115
		7.1.1	Scale and scaling-up	115
		7.1.2	Why is pixel size important?	116
	7.2	Geostatistical theory and method		118
		7.2.1	Regularization	118
		7.2.2	Selecting an optimum pixel size	121
	7.3	Remotely sensed imagery		123
	7.4	Examples		124
	7.5	Summary		132
	7.6	Acknowledgment		132

8 Conditional Simulation: An alternative to estimation for achieving mapping objectives — Jennifer L. Dungan   135

	8.1	Contrasting simulation with estimation		137
	8.2	Geostatistical basis of conditional simulation		140
	8.3	Algorithms for conditional simulation		142
		8.3.1	Adding variability back in	142
		8.3.2	Using Bayes Theorem	144
		8.3.3	Optimization	146
	8.4	Adding an ancillary variable		147
	8.5	Describing uncertainty		149
	8.6	The future of simulation in remote sensing		151
	8.7	Acknowledgements		152

9 Supervised image classification — Ben Gorte   153

	9.1	Classification		153
		9.1.1	Pattern recognition	154
		9.1.2	Statistical pattern recognition	155
	9.2	Probability estimation refinements		157
		9.2.1	Local probabilities	157
		9.2.2	Local *a priori* probability estimation	158
		9.2.3	Local probability density estimation	158
		9.2.4	Non-parametric estimation	159
	9.3	Classification uncertainty		162

10 Unsupervised class detection by adaptive sampling and density es-

timation — Cees H.M. van Kemenade, Han La Poutré and Robert J. Mokken **165**
    10.1 Unsupervised non-parametric classification 165
    10.2 Outline of the method 167
    10.3 Adaptive sampling of pixels 167
        10.3.1 Theoretical analysis of a one-class problem 168
        10.3.2 Discussion of a multi-class problem 170
    10.4 Density estimation 171
    10.5 Curse of dimensionality for density estimation 174
    10.6 Density based sampling for Remote Sensing 176
        10.6.1 Density estimation in Remote Sensing 176
        10.6.2 Using local density estimates during sampling 177
    10.7 Hierarchical clustering 179
        10.7.1 Water-level model 179
        10.7.2 Algorithm 179
    10.8 Remote sensing application 181
    10.9 Conclusions 183

11 Image classification through spectral unmixing — Freek van der Meer **185**
    11.1 Mixture modeling 186
    11.2 Spectral unmixing 187
    11.3 End-member selection 189
    11.4 Application to the Enschede image 190

III   195

12 Accuracy assessment of spatial information — Andrew K. Skidmore **197**
    12.1 Error inherent in raster and vector spatial data 197
        12.1.1 Raster Image errors 197
        12.1.2 Error In Vector Spatial Data 198
    12.2 Methods of quantifying errors 199
        12.2.1 The error matrix 199
        12.2.2 Sampling design 200
        12.2.3 Number of samples for accuracy statements 201
        12.2.4 Sample unit area 201
        12.2.5 Measures of mapping accuracy generated from an error matrix 203
        12.2.6 Change detection - testing for a significant difference between maps 205
        12.2.7 Methods of quantifying errors in a vector data layer 206
    12.3 Error accumulation in GIS overlay 207
        12.3.1 Reliability diagrams for each layer 207
        12.3.2 Modeling error accumulation 208

		12.3.3 Bayesian overlaying	209
12.4	Conclusions		209

13 Spatial sampling schemes for remote sensing — Jaap de Gruijter 211
13.1 Designing a sampling scheme	211
13.1.1 Towards better planning	211
13.1.2 A guiding principle in designing sampling schemes	213
13.1.3 Practical issues	214
13.1.4 Scientific issues	215
13.1.5 Statistical issues	215
13.2 Design-based and model-based approach	217
13.3 Design-based strategies	219
13.3.1 Scope of design-based strategies	220
13.3.2 Simple Random Sampling (SRS)	221
13.3.3 Stratified Sampling (StS)	223
13.3.4 Two-stage Sampling (TsS)	226
13.3.5 Cluster Sampling (ClS)	230
13.3.6 Systematic Sampling (SyS)	233
13.3.7 Advanced design-based strategies	235
13.3.8 Model-based prediction of design-based sampling variances	238
13.4 Model-based strategies	240

14 Remote sensing and decision support systems — Ali Sharifi   243
14.1 Decision Making Process (DMP)	244
14.2 Spatial Decision Making	246
14.2.1 A systematic approach for solving spatial problems	246
14.2.2 Methods and techniques to support spatial decisions	251
14.2.3 Uncertainty in decision making process	253
14.3 Decision Support Systems	255
14.3.1 GIS and Decision Support Systems	257

Bibliography   284

Introduction

Spatial statistics is a relatively new branch of science that is rapidly emerging. It addresses quantitative problems that have a spatial context. Spatial statistics is primarily a science it itself, with close links to other fields of statistics, in particular sampling, regression theory and time series analysis, and with a large set of applications in various applied fields of science. As a science, spatial statistics dates back to presumably the start of the 20th century, when the first papers emerged. But it started to really flourish with works at the Paris School of Mines in the nineteen sixties and seventies, in particular of G. Matheron and J. Serra. It has found many applications and extensions since. Applications started in mining, but nowadays can be detected in agriculture, geology, soils, water, the environment, the atmosphere, economy and geography, to name just a few.

This book is a collection of papers addressing issues of spatial statistics emerging in remote sensing. Remote sensing is a modern way of data collecting, that has seen an enormous growth since launching of modern satellites. Remote Sensing is a term meaning literally the acquisition of physical data on an object without contact. It is now primarily used to data collection by satellites or airplanes, although new studies involve analysis of images by cameras at a much more detailed level. The procedures in this book can probably be applied as well to such studies, but our primary focus has been on satellite and airborne data. These data are normally available as pixels, i.e. as binary values that somehow correspond to an area in the real world. Values are obtained by observing the reflectance of solar radiation on the earth cover. They are hence influenced by vegetation type, presence of buildings and water and the geological composition of the earth surface. Moreover, pixels arrive in different spectral bands, that are sensitive to different parts of the electro-magnetic spectrum. A property of satellite pixels is their spatial resolution, being the correspondence between an area in the field and the smallest unit at a satellite image.

In this book we will address issues of spatial statistics that are of relevance for remote sensing. For that purpose the book is divided into three parts.

Part I describes prerequisites for doing a statistical analysis of spatial data. It starts in chapter 1 with a description of the example data set that has been used throughout the book. In chapter 2 some basic elements of statistics are addressed that come back at various chapters in the remainder of the book. In chapter 3 physical aspects of remote sensing are discussed that deal with collecting data and the inherent uncertainty that exists in these data. In chapter 4 the relation between remote sensing and geographical information systems is discussed. Much geographical information is available on the same area where data are collected, but combining it with satellite data is no trivial matter. Geographical information systems are nowadays widely spread and are particularly useful for the purpose of combining layers of information. The first part therefore serves as a background for spatial statistics for remote sensing.

Part II is the core of the book. It discusses the issues that nowadays are of primary importance for a spatial statistical analysis of remote sensing data. Chapter 5 describes the role of spatial variation. Spatial variation is obviously present in remote sensing data, as pixels close to each other are more likely to be similar than pixels at a larger distance. Common geostatistical tools, such as the variogram are discussed here. Chapter 6 addresses spatial interpolation. Modern interpolation procedures are in particular relevant to predict values that are obscured by clouds, or to predict values of vegetation, soil or geology variables that are highly correlated with satellite images. In chapter 7 issues of scale are addressed. Remote sensing data are available at a single resolution, that always has a positive support size, whereas ideally it should have a point support. Therefore, much variation is still present *within* data. Also, interest of practical applications may focus on questions with a larger or a smaller support. This gives clues to the optimal pixel size. Changes in scale are important, and some new ideas are considered in this chapter. Interpolation usually causes smoothing of the data. Therefore, chapter 8 focuses on spatial simulation, a modern geostatistical procedure to generate a spatial field with the same variation and spatial variation as the original data. Next, three chapters follow on classification. Satellite data usually come in various bands (from one to a few hundred), thus classification with a clear physical meaning is important. Chapter 9 is on supervised classification, presupposing a set of classes to which any observed set of pixels is assigned. Chapter 10 is on unsupervised classification, i.e. classification that does not rely on a set of classes, but makes a free classification of a satellite image. Chapter 11 is on spectral unmixing, that could play a role in both supervised and unsupervised classification.

Part III addresses some newer elements of remote sensing, that are currently researched. Chapter 12 addresses accuracy of spatial information. Inaccuracy is inherent in all types of spatial information, but important is its degree to which it matters in actual problems. Chapter 13 considers spatial sampling of satellite images. A careful judgment of where to collect data is critical in many remote sensing studies, and remote sensing can be used to direct spatial sampling to enhance the quality of collected data. Chapter 14 focuses on decision support by addressing the question of where to use remote sensing data for in practical applications. Remote sensing is placed in the general framework of decision making, and its tools are combined.

To make the book coherent, we have applied a single database throughout. All authors have been asked to include these data into the different chapters. In many instances this was done, although some authors added also examples by themselves, whereas other authors did not consider it harmful to leave the chapter without a case study application.

Alfred Stein
Freek van der Meer
Ben Gorte

PART I

Chapter 1

Description of the data used in this book

Ben Gorte

Earth observation satellite sensors measure electro-magnetic radiation emitted or reflected by the earth surface. Measured radiation depends on local earth surface characteristics. The relationship between measurement values and land cover allows to extract terrain information from image data.

Active sensors send radiation pulses in the microwave range of the electro-magnetic spectrum to the earth and measure the returned amounts in successive time-intervals, according to radar principles. *Passive* sensors measure thermal infrared radiation which is emitted by the earth surface, or visible and near-infrared radiation which originates from the sun and is reflected by the earth surface (Fig. 1.1).

Several chapters in this book contain experiments and case studies using a *Landsat Thematic Mapper* image of the Twente region in the Netherlands [161]. The Thematic Mapper sensor is applied in the Landsat Earth observation satellite program. It is a passive sensor that measures reflected sunlight in the visible, near-infrared and mid-infrared spectral ranges, as well as thermal infrared emitted by the earth (see chapter 2 for more details).

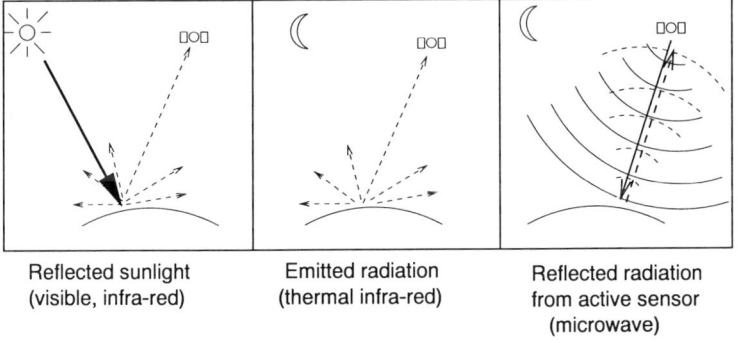

Figure 1.1: Sensor types

1.1 The Landsat Program

The idea of a civilian Earth resource satellite was conceived in mid 1960's and resulted in the Landsat program. The first Landsat satellite (at that time referred to as ERTS-A) was launched by NASA in 1972, followed by Landsat 2 (or ERTS-B), 3, 4 and 5 in 1975, 1978, 1982 and 1984 respectively, of which Landsat 5 is still operating. In the meantime, Landsat 6 was launched in 1993, but did not become operational, whereas Landsat 7 was launched in 1999.

NASA was responsible for operating the Landsats through the early 1980's. In January 1983, operations of the Landsat system were transferred to the National Oceanic and Atmospheric Administration (NOAA). In October 1985, the Landsat system was commercialized. After that date, all Landsat commercial rights became the property of Earth Observation Satellite Company (EOSAT) with exclusive sales rights to all thematic mapper (TM) data. Since 1992, the program is jointly managed by NASA and the US Department of Defence (DOD), to ensure that the design of Landsat 7 will meet the needs of both civilian and defense usage [244, 288].

1.2 Image radiometry

Multi-spectral sensors measure reflection in different parts of the electro-magnetic spectrum separately, but at the same time. The number of bands and their locations in the spectrum, expressed as wavelengths or wavelength intervals, specify the *spectral resolution* of a sensor. The Multi-Spectral Scanner (MSS), the primary sensor on board of the Landsat 1, 2 and 3 satellites operates in four bands in the visible and near-infrared range of the spectrum. The Thematic Mapper (TM) sensor in Landsat 4 and 5 [1] measures in seven spectral bands (Table 1.1). The characteristics of the MSS and TM bands were selected to maximize their capabilities for detecting and monitoring different types of Earth resources.

Typically, TM Bands 4, 3, and 2 can be combined to make false-color composite images where band 4 represents red, band 3 green, and band 2 blue (Fig. 1.2). This band combination makes vegetation appear as shades of red, brighter reds indicating more vigorously growing vegetation. Soils with no or sparse vegetation will range from white (sands) to greens or browns depending on moisture and organic matter content. Water bodies will appear blue. Deep, clear water will be dark blue to black, while sediment-laden or shallow waters will appear lighter [288]. Urban areas give rather high reflections in all visible bands and will be blue-gray in color composites.

Measurement principles vary according to sensor type, but after pre-preprocessing each measurement corresponds to a location in the terrain and is presented as a pixel in an image. A sensible geometrical correspondence exists between terrain locations and image pixels, such that the image can be regarded as a projection of the earth surface on an image plane. The distance between locations that correspond to adjacent image pixels is a sensor characteristic, which determines the *spatial resolution* of an image sensor.

Landsat Thematic Mapper measures reflection in scan-lines that run perpendicular to the ground track of the satellite. An oscillating mirror design acquires data

---

1. Landsat 4 and 5 also had MSS scanners. The one in Landsat 5 was switched off in 1992.

Figure 1.2: Landsat Thematic Mapper Image of Twente study area (band 4).

of 16 scan-lines within each sweep in both directions, since there are 16 detectors for each or the six reflected spectral bands (four for the thermal band). During one mirror sweep the satellite advances over a distance that corresponds to 16 scan-lines (474 m), such that eventually the entire terrain is covered. This geometry results in an along-track ground resolution of 29.625 m (118.5 m for the thermal band). The length of each scan-line, *i.e.* the across-track width of a TM image is 185 km, divided into approximately 6000 pixels, which yields an across-track ground resolution of approximately 30.8 m. At receiving stations, TM imagery is usually resampled to 28.5 × 28.5 m ground resolution.

It is quite common to pretend that the terrain is subdivided in rectangles or squares, the *terrain elements*, such that a pixel's measurement value is representative for the entire terrain element. However:

- The reflection may be not uniform within the terrain element, for instance at the boundary between two land cover types, or when details are smaller than the resolution.
- The satellite measures a reflection in an area called *instantaneous field of view (IFOV)*, which is rather circular than square. The measured value is a weighted average of the different reflections within that circular area,

where the weight is larger in the center than towards the outside of the circle. Sensors are designed with such a spatial resolution that the circular IFOV's slightly overlap each other. This ensures that the terrain is entirely covered by measurements, without creating too much data redundancy.

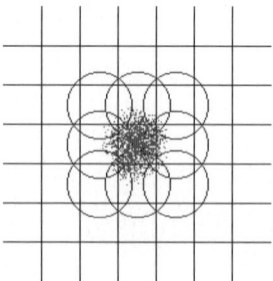

Figure 1.3: A grid of terrain elements and instantaneous field of view, shown as circles

1.3 Image geometry

Due to earth curvature, earth rotation, orbit parameters, satellite movements, terrain relief etc., the projection of the earth surface into a satellite image is not a cartographic one. To make an image coincide pixel by pixel with a raster map in a Geographic Information System (GIS), such that pixels with the same row and column position correspond to the same location in the terrain, extensive geometrical corrections have to be performed.

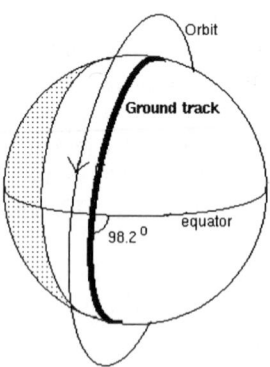

Figure 1.4: Sun synchronous, geostationary orbit

The Landsat satellites operate in near-polar, sun-synchronous orbits (Fig. 1.4). The altitude of the orbit of the Landsat 4 and 5 satellites (with Thematic Mapper) is

705 km. Each orbit takes approximately 99 minutes, with just over 14.5 orbits being completed in one day. The orbit being sun-synchronous implies that the satellite, when traveling from North to South, passes the equator always at the same local time, 9:45 am for Landsat 5. Between two successive orbits, the earth rotates over approximately $\frac{1}{14.5}360° = 24.8°$, which corresponds to 2752 km at the equator. With an across-track image width of only 185 km it is clear that each day only a small portion of the earth surface can be scanned. The orbit is chosen in such a way that every 16 days the whole earth surface is covered.

To keep an orbit sun-synchronous it has to be *near polar* instead of exactly polar. As a result Landsat satellites do not cross the equator perpendicularly, but with an *inclination angle* of 98.2°. When images are displayed scan-line by scan-line on a computer screen, the resulting picture is not exactly north-oriented. Image georeferencing, therefore, includes rotation to compensate for the near-polarness of the orbit. Two versions of the same image were available to the authors of this book, with and without georefencing The georeferenced one was made using an affine transformation and bilinear interpolation resampling to 30m × 30m resolution.

1.4 Study area

The area in the eastern part of the province of Overijssel in the Netherlands, the so-called Twente region, containing the towns Enschede and Hengelo, is predominantly rural. The predominant vegetation types are grassland, agricultural crops (mainly maize), woods and heather. The image area of approximately 26 by 22 km^2 also contains residential and industrial areas. The soil in the area is generally sandy. The area is flat, but with elevations between 20 and 80 m it is not as flat as most other parts of the Netherlands, The hydrography consists of some small rivers that are hardly visible in the image. The *Twentekanaal*, a channel entering the image from the west and ending in Enschede (slightly south to the center) appears clearly, as well as some artificial lakes, which remained after sand had been removed, to be used for highway construction. The main highway in the image curves through the image area from west to east, slightly north of the image center. Most of the eastern part, however, was still under construction when the image was recorded. The highway section going south to Enschede was already completed and is visible.

The Landsat TM image was recorded on 29th May 1992 by Landsat 5 (Fig. 1.2). It was supplied by EFTAS Nederland V.O.F., Oldenzaal, the Netherlands.

Table 1.1: Thematic Mapper Spectral Bands (from Lillesand and Kiefer [244])

Band	Wavelength (μm)	Spectral Name	Principal Applications
1	0.45 – 0.52	Blue	Designed for water body penetration, making it useful for coastal water mapping. Also useful for soil/vegetation discrimination and forest type mapping.
2	0.52 – 0.60	Green	Designed to measure green reflectance peak of vegetation for vegetation discrimination and vigor assessment.
3	0.63 – 0.69	Red	Designed to sense in a chlorophyll absorption region aiding in plant species differentiation.
4	0.76 – 0.90	Near infrared	Useful for determining vegetation types, vigor, and biomass content, for delineating water bodies, and for soil moisture discrimination.
5	1.55 – 1.75	Mid-infrared	Indicative of vegetation moisture content and soil moisture. Also useful to differentiate snow from clouds.
6	10.4 – 12.5	Thermal Infrared	Useful in vegetation stress analysis, soil moisture discrimination, and thermal mapping applications.
7	2.08 – 2.35	Mid-infrared	Useful for discrimination of mineral and rock types. Also sensitive to vegetation moisture content.

Chapter 2

Some basic elements of statistics

Alfred Stein

In statistics we are primarily concerned with a quantitative analysis of data. This introductory chapter gives a short summary of some basic statistical tools. In particular those procedures have been included that apply to spatially varying pixels and hence are of use for remote sensing data.

Variability (Oxford: the fact of quality of being variable in some respect) in space is defined as the phenomenon that a property changes in space. If at a location a variable is observed, this variable may be observed to deviate from the previous observation at a small distance from this location. The deviations may increase when distances increase. Description of variability (how large are the deviations if the distance increases) may be important for a process based interpretation. Relatively large variation within small distances indicate that the variable is subject to local influences. On the other hand, gradual changes indicate an influence at a more global scale, e.g. extending over the size of many pixels.

The study of variability is in principle independent of any discipline, and even from earth sciences. In remote sensing and earth sciences, however, a quantitative approach using GIS and geostatistics has now for a large part been developed.

Statistical (and geostatistical) procedures may be helpful in a number of stages of interpreting and evaluating data showing spatial variation. In many studies they have proven to be indispensable, especially when the number of available data increases as is the case for remote sensing images. Without having the intention to being complete, one may think of the following aspects:

1. The type and the amount of the variation. Some examples are: which band shows the most variation within a region? Is every band varying at the same scale? What is the relation between spatial variation of a satellite image and aspects of the earth surface it may represent, like vegetation, erosion, soil development and human influences in the past [350]?
2. To predict a pixel at an unvisited location. In particular it has been useful to predict the values of pixels below clouds, and to predict values related to pixel values [5].

3. To determine probabilities that a spatial variable related to a pixel value exceeds a critical value (see Chapter 6).
4. To develop a sampling scheme on the basis of pixel values. Typical questions include: how many observations does one need? Which configuration of the data is most appropriate given the objective of a particular inventarisation study (see Chapter 13)?
5. To use a satellite image in model calculations, for example to calculate water restricted yield of a particular crop.

Pixels are observed from satellite images and usually show variation. This variation is related to variation in reflectance at the earth surface, as it is influenced by for example variation in specific soil properties and in vegetation (Chapter 3). It is common to describe variation by variables, each variable related to some property. A variable therefore can take different values, on the basis usually of the properties at the earth surface. We distinguish two types of variables.

Continuous variables, i.e. variables that take values at a continuous scale. An example is given by pixel values, although these are currently only available as bytes and intermediate values are never recorded. But a number like 231.1, for example, can still be interpreted as a realistic pixel value between 231 and 232, being closest to 231. Other continuous variables include the content of a heavy metal or of nitrate, temperature and rainfall. One could distinguish ratio and interval variables, where ratio variables have a proper origin (rainfall) and interval variables that do not have this origin (temperature, pixel values), but this distinction will not be maintained throughout this chapter.

Discrete variables, i.e. variables that take only a limited number of discrete values. First there are ordinal variables, that take a value, but where the differences between succeeding values is not quantified. For example, land suitability classes could run from 1 (unsuitable) to 5 (highly suitable) where a difference between classes 1 and 2 is undetermined relative to that between classes 2 and 3. Second there are nominal variables, such as land use and historical records. Instances of these variables can be labeled and can therefore be given a name.

The aim of this chapter is to give an introduction to some basic elements of statistics. Many of these will come back in the succeeding chapters. It therefore serves as a basis for spatial statistics. The chapter is illustrated with a subset of the data set as described in chapter 1.

2.1 Population vs. sample

Here we will describe the difference between a population and a sample. A population is what we want to measure. In applications on remote sensing this can be an image, a set of images, an individual pixel, etc. The population is given by a capital, for which we usually take the Y. As Y is unknown, it is influenced by random influences and it follows a probability distribution: some values are more likely to occur than other values. The probability distribution is a measure for this. Having

Chapter 2: Some basic elements of statistics 11

access to the probability distribution of Y allows us to say that Y is contained in a probability interval, exceeds some threshold value with a certain probability or is below a threshold with another probability. But in general we will have to estimate the distribution from the data. For the moment, though, we assume that the distribution is given and known, and we consider some properties of it.

The cumulative distribution function (cdf) of a variable specifies the probability that Y takes a value smaller than some specified value y. It is denoted by $F_Y(y)$. The cdf has to obey some properties. For one, it must be continuous from the right. Secondly, it takes values between 0 and 1, with 0 being the limit for $y \to -\infty$ and 1 for $y \to +\infty$. Also, the cdf must be non-decreasing: if $y_2 > y_1$, then the probability that Y is smaller than y_2 is larger than or equal to the probability that Y is smaller than y_1. For many distribution functions the function $F_Y(y)$ has a derivative, the probability density function (pdf) $f_Y(y)$. The pdf has limiting value 0 for $y \to -\infty$ and for $y \to +\infty$.

Important properties of s distribution are its centered moments.

- The expectation, or the expected value, is denoted by $E[Y]$, or μ. It is defined as

$$E[Y] = \mu = \int y \cdot f_Y(y) dy,$$

the integral being taken over all values of Y for which $f_Y(y)$ is defined. The expectation is a measure for the central tendency of a distribution.

- The variance of Y is a measure of the spread. It is denoted by $\sigma^2 = E[Y-\mu]^2$ and is defined as

$$E[Y - \mu]^2 = \sigma^2 = \int (y - \mu)^2 \cdot f_Y(y) dy.$$

A larger value of σ^2 corresponds to a larger spread, that is: highly deviating values are likely to occur. On the other hand, a low value of σ^2 corresponds to a small spread: only values close to μ are likely to occur. The expectation and the variance are the first and the second moment of the distribution of Y.

- Other common measures, the so-called 3rd and 4th moment, are the skewness and kurtosis, and are defined as

$$E[Y - \mu]^3 = \text{skewness} = \int (\frac{y - \mu}{\sigma})^3 \cdot f_Y(y) dy.$$

$$E[Y - \mu]^4 = \text{kurtosis} = \int (\frac{y - \mu}{\sigma})^4 \cdot f_Y(y) dy$$

respectively. Skewness indicates whether a pdf is symmetric (skewness = 0), skewed to the right (skewness > 0, possibly caused by some very high observations) or to the left (skewness < 0). Kurtosis indicates flatness of the distribution, as compared to the normal distribution (for which the kurtosis is equal to 2), to be discussed below.

A special probability function is the so-called normal (or Gaussian) distribution. The cdf and the pdf are fully characterized by the first two moments μ and σ^2. The normal probability density function is defined as

$$\phi(y) = \frac{1}{\sqrt{2\pi\sigma^2}} e^{-\frac{(y-\mu)^2}{2\sigma^2}}.$$

The normal cumulative distribution function is defined as

$$\Phi(x) = \int_{-\infty}^{x} \phi(y) dy.$$

Notice that no analytic expression exists for $\Phi(x)$, and that tables have to be used instead. A special case is the so-called standard normal distribution with $\mu = 0$ and $\sigma^2 = 1$. The normal distribution is fully characterized by the mean and the standard deviation (Fig. 2.1).

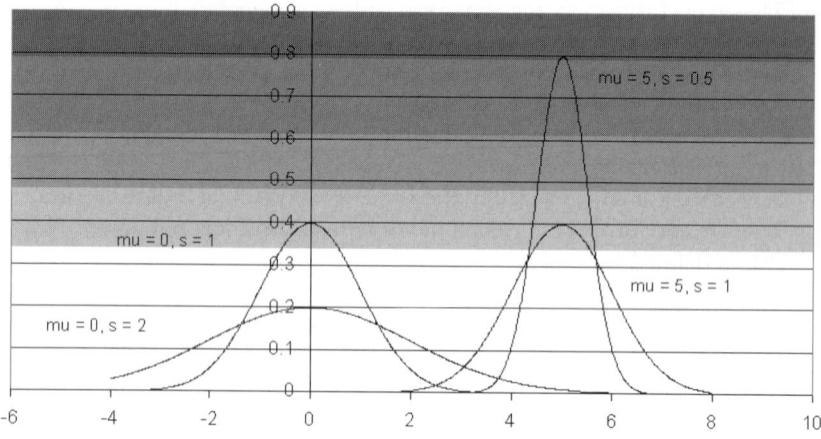

Figure 2.1: The normal distribution for various values of μ and σ^2.

Inference from observations derives the moments of the distribution. For that purpose descriptive statistics are calculated. Descriptive statistics characterize parameters of continuous environmental variables, such as mean, variance, median, minimum, maximum, skewness and kurtosis. Consider n observations, $y_1, ..., y_n$, i.e. the sample, of size n. i.e. n pixel values within an image. These observations are realizations from the population. On the basis of the observations inference about population parameters is made. We distinguish between parameters of location (such as minimum, maximum, mean and median) and other descriptive statistics, such as variance, standard deviation, variance of the mean, skewness and kurtosis. Minimum and maximum are obviously estimated. Mean and median are estimated as

$$\text{mean} = m_Y = \bar{y} = \sum_{i=1}^{n} y_i$$

$$\text{median} = 50^{th} \text{percentile}$$

and are both estimators for the population mean μ. For normally distributed data, mean and median are the same. In many studies, however, they differ, indicating that the data are not normally distributed, possibly as they contain some (isolated) extreme values.

Estimates for measures to characterize the spread in population values are the variance:

$$\text{Variance} = s_Y^2 = \frac{1}{n-1} \sum_{i=1}^{n} (y_i - \bar{y})^2$$

and its square root, the standard deviation

$$\text{Standard deviation} = s_Y = \sqrt{\frac{1}{n-1} \sum_{i=1}^{n} (y_i - \bar{y})^2}.$$

They are measures for the spread: data can have either a large or a small spread. The advantage of using the standard deviation in stead of the variance is that it is expressed in the same measurement unit as the original observations, whereas the variance is expressed in the squared units.

Related to the standard deviation is the standard error, for mutually independent data this is:

$$\text{Standard error} = s_e = \frac{s_Y}{\sqrt{n}} = \sqrt{\frac{1}{n \cdot (n-1)} \sum_{i=1}^{n} (y_i - \bar{y})^2}.$$

The standard error is a measure of the precision of the mean. Note that it is inversely proportional to the square root of n, the number of observations: the more data are collected the more precise the mean is being determined.

Skewness is calculated as

$$\text{Skewness} = \frac{\sum (y_i - \bar{y})^3}{s_Y^3} \cdot \frac{n}{(n-1)(n-2)}.$$

For a normally distributed variable this value is equal to 0. A negative value indicates a large tail to the left (often accompanied by a small mean as compared to the median), whereas a positive value indicates a large tail to the right. Finally, the kurtosis is calculated as

$$\text{Kurtosis} = \frac{n(n+1) \sum (y_i - \bar{y})^4 - 3(n-1)s_Y^4}{(n-1)(n-2)(n-3)s_Y^4}.$$

For a normal distribution it is equal to $2 \cdot s_Y$.

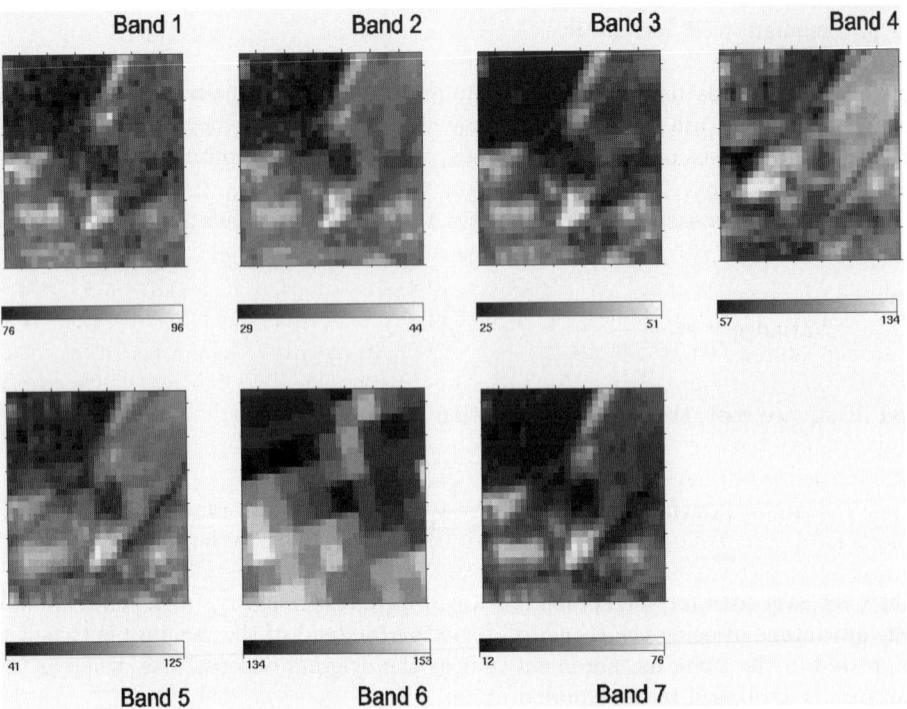

Figure 2.2: Display of the seven bands for a small region.

Example

We will consider a relatively small subarea (33 × 33 pixels) from the Twente data file (fig. 2.2). Obviously, the images show some relation to each other, although some images show more spread than other images. Also, the image for band 6 appears to show a somewhat coarser resolution. We consider descriptive statistics for the seven bands (table 2.1).

	n	Minimum	Maximum	Mean	Std.dev.	Skewness	Kurtosis
Band 1	1089	76	96	82.3	3.91	0.88	0.54
Band 2	1089	29	44	34.4	2.89	0.19	-0.24
Band 3	1089	25	51	32.4	5.23	0.91	0.29
Band 4	1089	57	134	89.6	16.49	0.29	-0.85
Band 5	1089	41	125	73.9	16.18	0.01	-0.51
Band 6	1089	134	153	141.1	5.02	0.37	-1.06
Band 7	1089	12	76	28.7	12.35	1.26	1.33

Table 2.1: Descriptive statistics for the 7 bands in the first sub-area.

Each band contains the same number of pixels ($n = 1089$). We notice that in this subregion different minimum and maximum values occur for the 7 bands, and that no band shows a value equal to 0. For example, band 1 ranges from 76 to 96, etc. Mean values differ between the 7 bands as do the standard deviations within each band: it is lowest for band 2 (2.89) and highest for band 4 (16.49). This implies that for band 4 much variation must be visible, whereas this is lowest for band 2. Upon inspection of the skewness we notice, that only band 5 appears to be non-skewed, whereas all other bands show a positive skewness: there are outliers to the right for each of these bands. Also the kurtosis indicates that the data are not-normal: normal data would have a kurtosis equal to 2. Therefore, each distribution is flatter than the standard normal distribution. This is confirmed when we consider the histograms. Histograms plot the number of measured pixels within different range classes (Fig. 2.3). For band 1 few (< 10) pixels are observed in the first range class labeled with the number 76, ranging from 75 to 77. Many more pixels (> 150) occur within the second range class, ranging from 77 to 79. Similar remarks apply to the other bands and all range classes. On top of the histograms the normal pdf is plotted, that is the density with calculated mean and standard deviation. A comparison between the curve and the histogram tells us whether the data are approximately normally distributed. For none of the bands there is strong evidence for normality, as all bands are skewed to the right, i.e. they have more values exceeding the mean than could be expected on the basis of normality. In practical studies, therefore, one may encounter deviations from the normal distribution.

2.2 Covariance and correlation

Descriptive statistics defined above are for single variables only. In several situations, we also need to have information about the relation between two (or more) variables, in particular when we wish to predict one from the other (see Chapter 6). Covariance and correlation quantify these relations. So assume that there are p variables: $Y_1, ..., Y_p$. The covariance between variables i and j with population means μ_i and μ_j, respectively, is defined as

$$Cov_{ij} = E[Y_i \cdot Y_j] - \mu_i \cdot \mu_j.$$

A positive covariance expresses a positive relation between Y_i and Y_j, i.e. an increase in Y_i corresponds to an increase in Y_j, whereas a negative covariance expresses an increase in Y_i corresponding to a decrease in Y_j. The covariance depends upon the variances of the two variables. The correlation removes this dependence, being defined as the covariance divided by the product of the standard deviations of the two variables:

$$\rho_{ij} = \frac{Cov_{ij}}{\sqrt{Var(Y_i) \cdot Var(Y_j)}}.$$

The correlation coefficient takes values between -1 and 1, where a value of -1 indicates a perfect negative relation, and a value of +1 a perfect positive relation. A value equal to 0 indicates absence of relation. As an example we calculated covariances and correlations between the 7 bands in the first test image, containing 1089

16   Alfred Stein

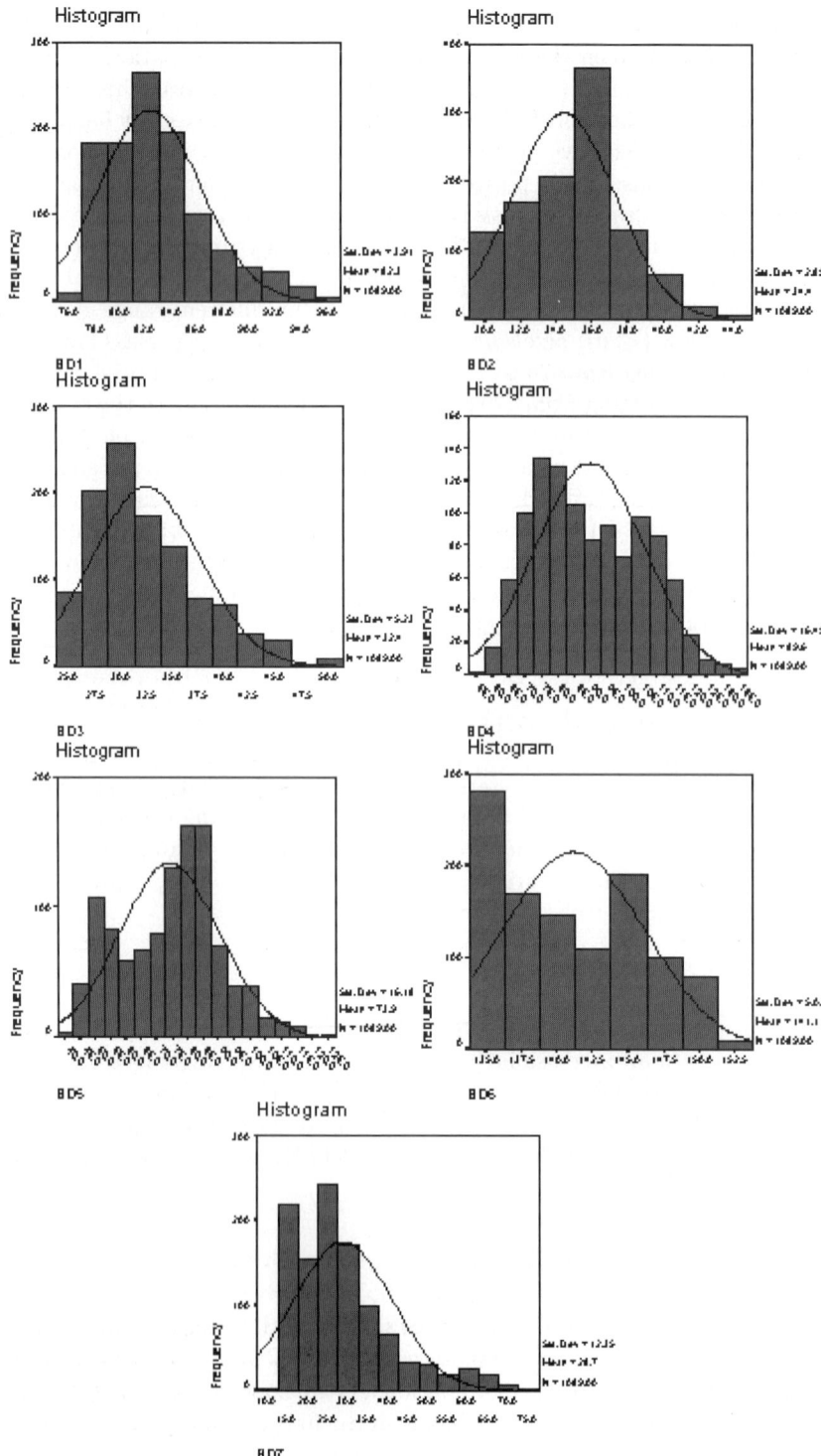

Figure 2.3: Histograms for each of the seven bands in the first sub-area.

Chapter 2: Some basic elements of statistics 17

pixels. Table 2.2 shows the covariances (upper right triangle), the variances (on the diagonal) and the correlations (lower left triangle).

	Band 1	Band 2	Band 3	Band 4	Band 5	Band 6	Band 7
Band 1	**15.313**	10.177	19.086	-15.466	49.872	14.104	44.074
Band 2	0.901*	**8.329**	13.949	0.505	41.269	10.130	30.922
Band 3	0.933*	0.925*	**27.307**	-23.116	67.763	19.097	58.903
Band 4	-0.240*	0.011	-0.268*	**271.896**	52.380	17.104	-51.798
Band 5	0.788*	0.884*	0.801*	0.196*	**261.793**	49.935	169.859
Band 6	0.718*	0.699*	0.728*	-0.207*	0.614*	**25.227**	44.169
Band 7	0.912*	0.867*	0.912*	-0.254*	0.850*	0.712*	**152.644**

Table 2.2: Correlations (lower left block), variances (diagonal) and covariances (upper right block) between the 7 bands within the sub-area.

2.2.1 Significance of correlation

In many instances we are interested in checking whether a correlation is significantly different from a particular value ρ_0 (most commonly $\rho_0 = 0$). This is usually formulated as a hypothesis H_0, which is tested against an alternative hypothesis H_a. In particular, H_0 can be formulated here as no difference between the observed correlation r and some hypothesized correlation ρ_0. To test the hypothesis a test statistic is formulated. The larger the difference between r and ρ_0 the larger will be the value of the test statistic. Sample size, however, is critical as well, as more confidence can be put on a correlation of 0.4 derived from 100 observations than from just 4 observations. The sampling distribution of the correlation coefficient r is generally not normal. To obtain normality, use can be made of Fisher's z-transformation, given by

$$Z_f = 1.1513 \cdot \log_{10} \frac{1+r}{1-r}.$$

Now Z_f is normally distributed with a mean equal to μ_f and a standard deviation equal to σ_f, where:

$$\mu_f = 1.1513 \cdot \log_{10} \frac{1+r}{1-r} \quad \text{and} \quad \sigma_f = \sqrt{\frac{1}{n-3}}$$

This leads to the standardized value z_f:

$$z_f = \frac{Z_f - \mu_f}{\sigma_f}.$$

The significance of z_f can be tested in a common way from the z tables (Table 2.3 and Fig. 2.4). In the majority of the cases, though, H_0 specifies whether r is equal to 0 and we wish to test the hypothesis of no correlation. But when $\rho_0 = 0$ also $\mu_f = 0$ and the significance of r depends on its own magnitude and the sample size only. A table to check this is given, and this only depends upon the sample size and the level of significance.

2.2.2 Example

The correlation coefficient between bands 4 and 5 is equal to 0.196. The number of data is equal to 1089, hence the critical value at an $\alpha = 0.01$ level, equal to 0.182 (for $n = 200$) is exceeded, and therefore this correlation coefficient is significant at the 0.01 level.

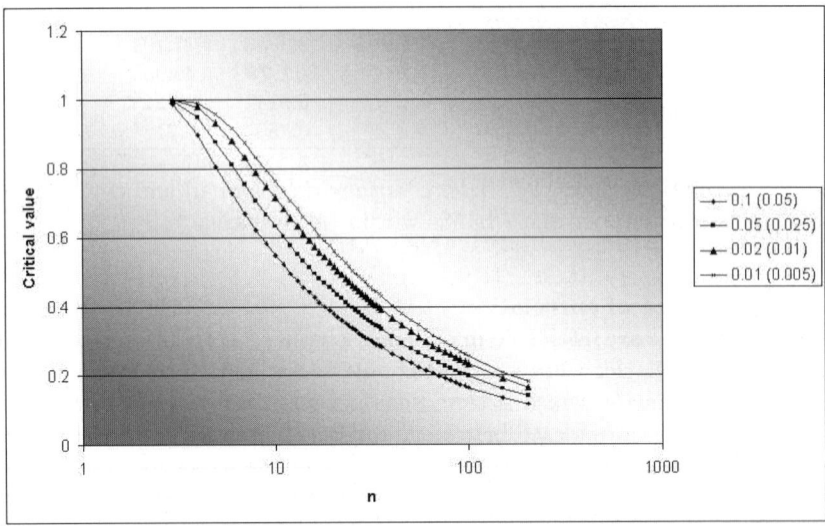

Figure 2.4: Critical values for the correlation coefficient for different values of n and different confidence levels.

2.3 Likelihood

The concept of likelihood is important for remote sensing studies. We will encounter it in the classification chapters below. Its background is that the probability model, the set of statistical hypotheses and the data form a triplet [118]. We consider $\Pr(R|H)$, the probability of obtaining results R, given hypothesis H according to the probability model. The data are implicit in this formulation. The probability is defined for any possible result given any hypothesis. It may be regarded as a function of both R and H, i.e. we can fix a hypothesis and consider the probability of any result, or we can fix the results and consider all possible hypotheses that lead to these results. This brings us to the concept of likelihood: the likelihood $L(H|R)$ of the hypothesis H given data R. A specific model for the likelihood is proportional to $\Pr(R|H)$, the constant of proportionality being arbitrary. Notice the difference with probability. In probability statements, R is variable and H is constant, but in likelihood H is the variable for constant R. The arbitrary constant of proportionality enables us to use the same definition of likelihood for discrete and continuous variables.

Example

We consider a binomial model for the occurrence of two forms of land use LU_1 and LU_2 in a set of two pixels. Suppose that there are two sets of data: R_1, a set of two pixels with LU_1 and LU_2 and a set R_2 a set of two pixels both with LU_1. A set of two pixels is drawn at random, i.e. either R_1 or R_2 shows up. Let p be the probability that LU_1 occurs. We consider two hypotheses: $H_1 : p = \frac{1}{4}$ and $H_2 : p = \frac{1}{2}$. Then the four probabilities $Pr(R_1|H_1)$, $Pr(R_2|H_1)$, $P(R_1|H_2)$ and $P(R_2|H_2)$ are as follows:

Hypothesis		LU_1 and LU_2	Both LU_1
	$H_1 : p = \frac{1}{4}$	$\frac{3}{8}$	$\frac{1}{16}$
	$H_2 : p = \frac{1}{2}$	$\frac{1}{2}$	$\frac{1}{4}$

These numbers are probabilities and if the hypothesis $p = \frac{1}{2}$ is valid, the probability of getting (LU_1 and LU_2) or (Both LU_1) equals $\frac{3}{4}$. This is also termed the addition axiom for each hypothesis. On the other hand, we may not use the addition axiom over different hypotheses (the probability that $p = \frac{1}{4}$ or $p = \frac{1}{2}$ is not necessarily equal to $\frac{1}{2} + \frac{3}{8} = \frac{7}{8}$). Also, we may not invert the probability statements to conclude, for example, that $Pr(H_1|R_1) = \frac{3}{8}$. Applying the definition of likelihood, the likelihoods of the hypotheses given the data are as follows:

Hypothesis		LU_1 and LU_2	Both LU_1
	$H_1 : p = \frac{1}{4}$	$k_1 \cdot \frac{3}{8}$	$k_2 \cdot \frac{1}{16}$
	$H_2 : p = \frac{1}{2}$	$k_1 \cdot \frac{1}{2}$	$k_2 \cdot \frac{1}{4}$

where k_1 and k_2 are arbitrary constants. They remind us that with likelihoods the hypotheses are the variables and the data are fixed. We cannot compare $L(H_1|R_1)$ with $L(H_2|R_2)$ but we can state that on the data R_1 the likelihood of the hypothesis H_1 is $\frac{3}{4}$ the likelihood of the hypothesis H_2, whereas on data R_2 the likelihood of H_1 is $\frac{1}{4}$ that of H_2.

Maximum likelihood estimation

Likelihoods are in particular useful and regularly applied when the probability model is a Normal distribution. They may give transparent expression and relatively easy ways of estimation. Suppose therefore that the pixels in an image follow a Normal distribution. That is, if μ is the true mean and σ^2 the theoretical variance of the pixels within a class C the probability that a single pixel lying in the interval $(x, x + dx)$ equals

$$dF = \frac{1}{\sqrt{2\pi\sigma^2}} e^{-\frac{(x-\mu)^2}{\sigma^2}} dx$$

and hence the probability of obtaining a set of n observations in the interval $(x_1, x_1 + dx_1)$ $(x_2, x_2 + dx_2)$, ... , $(x_n, x_n + dx_n)$ equals

$$dF = \left(\frac{1}{\sqrt{2\pi\sigma^2}}\right)^n e^{-\sum_{i=1}^{n} \frac{(x_i-\mu)^2}{\sigma^2}} dx_1...dx_n.$$

The likelihood of (μ, σ^2) given the sample is thus

$$L(\mu, \sigma^2) = k \cdot e^{-\sum_{i=1}^{n} \frac{(x_i-\mu)^2}{\sigma^2}}.$$

Now

$$\sum_{i=1}^{n}(x_i - \mu)^2 = \sum_{i=1}^{n}(x_i - \bar{x})^2 + n(\bar{x} - \mu)^2$$

where $\bar{x} = \frac{1}{n}\sum_{i=1}^{n} x_i$, the sample mean. Notice that the first term of this expression is n times the sample variance. Therefore, the likelihood may be expressed in terms of the sample mean and the variance s^2.

In many circumstances, an analytical solution is not available and we have to turn to iterative procedures. Let us consider a vector of parameters for which we wish to find a maximum likelihood estimate θ; in the example $\theta = (\mu, \sigma^2)$. Let the solution for obtained at the ith step be denoted by $\theta^{(i)}$ and the corresponding likelihood by $L(\theta^{(i)})$. The estimate $\theta^{(i+1)}$ at the next step is obtained as

$$\theta^{(i+1)} = \theta^{(i)} - \rho^{(i)} R^{(i)} s^{(i)}.$$

Here, $R^{(i)}$ is the inverse of the second derivative of L to the different elements of θ, estimated at the ith iteration, i.e. the (a,b) element of $R^{(i)}$ is given by

$$R_{ab}^{(i)} = E\left[\frac{\partial^2 L(\theta^{(i)})}{\partial \theta_a^{(i)} \partial \theta_b^{(i)}}\right]^{-1}.$$

In the method of scoring (Gauss-Newton) $R^{(i)}$ is approximated by the inverse of the Fisher information matrix at the ith iteration, i.e. $R^{(i)} = \left[M^{(i)}\right]^{-1}$. The (a,b) element of the Fisher information matrix is

$$M_{ab}^{(i)} = E\left[\frac{\partial L(\theta^{(i)})}{\partial \theta_a^{(i)}} \frac{\partial L(\theta^{(i)})}{\partial \theta_b^{(i)}}\right],$$

based on Gaussian moments [232]. Notice that in this case only first derivatives are necessary.

Further, $\rho^{(i)}$ is a scalar, which value is chosen such that $L(\theta^{(i+1)}) < L(\theta^{(i)})$ and $s^{(i)}$ is the vector with elements equal to the derivative to the parameters, evaluated at the ith iteration step:

$$s^{(i)} = \frac{\partial L}{\partial \theta}|_{\theta=\theta^{(i)}}.$$

This procedure maximizes the likelihood of the parameter vector .

A key statistic for model identification is Akaike's information criterion, defined as:

$$AIC = L(\theta^{(i)}) + 2p,$$

where p is the number of free estimated parameters, i.e. the number of elements of the vector θ. For small numbers of data the corrected Akaike's Information Criterion

(AIC$_C$) may be used [198]:

$$AIC_C = L(\theta^{(i)}) - n + n * \left(\frac{1 + \frac{p}{n}}{1 - \frac{p+2}{n}} \right).$$

The aim is to minimize the AIC (or the AIC$_C$): therefore the model is selected that shows the lowest of these values. Often, the likelihood function is calculated assuming a Gaussian distribution. Actual data may show a different distribution though. Hence the AIC can be used as a measure for model identification subject to the restriction of Gausianity.

2.4 Regression and prediction

In regression the variation of one variable is explained by variation of one or more other variables. The variable for which the variation has to be explained is also called the dependent variable. The others are the explanatory variables. For example, variation in remote sensing band 4 is likely to depend on hydrologic conditions (moisture availability) in an area. Therefore, variation of band 4 may be explained by variation in moisture availability. It can be important to quantify this effect by means a linear model, to see which part of the total variation is explained by the explanatory variable and to predict moisture availability values by using band 4 values and with the model.

So, let there be two variables, Y and Z, and assume that Y is the dependent variable (for example: moisture content) and Z the explanatory variable (for example: the pixel values in band 4). The linear regression from Y on Z is then defined as fitting a line $a + b \cdot Z$ through the data Y_i. Estimates for the coefficients a and b are obtained as:

$$a = m_Y - b \cdot m_Z$$
$$b = \frac{\sum_{i=1}^{n}(Z_i - m_Z)(Y_i - m_Y)}{\sum_{i=1}^{n}(Z_i - m_Z)^2}.$$

Note that regression is different from correlation: regressing Y on Z is different from regressing Z on Y, whereas correlation is symmetric.

Including observations of the explanatory variable in the regression equation yields estimated values $\widehat{Y}_i$ of the dependent variable. These will generally differ from the observations Y_i.

To determine the quality of the fit, use can be made of the residuals between observations and estimated values, $\widehat{Y}_i - Y_i$. To do so, the coefficient of determination R^2 is applied:

$$R^2 = \frac{\sum_{i=1}^{n}(\widehat{Y}_i - m_Y)^2}{\sum_{i=1}^{n}(Y_i - m_Y)^2}.$$

The larger the value of R^2, the better the regression model explains the total variation: if $R^2 = 0$, no variation is explained by the regression model, whereas if

$R^2 = 1$, all variation is explained by the regression model, i.e., all observations actually lay on the regression line. The percentage of total variation explained by the model is equal to $R^2 \times 100\%$.

Example
In the following table we sketch for sub-area 1 a regression analysis of band 3 on band 4, i.e. the variation in band 3 explained by means of variation in band 4. We distinguish as well 4 classes.

	Model	R^2
All data	Band 3 = 40.0−0.085·Band 4	0.072
Class 1	Band 3 = 55.5−0.223·Band 4	0.493
Class 2	Band 3 = 28.9+0.162·Band 4	0.040
Class 4	Band 3 = 57.9−0.289·Band 4	0.481
Class 6	Band 3 = 27.2+0.013·Band 4	0.002

If we neglect the presence of classes, the model is not very useful, showing a negative relation though between bands 3 and 4. This problem is partly overcome by using the available classification, where we notice that for classes 1 and 4 the model is rather good, explaining 49% and 48% respectively of the total variation in band 3, showing a negative relation between the two bands (Fig. 2.5). In classes 2 and 6, however, the model is not very useful at all, with a very low R^2 value. Notice, though that for class 2 this is based on 4 observations only.

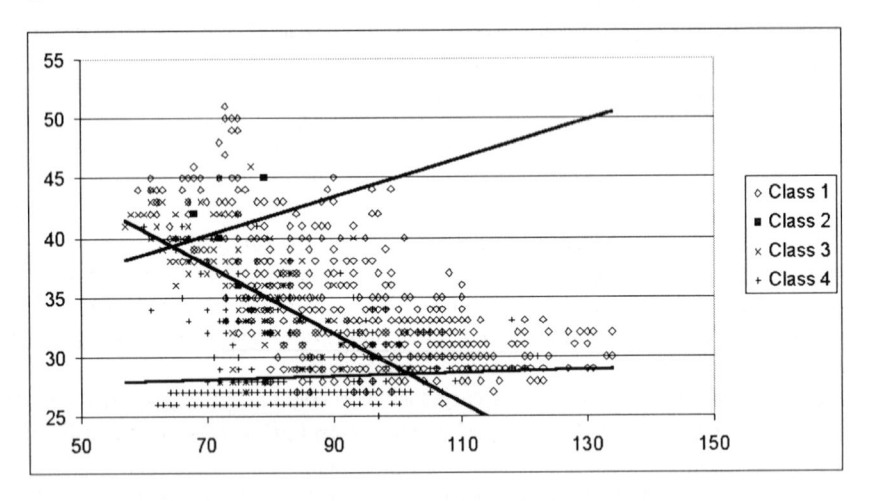

Figure 2.5: Regression of band 3 to band 4 within different classes. The regression line for class 2 is not drawn

2.5 Estimation and Prediction

A central activity in statistical studies is to use a model derived with for example regression analysis to make predictions. As an example, suppose that observations on band 5 are related to the moisture content of the topsoil. Based upon the satellite image the moisture content is to be predicted throughout the region. A regression model formulates the expectation. Here a distinction is made between estimating the expectation in an unobserved point (i.e. formulating a regression model) and predicting the most likely value in the same point. Predicting is equivalent to estimating a individual observation. Predicting is therefore much more risky than estimating the expectation. Also, a measure of the quality of this prediction is required, such as the variance or the standard deviation.

If all observations have the same variance, the best prediction and the estimate of its expectation for a new point have the same value. If the expectation is estimated, however, uncertainty reduces to uncertainty in determining μ. When predicting, additional uncertainty has to be dealt with, because an individual observation has to be predicted. Apart from numerical differences prediction of a random variable and estimation of its expectation are fundamentally different. From now on we deal mainly with prediction, that is in fact estimation of a stochastic effect.

We will formulate prediction as follows. The pixel values are related to the spatial location x, possibly the center of the grid cell. Let us first estimate the expectation of $Y(x)$ in an unvisited location with pixel value equal to x_0. The first approach would be to estimate the average value m and assigning this value to $x = x_0$, being a prediction for the value of $Y(x)$ in $x = x_0$. For independent data, the variance of the estimation error equals σ^2/n, where σ^2 is the variance of Y. In practice we use an estimate s^2 for σ^2. Also, if we take a linear model with estimated coefficients a and b: $y = a + b \cdot x$, the expected value in a pixel with value x_0 equals $a + b \cdot x_0$. If we estimate the expectation, an error is being made. Variance of this error, the so-called estimation error variance equals

$$Var[Y(x_0) - \widehat{Y}(x_0)] = \begin{pmatrix} 1 & x_0 \end{pmatrix}' (X'X)^{-1} \begin{pmatrix} 1 \\ x_0 \end{pmatrix} \cdot \sigma^2,$$

where X equals the design matrix, i.e. the matrix with the first column equal to 1 and the second column equal to the first coordinate of the pixel locations with an observation, the 3rd column equal to the 2nd coordinate of the pixel locations with an observation. To estimate the variance the population variance σ^2 is replaced by the sample variance s^2.

The next stage is to predict the value, i.e. the most likely individual value. If in a simple model the expectation is equal to the mean independent data this value is again the best linear unbiased predictor. The variance of the prediction error differs from the variance of the estimation error, being equal to $(1 + \frac{1}{n})\sigma^2$. It is equal to the sum of the variance of the estimation error and the variance σ^2 of Y. If data show a linear relation, then the prediction error variance changes in a similar way as for the estimation error variance and becomes equal to

$$Var[Y(x_0) - \widehat{Y}(x_0)] = \left[1 + \begin{pmatrix} 1 & x_0 \end{pmatrix}' (X'X)^{-1} \begin{pmatrix} 1 \\ x_0 \end{pmatrix}\right] \cdot \sigma^2,$$

where the symbols have the same meaning as before. Notice, that if X becomes equal to a column vector with elements equal to 1, i.e. in the simple model described above, then $X'X$ is equal to n and the same prediction error emerges as before.

Example
We turn to predictions made for data in classes 1 and 2 of the sub-area (Fig. 2.6). Below, the data, the fitted lines and the confidence lines are given. We notice relatively narrow (and straight) boundaries for data in class 1, due to the good model and the large number of data. For class 2, though, the confidence bounds are curved, and in particular when data have to be predicted outside the data range the prediction interval becomes rather wide.

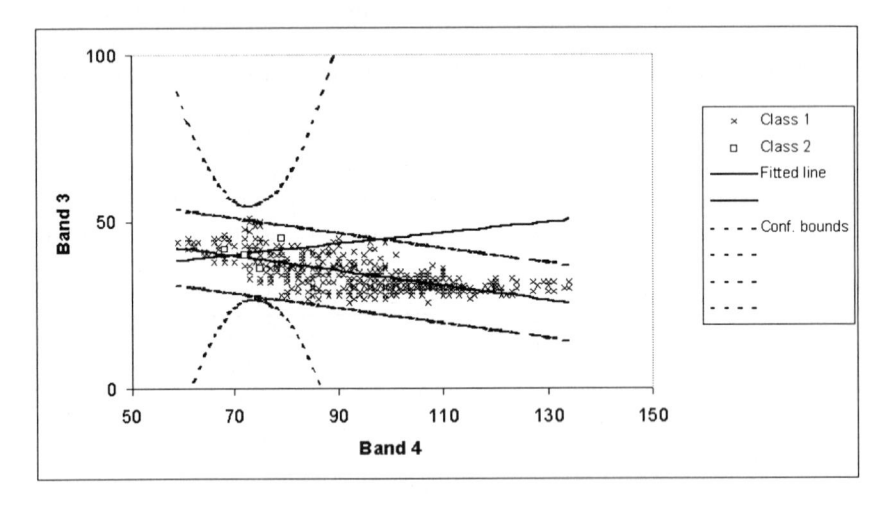

Figure 2.6: Predictions and estimations of band 3 by band 4 data, including confidence bounds

The procedures of summarizing the data by means of descriptive statistics are generally adequate to investigate linear and causal relations. However, they do not give any insight as concerns the spatial relations. Of course, one could choose the coordinates as explanatory variables. But even then only a global indication is available as to how much variation can be explained by a coordinate, and no insight is given in where the variation occurs. Moreover, for spatial data we will often be confronted with spatial dependencies: observations close to each other are more similar than observations at a larger distance. To analyze these data we need spatial statistical procedures.

Table 2.3: Significance of correlation coefficients for different numbers of observations (n) and different two-sided significance levels (α). One-sided intervals are given in brackets.

	α			
n	0.1 (0.05)	0.05 (0.025)	0.02 (0.01)	0.01 (0.005)
3	0.988	0.997	1.000	1.000
4	0.900	0.905	0.980	0.990
5	0.805	0.878	0.934	0.959
6	0.729	0.811	0.882	0.917
7	0.669	0.754	0.833	0.875
8	0.621	0.707	0.789	0.834
9	0.582	0.666	0.750	0.798
10	0.549	0.632	0.715	0.765
11	0.521	0.602	0.685	0.735
12	0.497	0.576	0.658	0.708
13	0.476	0.553	0.634	0.684
14	0.458	0.532	0.612	0.661
15	0.441	0.514	0.592	0.641
16	0.426	0.497	0.574	0.623
17	0.412	0.482	0.558	0.606
18	0.400	0.468	0.543	0.590
19	0.389	0.456	0.529	0.575
20	0.378	0.444	0.516	0.561
21	0.369	0.433	0.503	0.549
22	0.360	0.423	0.492	0.537
23	0.352	0.413	0.482	0.526
24	0.344	0.404	0.472	0.515
25	0.337	0.396	0.462	0.505
26	0.330	0.388	0.453	0.496
27	0.323	0.381	0.445	0.487
28	0.317	0.374	0.437	0.479
29	0.311	0.367	0.430	0.471
30	0.306	0.361	0.423	0.463
31	0.301	0.355	0.416	0.456
32	0.296	0.349	0.409	0.449
33	0.291	0.344	0.403	0.443
34	0.287	0.339	0.397	0.436
35	0.283	0.334	0.391	0.430
40	0.264	0.312	0.367	0.403
45	0.249	0.294	0.346	0.380
50	0.235	0.279	0.328	0.361
55	0.224	0.266	0.313	0.345
60	0.214	0.254	0.300	0.330
65	0.206	0.244	0.288	0.317
70	0.198	0.235	0.278	0.306
75	0.191	0.227	0.268	0.296
80	0.185	0.220	0.260	0.286
85	0.180	0.213	0.252	0.278
90	0.174	0.207	0.245	0.270
95	0.170	0.202	0.238	0.263
100	0.165	0.197	0.232	0.256
150	0.135	0.160	0.190	0.210
200	0.117	0.139	0.164	0.182

Chapter 3

Physical principles of optical remote sensing

Freek van der Meer

When light interacts with a material, light of a certain wavelength is preferentially absorbed while at other wavelengths light is transmitted in the substance. Reflectance, defined as the ratio of the intensity of light reflected from a sample to the intensity of the light incident on it, is measured by reflection spectrophotometers which are composed of a light source and a prism to separate light into different wavelengths. This light beam interacts with the sample and the intensity of reflected light at various wavelengths is measured by a detector relative to a reference standard of known reflectance. Thus a continuous reflectance spectrum of the sample is obtained in the visible- and near-infrared wavelength region. In this chapter, the principles of optical remote sensing are outlined through deriving the physics of radiation, reflectance, scattering and atmospheric attenuation [319]. These are compared with the sensor characteristics of the LANDSAT Thematic Mapper system; the system of which data is used in this book. Much of this discussion is based on [301].

3.1 Electromagnetic Radiation

Electromagnetic radiation (EMR) can be regarded as energy in the form of linked electric and magnetic force fields transmitted in discrete packages known as photons or quanta following Einstein's relationship

$$E = mc^2,$$

where E is energy, m is mass and c is the velocity of the EMR in a vacuum. The photons propagate as waves involving vibrations at right angle to the direction of movement. The distance between the wave crests is the wavelength (λ) and the number of vibrations passing a point in one second is the frequency (ν) where $\lambda\nu = c$. The frequency or wavelength of EMR is a function of the energy of the photons. Planck formulated this as

$$E = \nu h = \frac{ch}{\lambda},$$

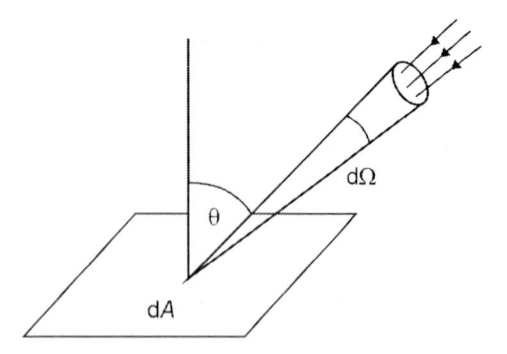

Figure 3.1: Geometry of radiance of the incoming radiation on a surface area, dA (after [301])

where h is Planck's constant $(6.62 \cdot 10^{-34} Js)$. This equation is known as Planck's law.

EMR is produced whenever the size or direction of an electric or magnetic field fluctuates with time. The source of energy for optical remote sensing is the sun which can be regarded a blackbody with a temperature of approximately 6000 K. A blackbody is any material capable of absorbing and re-emitting all electromagnetic energy which it receives. The total energy emitted by a blackbody (e.g. its total radiant emittance per unit area in Wm^{-2}) is proportional to the fourth power of its absolute temperature (T) as described in Stefan-Boltzmann's law

$$H = \sigma T^4,$$

where σ is Stefan-Boltzmann's constant $(5.7 \cdot 10^{-8} Wm^{-2}K^{-4})$. A blackbody emits EMR at any wavelength, however its absolute temperature determines the maximum amount of transmitted energy and at which wavelength this peak in the emittance curve occurs. Wien's displacement law relates the absolute temperature of the blackbody and its wavelength of maximum emittance (λ_m) as

$$\lambda_m = 2898/T.$$

The sun, a blackbody of 6000 K absolute temperature, has a peak in the emittance curve at a wavelength of $0.5\mu m$ and does not emit energy at wavelengths less than $0.1\ \mu m$ and larger than $100\mu m$.

3.2 Physics of Radiation and Interaction with Materials

Thermal radiation is emitted by all objects at temperatures above absolute zero. Consider an area dA and radiation arriving at a direction to the normal of dA (Fig. 3.1), but in the range of directions forming a solid angle of d steradian. Radiance, L in $Wm^{-2}sr^{-1}$, is defined as

$$d\Phi = L dA d\Omega \cos\theta.$$

The total power falling on dA from all directions is given by the integration over 2π steradian in case of no absorption nor scattering by

$$\Phi = dA \int_{2\pi} L\cos\theta d\Omega = EdA,$$

where E is called irradiance measured in Wm^{-2}. In the reverse case when E measures the total radiance leaving dA we refer to exitance, M, and the total power emitted by the source is radiant intensity, I, defined as

$$\Phi_{total} = Id\Omega.$$

We can also express these quantities in spectral format by introducing an interval of wavelength and integrating over d. Spectral radiance according to Planck's relationship is given by

$$L_\lambda = \frac{2hc^2}{\lambda^5}(e^{hc/\lambda kT} - 1)^{-1},$$

where k is Boltzmann's constant $(1.38 \cdot 10^{-23} JK^{-1})$, c is the speed of light (in a vacuum) and T is the temperature. The total outgoing radiance of a black body of temperature T is given by

$$L = \int_0^\infty L_\lambda d\lambda = \frac{2\pi^4 k^4}{15c^2 h^3}T^4,$$

where $\sigma = \frac{2\pi^4 k^4}{15c^2 h^3}$ is again Stefan-Boltzmann's constant $(5.67 \cdot 10^{-8} Wm^{-2}K^{-4})$. Wien's displacement law gives the relation between the wavelength at which the maximum radiation is reached, λ_{max}, and the temperature of the black body as

$$\lambda_{\max} = c_w/T,$$

where c_w is a constant (Wien's constant) of $2.898 \cdot 10^{-3}$ km (Fig. 3.2).

3.3 Surface Scattering

When radiation interacts with a surface, it is partly absorbed into the substance, and partly scattered or reflected by the object. Consider a collimated beam of radiation incident on a surface at an incidence angle θ_o. The irradiance E is given by $F\cos\theta$ and scattered into a solid angle $d\Omega$ in a direction θ_1. The outgoing radiance of the surface as a result of this illumination is L_1 in the direction (θ_1, ϕ_1) where ϕ_1 is the azimuthal angle. The bidirectional reflectance distribution function (BRDF)R (in sr^{-1}) is defined as

$$R = L_1/E.$$

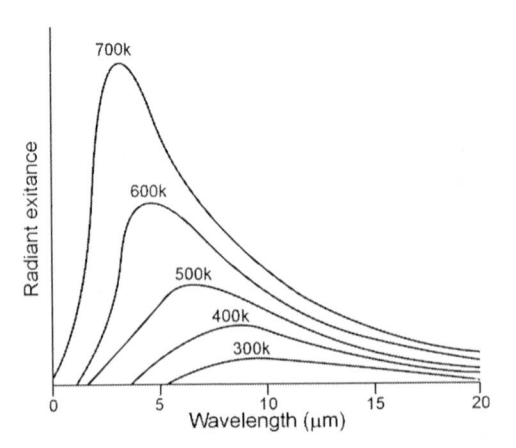

Figure 3.2: Energy from perfect radiators (black bodies) of different temperatures as a function of wavelength (after [61]).

R is a function of the incident and scattered directions and can thus be noted as $R(\theta_o, \phi_o, \theta_1, \phi_1)$. The reflectivity of the surface, r (also known as albedo), is the ratio of the total power scattered to the total power incident as

$$r(\Theta_o, \Phi_o) = \int_{\Theta=0}^{\pi/2} \int_{\phi=0}^{2\pi} R \cos \Theta_1 \sin \Theta_1 d\Theta_1 d\phi_1.$$

Two extreme cases of scattering surface can be defined, the perfectly smooth surface (specular surface) and the perfectly rough surface (the Lambertian surface; Fig. 3.3). A perfect Lambertian surface will scatter all the radiation incident upon it so that the radiant exitance M is equal to the irradiance E and the albedo is unity. A measure of roughness of a surface is given by the Rayleigh criterion. For a surface to be smooth according the Rayleigh criterion it should satisfy

$$\Delta h \cos \Theta_o / \lambda < 1/8,$$

where Δh is the surface irregularity of height and the wavelength considered.

3.4 Reflectance properties of Earth surface materials
3.4.1 Minerals and Rocks

Reflectance spectra have been used for many years to obtain compositional information of the Earth surface. The following discussion is based on [301]. Spectral reflectance in visible and near-infrared offers a rapid and inexpensive technique for determining the mineralogy of samples and obtaining information on chemical composition. Electronic transition and charge transfer processes (e.g., changes in energy states of electrons bound to atoms or molecules) associated with transition metal

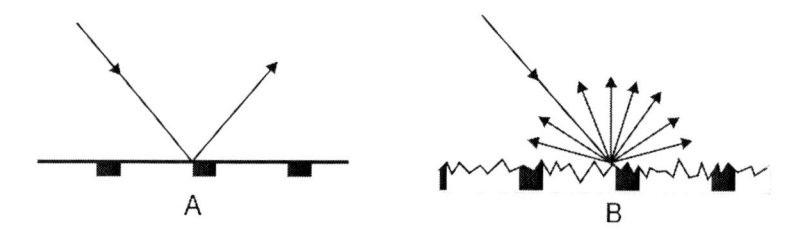

Figure 3.3: Example of (a) specular scattering and (b) diffuse or Lambertian scattering behavior.

ions such as Fe, Ti, Cr, etc., determine largely the position of diagnostic absorption features in the visible- and near-infrared wavelength region of the spectra of minerals [56, 2, 3]. In addition, vibrational processes in H_2O and OH^- (e.g., small displacements of the atoms about their resting positions) produce fundamental overtone absorptions [189, 185]. Electronic transitions produce broad absorption features that require higher energy levels than do vibrational processes, and therefore take place at shorter wavelengths [189, 154]. The position, shape, depth, width, and asymmetry of these absorption features are controlled by the particular crystal structure in which the absorbing species is contained and by the chemical structure of the mineral. Thus, variables characterizing absorption features can be directly related to the mineralogy of the sample.

The total energy of a molecule W_t is the sum of the electronic energy W_e the vibrational energy W_v and the rotational energy W_r as

$$W_t = W_e + W_v + W_r.$$

Changes in energy states of molecules due to changes in rotational energy levels do not occur in solids and will not further be treated here, I restrict this discussion to changes in electronic states and vibrational processes.

Harmonic vibration is described by

$$\omega_v = \sqrt{\frac{f(m_1 + m_2)}{m_1 m_2}},$$

where f is the restoring force constant (spring constant), ν the vibrational states $\nu=0,1,2,\ldots$ and m_1 and m_2 are the masses of the molecule. The possible energy states of the harmonic vibration are given by quantum mechanics as

$$W_v = (v + \frac{1}{2})\hbar\omega_v.$$

These define the spectral regions where absorption can occur as a result of vibrational processes.

Electrons in solid materials can occupy a discrete number of energy states $q = 1, 2, \ldots$ When photons are incident on a material they interact with the electrons and the absorption of a photon of a proper energy $h \cdot v$ may cause the transition of the electron to a higher state. Absorption of energy into the medium results in absorption features in reflectance spectra. The velocity of an electron is given by

$$v = \frac{qh}{2mL},$$

where m is the mass of the electron and $L = N \cdot l$ with l the internuclear bond length and N the number of electrons. Thus the total energy of the electron is the sum of the kinetic energy $mv^2/2$ and the potential energy (often set to zero). Now the energy states of the electron can be described by

$$W_q = mv^2/2 = \frac{q^2 h^2}{8mL^2}.$$

Each molecular orbital starting from the one with the lowest energy can accommodate only two electrons with antiparallel spin orientations (i.e., the Pauli exclusion principle). Therefore N electrons can maximally occupy $q = N/2$ states in the lowest energy state representing an energy of $W_{N/2}$ and the lowest empty one has $W_{(N/2)+1}$. We can now calculate that absorption of a photon of proper energy hv causes the transition of an electron from orbital q to $q = 1$ corresponding to an energy difference of

$$\Delta W = W_{(N/2)=1} - W_{N/2} = \frac{h^2}{8mL^2}(N+1).$$

Acknowledging that $hv = hc/\lambda$ and λ is the wavelength at which the absorption occurs we can find the longest wavelength at which radiation can be absorbed by the molecule due to the first transition as

$$\lambda_{\max} = \frac{8mc}{h} \frac{N^2 l^2}{N+1}.$$

To be able to interpret or predict the wavelengths at which absorptions due electronic transitions occur it is useful to know the energy level schemes of the molecules involved in the transitions.

The reflectance spectra of minerals are well known [189, 185, 186, 190, 191, 192, 193, 67, 84, 166] and several studies have been conducted to determine reflectance spectra of rocks [194, 195, 196, 197, 187, 188]. The reflectance characteristics of rocks can be simulated accurately by studying the compound effect of reflectance of minerals in a spectral mixture forming the rock. Reflectance spectra of minerals measured by different spectroradiometers with different spectral resolution are stored in spectral libraries that are available in digital format (e.g. [166, 67]). Examples of minerals exhibiting absorption features resulting from electronic transitions are shown in Fig. 3.4. Fig. 3.5 shows mineral absorption features due to vibrational transitions.

Chapter 3: Physical principles of optical R.S. 33

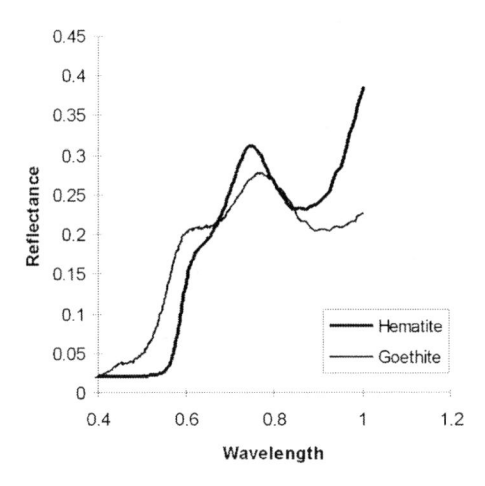

Figure 3.4: Spectral reflectance of minerals showing diagnostic iron absorption bands due to electronic transitions in the visible part of the spectrum.

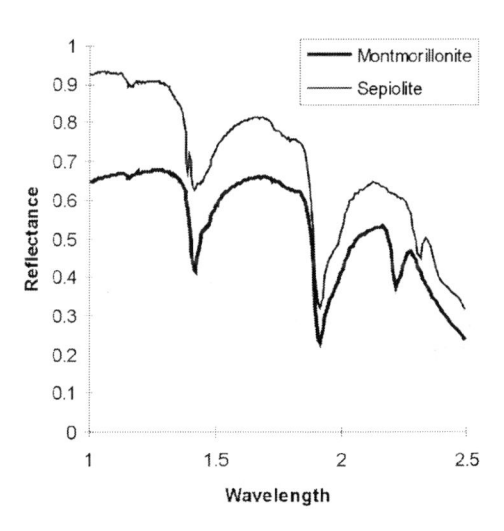

Figure 3.5: Spectral reflectance of minerals showing diagnostic H_2O absorption bands due to vibrational transitions in the shortwave infrared part of the spectrum.

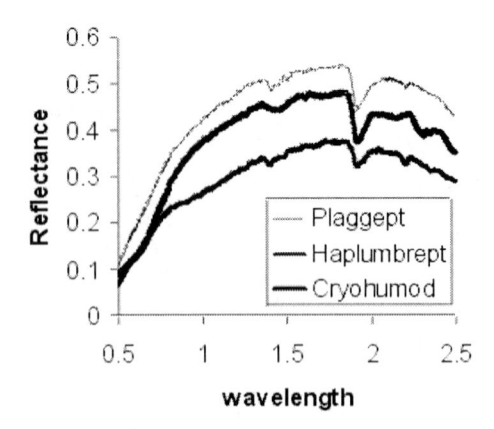

Figure 3.6: Spectral reflectance of three typical soils: a dark grey silty loam (plaggept), a brown sandy loam (haplumbrept) and a dark reddish organic rich silty loam (cryohumod).

3.4.2 Vegetation

Reflectance studies of vegetation (e.g. [147]) generally restrict to the green leaf part of the plants giving little attention to the non-green dry vegetation components. Reflectance properties of vegetation in the visible part of the spectrum are dominated by the absorption properties of photosynthetic pigments of which chlorophyll, having absorptions at 0.66 and 0.68 μm for chlorophyll a and b respectively, is the most important (Fig. 3.6). Changes in the chlorophyll concentration produce spectral shifts of the absorption edge near 0.7 μm : the red edge. This red edge shifts toward the blue part of the spectrum with loss of chlorophyll. The mid-infrared and short-wave infrared part of the vegetation spectrum is dominated by water and organic compounds of which cellulose, lignin, starch and protein [122]. Absorption features due to bound and unbound water occur near 1.4 and 1.9 μm and at 0.97, 1.20 and 1.77 μm. Cellulose has absorptions at 1.22, 1.48, 1.93, 2.28, 2.34, and 2.48 μm while lignin has absorption features at 1.45, 1.68, 1.93, 2.05-2.14, 2.27, 2.33, 2,38, and 2.50 μm [122]. Starch has absorption features at 0.99, 1.22, 1.45, 1.56, 1.70, 1.77, 1.93, 2.10, 1.32, and 2.48 μm [122]. The most abundant protein in leaves is a nitrogen bearing compound having absorption features at 1.50, 1.68, 1.74, 1.94, 2.05, 2.17, 2.29, 2.47 μm [122]. Dry plant materials lack the chlorophyll absorptions and intense water absorptions that are characteristic for green leaves and thus lack the intense absorption wing produced by high blue and UV absorptions. Dry plant materials have diagnostic ligno-cellulose absorption features at 2.09 and in the 2.30 μm region [122]. More details on reflectance properties of plant materials can be found in [122, 147, 394].

Chapter 3: Physical principles of optical R.S. 35

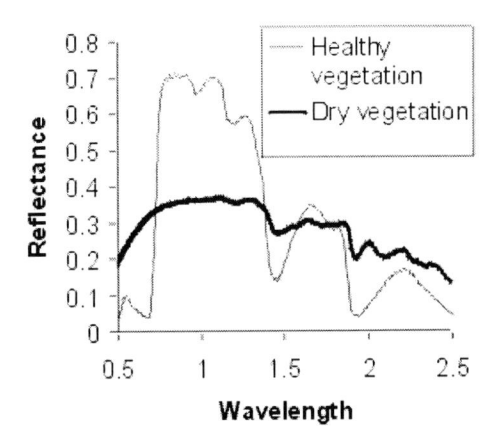

Figure 3.7: Spectral reflectance of healthy vegetation (grass) and dry vegetation (grass).

3.4.3 Soils

Spectral reflectance characteristics of soils (Fig. 3.7) are the result of their physical and chemical properties and are influenced largely by the compositional nature of soils in which main components are inorganic solids, organic matter, air and water. In the visible and near-infrared regions extending to 1.0 μm, electronic transitions related to iron are the main factor determining soil spectral reflectance. The majority of absorption features diagnostic for mineral composition occur in the short-wave infrared (SWIR) portion of the wavelength spectrum ranging from 2.0 to 2.5 μm. A strong fundamental OH^- vibration at 2.74 μm influences the spectral signature of hyrdoxyl-bearing minerals. Furthermore, diagnostic absorption features characteristic for layered silicates such as clays and micas and also of carbonates occur in the SWIR region. Organic matter influences the spectral reflectance properties of soils because amounts exceeding 2% are known to have a masking effect on spectral reflectance thus reducing the overall reflectivity of the soil and reducing (and sometimes completely obscuring) the diagnostic absorption features. Thus soils with a high (>20%) amount of organics appear dark throughout the 0.4 to 2.5 μm range. In contrast, less decomposed soils have higher reflectance in the near-infrared region and enhanced absorption features. Prominent absorption features near 1.4 and 1.9 μm due to bound and unbound water are typical for soil reflectance. Less prominent water absorption features can be found at 0.97, 1.20 and 1.77 μm. Increasing moisture content generally decreases the overall reflectance of the soil. A similar effect results from increasing the particle size resulting in a decrease in reflectivity and contrast between absorption features. Some studies on the spectral reflectance characteristics of soils and attempts to make classifications can be found in [33, 75, 351, 201, 331].

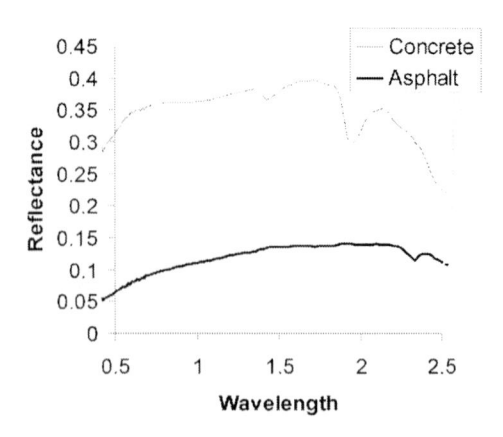

Figure 3.8: Spectral reflectance of building materials.

3.4.4 Man-made and other materials

It is impossible to provide a thorough review including an understanding of the physical nature of reflectance spectra of man-made structures and other materials covering the Earth surface. Where these structures are designed of natural materials which can be mono- or multi-mineralic the spectroscopic properties are directly related to these basic materials. For a large suite of chemical constituents no clear background information on spectral reflectance characteristics is available. In Fig. 3.8 spectral reflectance curves of concrete and asphalt are shown as examples of commonly used building materials.

Another important material that can be observed in remote sensing imagery is water. Water bodies have a different response to EMR than water bound up in molecules: they do not exhibit discrete absorption features. Water has a high transmittance for all visible wavelengths, but transmittance increases with decreasing wavelength. In deep and clear water nearly all radiation is absorbed, however suspended sediment, plankton and pigment cause increase of reflectance, whereas in the visible portion of the spectrum. In the near infrared, almost all energy is absorbed by water as is also the case in the SWIR.

3.4.5 Spectral re-sampling

The spectral resolution (i.e., the width of each channel in terms of wavelength covered for a spectrometer) and spectral sampling interval (i.e., the wavelength distance between subsequent channels) determine the accuracy of the spectral reflectance curves. Spectra shown so far in this chapter were measured on a laboratory spectrometer with 826 channels of less than 1 nm nominal in width over the 0.4 to 2.5 μm wavelength region. Most imaging devices acquire image data at coarser resolution. To be able to compare the expected spectral signatures in such data with the observed spectral reflectance curves from spectral libraries, re-sampling of the library spectra is necessary. Spectra re-sampling is the process of creating lower resolution spectral data sets (i.e., spectra) from higher resolution input spectra. This

Chapter 3: Physical principles of optical R.S.

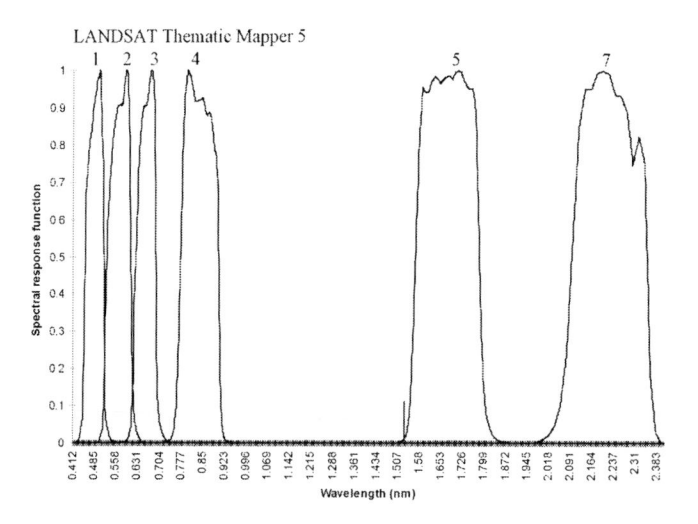

Figure 3.9: Spectral response functions for the LANDSAT 5 spectral bands.

procedure involves estimating the Gaussian spectral response function (SRF, the spectral response functions for the TM channels, shown in Fig. 3.9) of each to be simulated channel using the band center and width (at full-width-half-maximum, i.e., at 50% of the height of the curve). This SRF is of the type

$$R_i(\lambda) = \frac{1}{\sigma\sqrt{2\pi}} e^{-0.5(\lambda-\mu)^2/\sigma^2},$$

where μ is the band center, σ the standard deviation (equivalent to the FWHM) of the channel and λ the wavelength relative to and $R_i(\lambda)$ the spectral response. Note that the integral over the SRF is 1. The channels of the higher spectral resolution spectral data are integrated by dividing them into the derived SRF assuming that the initial SRF of the input data channels are a single bright source (i.e., a continuous input spectrum). This is done as follows

$$\rho_{res}(\lambda_i) = \frac{\int_{\lambda_1}^{\lambda_2} \rho(\lambda) R_i(\lambda) d\lambda}{\int_{\lambda_1}^{\lambda_2} R_i(\lambda) d\lambda},$$

where $\rho_{res}(\lambda_i)$ is the re-sampled spectrum using the continuous spectrum $\rho_{res}(\lambda)$ and the SRF $R_i(\lambda)$.

3.5 Atmospheric attenuation

Radiation passing through the atmosphere to the sensor interacts with particles and gasses in the atmosphere causing scattering by which energy is redirected and absorption by which energy is absorbed by molecules. The combined effect of scattering and absorption is called atmospheric attenuation. The main molecules involved in absorption in the visible and near-infrared portion of the wavelength spectrum are H_2O (causing absorption lines at 0.9, 1.1, 1.4, 1.9 and 2.7μm), O_2 (causing absorption lines at 0.8μm) and CO_2 (causing absorption lines at 2.7μm). Rayleigh scattering is caused by individual molecules and is defined on basis of the scattering cross-section, σ, which has the unit of an area and defined such that if a flux density F is incident on a single molecule the scattered power is F. in case of Rayleigh scattering is defined as

$$\sigma = 128\pi^5 a^6/3\lambda^4,$$

where a is the radius of the particle. Thus scattering is proportional to λ^{-4}. Blue light, therefore, is more strongly scattered than red light. In case of particle scattering this is defined as

$$\sigma_A = \frac{const}{\lambda^n},$$

where n ranges typically from 0.5 to 1.5. Rayleigh scattering occurs when particles are small compared to the wavelength, the opposite scattering type is referred to as non-selective scattering and occurs when particles are much larger than the wavelength of the radiation. Intermediate cases are often referred to with the term Mie scattering where σ is proportional to λ^{-1}. Besides molecular particles, also non-molecular particles cause atmospheric scattering. Typical examples are aerosols (e.g. small particles suspended in a gas, i.e. haze) with a particles size in the range of 0.01-1.0 μm, fog particles with sizes in the range of 1-10 μm, and cloud and rain particles with sizes in the range of 1-10 μm and 100-10000 μm respectively.

3.6 Calibration of LANDSAT TM to at-sensor spectral radiance

Given the effects of atmospheric attenuation and sensor characteristics, remotely sensed data need to be properly calibrated to derive radiance or reflectance at sensor or at the surface. Absolute radiance or reflectance calibration can be achieved through the use of radiative transfer models that model an ideal atmosphere. Relative measurements can be derived by applying calibration coefficients provided by the instrument manufacturers. These coefficients allow to translate the given raw digital numbers to at-sensor radiance and reflectance and are usually the result of an experiment in an integrating sphere, i.e. a pure reflecting environment in which sets of lambs allow to create radiation with known radiance levels which are in turn measured by the sensor. The relation between the radiance levels and the sensor registered digital values provide empirical coefficient for the correction. Conversion from the raw digital values of the LANDSAT Thematic Mapper to at-satellite spectral radiances is accomplished with the following equation

$$L_\lambda = Gain \times DN + Bias.$$

Table 3.1: TM post-Calibration dynamic ranges for Landsat TM data with L_{min} and L_{max}, in mW per cm² per steradian per μm.

	before 1/8/1983		before 1/1/1984		15/1/1984 - 1/10/1991	
band	L_{min}	L_{max}	L_{min}	L_{max}	L_{min}	L_{max}
TM1	-0.152	15.842	0.00	14.286	-0.15	15.21
TM2	-0.284	30.817	0.00	29.125	-0.28	29.68
TM3	-0.117	23.463	0.00	22.50	-0.12	20.43
TM4	-0.151	22.432	0.00	21.429	-0.15	20.62
TM5	-0.037	3.242	0.00	3.00	-0.037	2.719
TM6	0.20	1.564	0.484	1.240	0.1238	1.560
TM7	-0.015	1.700	0.00	1.593	-0.015	1.438

Table 3.2: Bandwidths for Landsat 4 and 5 (mum).

	Band1	Band2	Band3	Band4	Band5	Band6	Band7
Landsat 4	0.066	0.081	0.069	0.129	0.216	1.000	0.250
Landsat 5	0.066	0.082	0.067	0.128	0.217	1.000	0.252

For these same scenes the values of *Gain* (i.e., slope of the radiometric response function) and *Bias* (i.e., offset of the radiometric response function) are derived from the lower (L_{min}) and upper (L_{max}) limit (in units of $mW \cdot cm^2 \cdot sr \cdot \mu m^{-1}$) of the post-calibration dynamic range provided in Table 3.1 for a specific band and using the following equations (after [251]):

$$Gain = \frac{L_{\max} - L_{\min}}{255} \quad \text{and} \quad Bias = L_{\min}.$$

For TM scenes older than October 1, 1991 the values of *Gain* and *Bias* for a specific band can be determined by using the lower and upper limit of the post-calibration dynamic range and the following equation

$$Gain = \frac{L_{\max}}{254} - \frac{L_{\min}}{255} \quad \text{and} \quad Bias = L_{\min}.$$

To obtain radiance in units of milliwatts per centimeter squared per steradian per micrometer, the computed radiance value has to be divided by the detector bandwidth (Table 3.2). Subsequently, spectral radiance can be corrected for solar irradiance by converting the radiance values to at-satellite reflectance or planetary TM albedo, ρ_p, by

$$\rho_p = \frac{\pi L_\lambda d^2}{E_{sun}(\lambda) \cos \Theta_s},$$

where d is the earth-sun distance in astronomical units and $E_{sun}(\lambda)$ is the mean exo-atmospheric irradiance (Table 3.3; [252]) and $\cos \Theta_s$ the cosine of the solar zenith angle (in degrees). The earth-sun distance can be approximated by

$$d = 1 + 0.0167 \sin[2\pi(D - 93.5)/365],$$

where D is the day number of the year (see [124]). In Fig. 3.10 an example of a raw DN spectrum, a radiance spectrum and a reflectance spectrum for three pixels on relatively homogenous and pure areas of vegetation, town and water are displayed.

Table 3.3: TM Solar Exo-atmospheric Spectral Irradiances (in mW per cm^2 per μm; after [252]).

Band	Landsat 4	Landsat 5
TM1	195.8	195.7
TM2	182.8	182.9
TM3	155.9	155.7
TM4	104.5	104.7
TM5	21.91	21.93
TM6	—	—
TM7	7.457	7.452

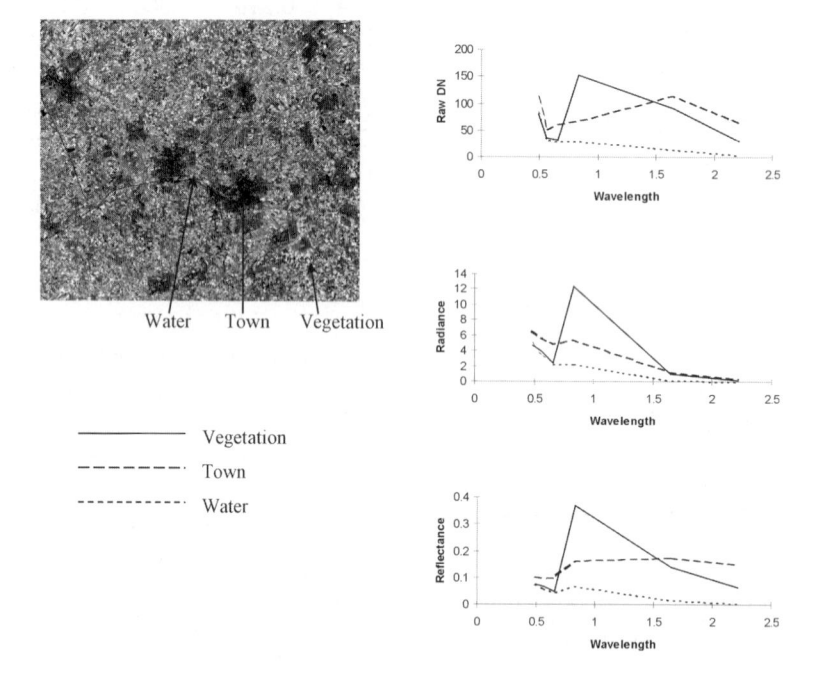

Figure 3.10: Three pixels over an area of relatively pure and homogenous cover for the classes vegetation, town and water and their respective raw DN spectra, radiance spectra and reflectance spectra.

Chapter 4

Remote sensing and geographical information systems

Sytze de Bruin and Martien Molenaar

Spatial information derived from remotely sensed imagery is often further analyzed within a geographical information system (GIS). A GIS can be defined as a computer-based system that allows input, management, retrieval, analysis, manipulation and presentation of geo-information [400]. Imaging remote sensing provides spatial data on electromagnetic radiation emitted or reflected by the Earth surface. Through analysis, including image processing and information extraction, these data may be converted into information relevant to a GIS. There are at least three reasons why GIS and remote sensing can benefit from their mutual integration:

1. Analysis of remotely sensed imagery may benefit from GIS-stored data. Examples of this can be found in other chapters of this book, and elsewhere (e.g. [199], [210], [162]).
2. Remotely sensed imagery can be the basis of up-to-date geo-information.
3. Integration of the information extraction process in GIS may help to keep track of errors and uncertainties inherent in the capture and manipulation of data [146].

The aim of this chapter is to introduce several aspects concerning the integration of GIS and remote sensing that are directly related to the analyses described in other chapters of this book. Its emphasis is on geo-information acquisition, or, more precisely, the two-way data flow between GIS and image processing software modules for image processing and geo-information extraction from remotely sensed imagery.

Acquisition of geo-information is always done with a particular view or model of real world phenomena in mind. This view affects how geographical data is modeled in the computer and the way in which it can be used in further analysis. Therefore we start with a brief section on data models. Treatment of this subject is limited to the level of conceptual data modeling [272, 273] and does not involve logical data schemes nor physical implementation of these on the computer.

4.1 Data modeling

4.1.1 Geographic data models of real world phenomena

A terrain description inevitably is a generalization and an abstraction of the real terrain it represents. Until recently, two fundamentally different data models allowed for representing geographical phenomena: the exact object model and the continuous field model [58]. The exact object model views the world as being composed of exactly definable and delimitable spatial entities. Each object has an identity, occupies space and has properties. Examples are buildings, runways, farm lots, railways, etc. The continuous field model, on the other hand, views geographic space as a smooth continuum. It assumes that every point in space can be characterized in terms of a set of attribute values measured at geometric coordinates in a Euclidean space [57]. Examples are elevation and slope in an undulating landscape, concentration of algal chlorophyll in surface water, green leaf area index in an agricultural field, etc.

Several studies (e.g. [264], [176], [99]) have shown that fuzzy set theory provides a way to represent geographical phenomena that do not properly fit into either of the above models. [406] first introduced the concept of fuzzy sets to deal with classes that do not have sharply defined boundaries. Fuzzy sets are characterized by a membership function that assigns to each element a grade of membership ranging from zero to one. Therefore, membership in a fuzzy set is not a matter of yes or no but of a varying degree. Consequently, an element can partially belong to multiple fuzzy sets.

Fuzzy set theory allows geographical phenomena to be modeled as objects that are not exactly definable. Their components belong to partially overlapping sets (classes) having diffuse boundaries in attribute space. Presence of spatial correlation among these components - in fact a necessity for any kind of mapping [223] - ensures that they form spatially contiguous fuzzy objects [59]. Examples of phenomena that have been modeled using fuzzy set theory are: climatic classes [264], polluted areas [176], soils [59], soil-landscapes [99], and vegetation [131].

4.1.2 GIS data structures

The nature of digital computers imposes that computerized geographical data are always stored in a discrete form. There are two basic data structures to store geographical data in the computer: the vector structure and the raster structure. A third structure, based on object orientation is not treated here separately, because in essence it recurs to the basic structures. Moreover, development of GIS based on object oriented structures is still essentially at the theoretical stage at present [60].

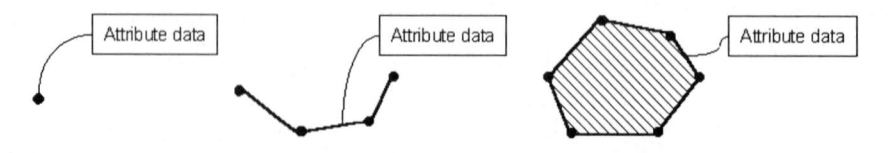

Figure 4.1: Point, line and polygon of the vector structure.

The vector structure (Fig. 4.1) uses points, lines and polygons to describe geographical phenomena. The geometry of these elementary units is explicitly and precisely defined in the database. Points are geometrically represented by an (x, y) coordinate pair, lines consist of a series of points connected by edges and polygons consist of one or more lines that together form a closed loop. The thematic attribute data of a vector unit reside in one or more related records.

The vector structure is very suitable to represent discrete geographical objects. It also lends itself to represent continuous fields and fuzzy objects (see Fig. 4.2). For example, a triangular irregular network (TIN) based on a Delauney triangulation of irregularly spaced points provides a vector data model of a continuous field [60].

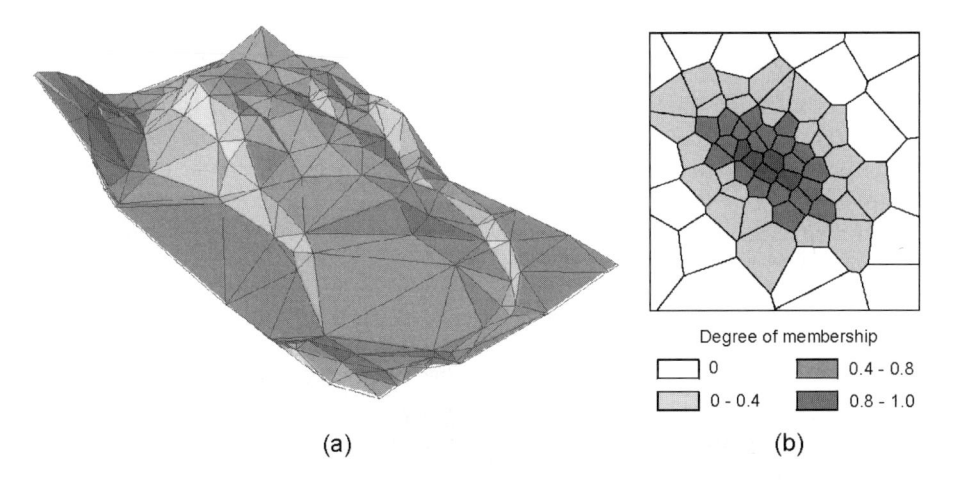

Figure 4.2: Vector representations of a continuous field (a) and a fuzzy object (b). Fig. 4.2(a) is a perspective view of a TIN-based digital elevation model. Fig. 4.2(b) shows Thiessen polygons that are shaded according to their degree of class membership.

The raster data structure comprises a matrix of n rows × m columns. Each element of the matrix holds an attribute value or a pointer to a record storing multiple attribute data of a geographic position. The raster structure has two possible interpretations (see Fig. 4.3): the point or lattice interpretation and the cell interpretation [125, 129, 273]. The former represents a surface using an array of mesh points at the intersections of regularly spaced grid lines. Each point contains an attribute value (e.g. elevation). Attribute values for locations between mesh points can be approximated by interpolation based on neighboring points (Fig. 4.3a). The cell interpretation corresponds to a rectangular tessellation of the surface. Each cell represents a rectangular area using a constant attribute value (Fig. 4.3b).

The spatial resolution of a raster refers to the step sizes in x (column) and y (row) direction. In case of a point raster these define the distances between mesh points in the terrain. In a cell raster they define the size of the sides of the cell. Given the coordinates of the raster origin, its spatial resolution and information on

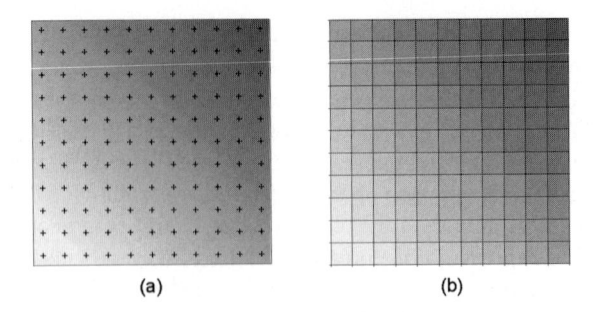

Figure 4.3: Point interpretation (a) and cell interpretation (b) of the raster structure.

projection, the geographic position of a raster element is referred to implicitly by means of the row and column indices.

Like the vector structure, the raster structure is capable of representing all three geographical data models described in section 2.1. Fig. 4.3 shows raster representations of a continuous field. Fig. 4.4 shows examples of cell rasters representing an exact object and a fuzzy object.

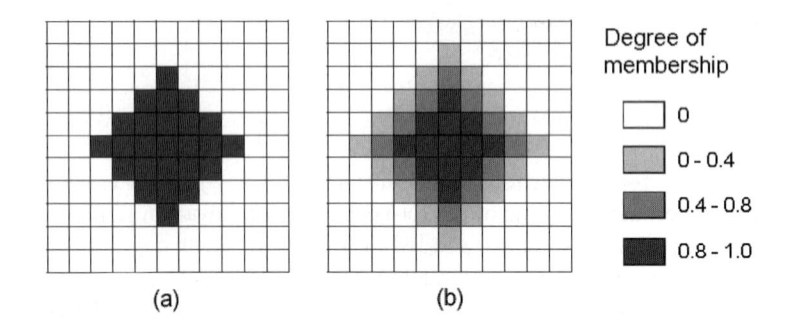

Figure 4.4: Cell raster representations of an exact object (a) and a fuzzy object (b).

The choice of using either the raster structure or the vector structure to model geographical information used to be an important conceptual and technical issue. At present, the data structures are no longer seen as mutually exclusive alternatives [60]. Table 4.1 summarizes how both structures enable representation of all three geographical data models. In addition, earlier problems regarding the quality of graphical output and data storage requirements of raster systems have largely been overcome with today's computer hardware and software. Many GIS now support both structures and allow for conversion between them [60]. Yet, if a GIS analysis involves multiple data sets these are usually required to be in the same structural form.

Table 4.1: Possible implementations of geographic data models in the vector structure and the raster structure.

Geogr. data model	Vector structure	Raster structure
Exact object	Direct	Assign object identifier to raster cells belonging to individual objects
Continuous field	Create a TIN by means of a Delauney triangulation of irregularly spaced sample points	Discretize continuous field into point raster or cell raster
Fuzzy object	Dissect space into components; assign fuzzy class memberships to components	Assign fuzzy class memberships to raster cells

4.2 Integrating GIS and remote sensing data

Digital remote sensing imagery conforms the continuous field data model implemented in a cell raster structure. The elementary unit of a remotely sensed image is the pixel (picture element). A recorded pixel value is primarily a function of the electromagnetic energy emitted or reflected by the section of Earth surface that corresponds to the sensor's instantaneous field of view (IFOV, see chapter 1). It is often assumed that the energy flux from the IFOV is equally integrated over adjacent, non-overlapping rectangular cells; the pixels' ground resolution cells. In practice, most sensors are center biased such that the energy from the center of the IFOV has most influence on the value recorded for a pixel [129]. The IFOV of a sensor can also be smaller or larger than the ground resolution cell. However, in a well designed sensor system the ground resolution cell will approximate the instantaneous field of view of the instrument [354].

The pixel is the usual vehicle for integrating data between GIS and remote sensing [129]. In the earlier used terminology, the pixel is equivalent to the raster cell. Yet, whereas remotely sensed imagery always conforms a continuous field model, geo-information may refer to phenomena according to either of the three geographical data models. Hence, the extraction of geo-information from remotely sensed imagery may involve transformation of the geographical data model. This will be elaborated on below.

Of importance is the pixel's ground resolution in comparison with spatial variability of the geographical phenomena of interest. In this context, [354] distinguished between $H-$ and $L-$resolution situations. The former corresponds to the case in which geographical objects are larger than the ground resolution cells. $L-$resolution refers to the opposite case. $L-$resolution obviously leads to a mixed pixel problem, where it is not possible to extract information about individual objects from the image. However, using a continuous field geographical model it may still be possible to extract relevant information about concentrations of substances or mixtures of object classes from the imagery.

4.2.1 Geometric transformation

Geographical information extracted from remotely sensed imagery is liable to be combined with other layers of geo-information. On the other hand, image processing and information extraction from imagery may be enhanced using layers with existing geographical data. A fundamental requirement for integrated processing of remotely sensed data and GIS data is that they be spatially registered, as in Fig. 4.5 [121].

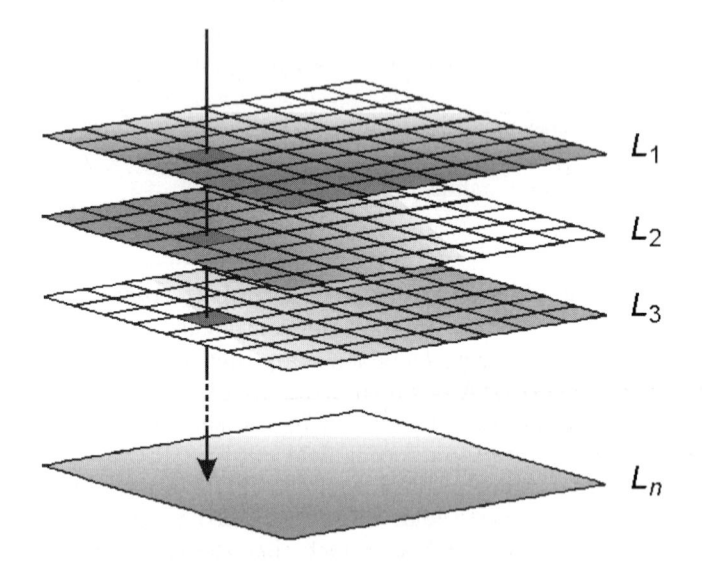

Figure 4.5: Mutually registered data layers

Registration of remotely sensed images and/or other spatial data sets is preceded by a referencing step in which a coordinate transformation function is calculated. The registration itself involves the actual geometric transformation of the input data set to the geometry of a reference data set. In case the reference data set is a geographic or geodetic coordinate system, the transformation is referred to as a rectification [121].

Two approaches exist for coordinate transformation: a deterministic approach and a statistical approach. A well-known application of the deterministic method is digital ortho-photo production by means of differential restitution of aerial photography (e.g. [29], [174]). Registration or rectification of imagery from satellite platforms is usually performed using a statistical method. The most commonly used method is to find a lower-order polynomial function, F, that maps the coordinate system of the master data set (X, Y) to the coordinates (x, y) of the slave data set:

$$(x, y) = F(X, Y)$$

[53, 121, 302]. The function, of order n, is of the general form

$$x = \sum_{i=0}^{n}\sum_{j=0}^{i} a_{ij} X^{i-j} Y^j \quad \text{and} \quad y = \sum_{i=0}^{n}\sum_{j=0}^{i} b_{ij} X^{i-j} Y^j.$$

For rectification of Landsat TM imagery it usually suffices to use a first order polynomial, so that

$$x = a_{00} + a_{10}X + a_{11}Y \quad \text{and} \quad y = b_{00} + b_{10}X + b_{11}Y.$$

The coefficients a_{ij} and b_{ij}, $i, j \in \{0, ..., n\}$ are found by means of least-squares estimation using a set of control points that have been identified in both data sets.

Registration of remotely sensed images or other raster data sets requires that the input data be resampled to determine the cell values of the registered matrix. Three techniques are commonly used for this purpose.

- Nearest neighbor assignment uses the value of the nearest cell center on the input raster. As it preserves the original cell values, it is the standard method for resampling nominal or ordinal data. This also applies if the transformed data await further processing, such as image classification. However, when performing an image classification, the training statistics are preferably derived from the untransformed image. Several pixels in the transformed image may have the value of one repeatedly sampled input pixel while other input pixels may not have been used at all.
- Bilinear interpolation assigns a distance-weighted average of the values of the four nearest input cells to the output cell. When applied to remotely sensed imagery, bilinear interpolation produces a smoother and more appealing resampled image compared with nearest neighbor assignment. It is the standard resampling technique for raster data if interpolation of input data is permitted, because it is not too computationally intensive.
- Cubic convolution determines the transferred value by fitting cubic polynomials through the 16 surrounding cells in the input raster. The result is not a weighted average of the input cells, so that the output cell value may in some cases lie outside the range of values of the input cells [125]. Cubic convolution is usually only applied to image data. It yields an output that is generally the most appealing in appearance. Similar to bilinear interpolation, it should not be used when further processing requires the original data values [121]. For a more detailed discussion of geometric transformation techniques we refer to [53], [121] and [302].

4.2.2 GIS data used in image processing and interpretation
Image processing

GIS-stored information can play an important role in the geometric transformation and radiometric correction of remotely sensed imagery. Here are some examples:

- Production of ortho imagery by means of differential restitution (e.g. [29], [174]) requires elevation data. A GIS may provide these data in the form of a digital elevation model (DEM).

- The coordinates of ground control points (e.g. road junctions) required for rectification of remotely sensed imagery can be extracted from a GIS storing infrastructural information [180].
- Slope and aspect data derived from a DEM can be used to correct for the effect of topography on the radiometric response of terrain features [72, 214, 268].

For an example of the use of GIS data in image enhancement we refer to [5]. The authors used stratification according to geological map delineations, and co-kriging with pixel values of different date images to replace clouded pixels in NOAA-AVHRR imagery.

Image classification

An obvious benefit of using GIS-stored information in image classification is in the selection of training areas. Through overlaying, existing information on, for example, land cover can be used to compile a training set for the classifier. Ideally, integration of GIS and remote sensing would allow to have instantaneous access to information that supports taking a decision on which areas to include in the final training set [180]. In this context, the error matrix (e.g., [79], [352]) and per-area image statistics and erroneous class omissions after classifying a preliminary training set could be of particular interest.

A GIS can also be the source of ancillary data to be used in an image classification. Classification of remotely sensed imagery is essentially a matter of deciding about a pixel's class label on the basis of incomplete knowledge. Incorporation of attributes derived from ancillary data sources may considerably contribute to the accuracy of the classification. This is particularly true if there is spectral overlap amongst classes, as illustrated in Fig. 4.6. Data concerning geology, topography, soils, vegetation or some other feature may improve discrimination of these classes. There are several methods to make use of ancillary data. These methods can be categorized into pre- and post-classification techniques and classifier operations, according to the ancillary data's function in the classification process [199] — see also chapters 9–11 in this volume.

Use of ancillary data preceding classification involves a stratification of the image according to some criterion or rule. The spectral characteristics of any sets of objects, such as vegetation types or land cover classes are likely to vary over distance. As variance increases, the likelihood of confusion between spectrally similar classes also increases. The purpose of stratification is to reduce the variation within strata and thus improve class distinction [199].

Post-classification techniques involve some sort of rule-based re-labeling of image classification results. Janssen et al. [209], for example, assumed that within a parcel only one type of land cover occurred. They further assumed that the majority of pixels within each parcel had been correctly classified in a per-pixel image classification. After overlaying the latter classification with a digital parcel map, all pixels within a given parcel were assigned the class label having the largest frequency in that parcel. The approach resulted in a considerably improved classification accuracy. This was mainly due to the correction of incorrectly classified mixed pixels near

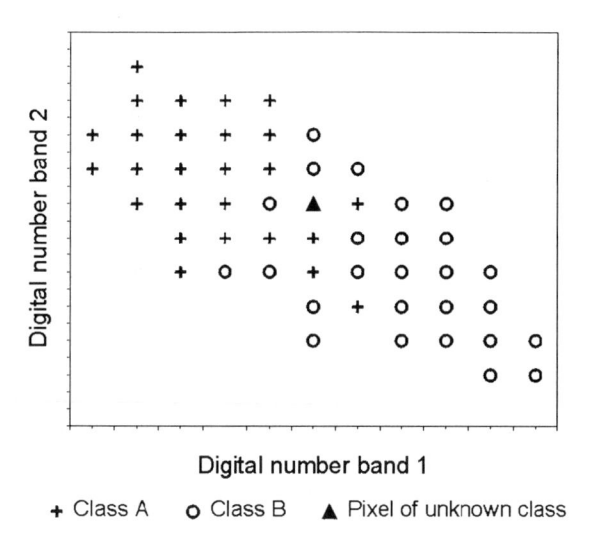

Figure 4.6: Overlap between classes A and B in a two band spectral space. Based on spectral data alone, the pixel of unknown class cannot unambiguously be assigned to either class.

parcel boundaries.

Strahler [353] showed how ancillary knowledge can be effectively incorporated into image classification through the use of modified prior probabilities in maximum likelihood (ML) classification (see Chapter 9).

In an experiment involving classification of natural vegetation, the use of prior probabilities contingent to elevation and aspect classes resulted in a considerably improved classification accuracy over that obtained with spectral data alone [353]. Janssen and Middelkoop [208] used a similar approach to incorporate historical land cover data stored in a GIS and knowledge on crop rotation schemes in the classification procedure. They also reported improved classification accuracy when prior probabilities were used. Gorte and Stein [162] developed an extension to ML classification that uses classification results to iteratively adjust spatially distributed prior probabilities. Their procedure does not require the priors to be known at the outset, but estimates them from the image to be classified.

4.2.3 Geographic information extraction from RS imagery
Exact objects

Image classification results in a field model of the terrain in the sense that posterior probabilities and class labels are treated as functions that take a value at any position in a plane [273]. This model does not match the often pursued objective of representing geographical objects, such as parcels and water bodies, in a GIS database. To achieve the latter it is necessary to subdivide an image into different segments that correspond to meaningful objects in the terrain. These objects can then be subjected to further analysis or interpretation [386]. For example, they can be assigned a class label on the basis of image classification results.

Sometimes object geometry is already available prior to image analysis. In that case, the analysis only has to provide the thematic part of object description. An example is the study in section 4.3.2 [209], in which each parcel was assigned the class label having the largest number of pixels in that parcel. Otherwise object geometry has to be derived from the image itself by a technique called image segmentation.

The prevalent image segmentation methods are either edge oriented or region oriented. Edge oriented segmentation methods attempt to detect edges between local image areas with different characteristics. They usually follow a two stage approach. First, every pixel is assigned a gradient value based on a local neighborhood operation. Next, the gradient image resulting from the first step is converted into thin binary-valued object boundaries [386]. The second step can considerably benefit from prior knowledge about object geometry. Janssen et al. [210] used edge detection to subdivide digitally available farm lots into elementary fields. They used the rule that in the study area field boundaries intersect perpendicular to the boundaries of lots or other fields. Each field thus delineated was assigned the class label corresponding to the majority class that resulted from a per-pixel classification.

Region oriented segmentation identifies areas of the image with homogeneous characteristics. A distinction is made between region growing and split and merge algorithms. Region growing algorithms grow segments from suitable initial pixels; ideally one pixel per segment. These are iteratively augmented with neighboring pixels that satisfy some homogeneity criterion. The process stops when all image pixels have been assigned to a segment [386]. Split and merge algorithms start by subdividing the image into squares of fixed size. These usually correspond to leaves at a certain level in a quadtree. Recursively, leaves are tested for homogeneity: Heterogeneous leaves are subdivided into four lower level ones; homogeneous leaves can be merged with three neighbors to form a leaf at a higher level. Recursion stops when no further subdivisions or junctions are possible [161].

Region oriented segmentation generally suffers from order-dependency. This means that the geometry of resulting segments depends on the order in which the image is processed. Gorte [161] developed a region merging algorithm that avoids order-dependency by slowly relaxing spectral homogeneity criteria. The method's intermediate result is a quadtree-based segmentation pyramid. Subsequently, relevant terrain objects are extracted on the basis of class homogeneity criteria, starting from the coarsest segmentation in the pyramid. The approach distinguishes between pure and mixed objects. The former unambiguously belong to one class; pixels within mixed objects belong to a mixture of classes [161].

An obvious approach to avoid complex image segmentation procedures is to perform a connected component labeling after classification of the image. Mutually adjacent pixels belonging to the same class are then merged to form the image segments. Of course, this method will not improve classification accuracy. Pixels located at an object boundary have spectral properties somewhere intermediate between the responses of the two neighboring objects. Such pixels may be incorrectly classified into a third class that most closely resembles their spectral response. Connected component labeling will not correct such classification errors.

Continuous fields

The extraction of exact objects from remotely sensed imagery always requires high resolution imagery in the sense that terrain objects are represented by several pixels. Otherwise, i.e., if objects are smaller than the pixel, it will be necessary to resort to a continuous field geographical model (cf. [354]). By doing so it may be possible to extract information about mixtures of object classes (Fig. 4.7a) or concentrations of substances (Fig. 4.7b) from the imagery.

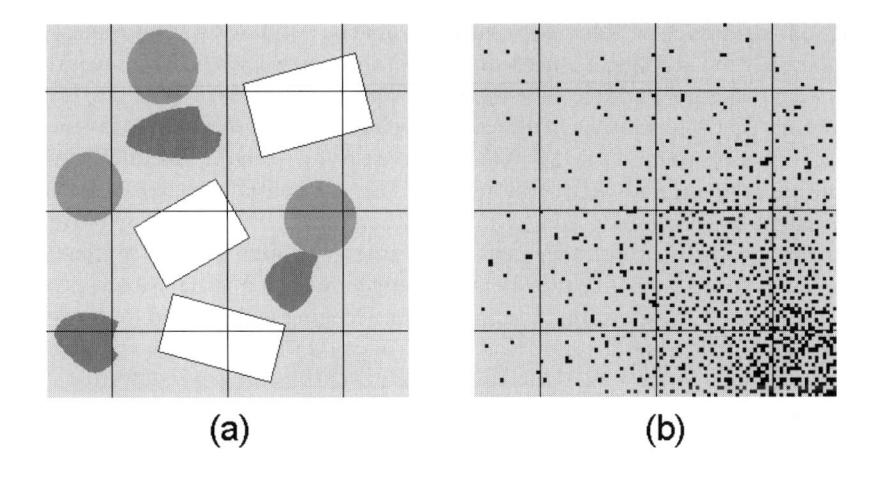

Figure 4.7: Schematic representations of sub-pixel objects.

Mixtures of discrete classes. A mixed pixel may represent a ground resolution cell that comprises more than one discrete object class (Fig. 4.7a). Consider for example the use of NOAA-AVHRR imagery (resolution 1.1 km) for agricultural surveys. Conventional classification will force the allocation of a mixed pixel to one class, which need not even be one of the component classes of the ground resolution cell [134]. Consequently, conventional classification does not result in accurate terrain descriptions in situations where mixed pixels abound. It would be more appropriate to estimate the relative area proportions of different object classes within the ground resolution cell, i.e., to unmix the pixel into its component fractions. This spectral unmixing can be accomplished in several ways (see chapter 11). Note that the success of spectral unmixing strongly depends on the separability of the component classes' spectral signatures [321, 342].

Concentrations and other continuously variable properties. As the size of objects becomes increasingly small compared to the size of the resolution cell, the applicability of the continuous field geographical model becomes increasingly evident

(Fig. 4.7b). Instead of observing individual objects it becomes more appropriate to consider concentrations, which vary continuously in space. For example, it makes no sense to study individual suspended sediment particles in coastal waters using remote sensing. However, it makes a lot of sense to use remote sensing to estimate suspended sediment concentrations (e.g., [304]).

Ritchie and Cooper [304] used least-squares estimation to derive an regression equation relating sparsely sampled suspended sediment concentrations to radiometrically and geometrically corrected Landsat MSS data. Similar approaches were used by [87], [241] and others. Once a suitable regression equation has been found it can be used to estimate the response variable from the imagery, for which area coverage is complete.

A serious drawback of using regression for spatial estimation from remote sensing imagery is that it takes no account of the autocorrelation in a scene. At each location the response variable is estimated from spectral data in a single pixel, thus ignoring any information in the neighboring sites. Therefore, regression does not make full use of the data [24]. On the contrary, co-kriging assumes that geographical phenomena and their spectral response are spatially correlated, both to themselves (auto correlated) and to one another (cross correlated). Co-kriging is most advantageous if the primary variable of interest is less densely sampled than other co-variables. This is likely to be the case where the co-variables are provided by remotely sensed imagery. The technique of co-kriging is described extensively in other chapters of this book.

Fuzzy objects

Besides the interpretation of a mixed pixel illustrated in Fig. 4.7, a mixed pixel may also represent a ground resolution cell in which two or more classes of mixed composition gradually inter-grade. This representation corresponds with the fuzzy object geographical model, which recognizes that objects may not be exactly definable. Their components belong to partially overlapping classes having diffuse boundaries in attribute space. For example, in lowland heath one usually does not observe a sudden change in land cover from dry heath to bog but instead views a gradual change along a moisture gradient from dry through wet heath to bog. Along the continuum the composition of the vegetative cover contains variable proportions of different species. The botanical composition at some locations may exactly correspond to the central concept of one class. In zones where classes inter-grade, locations exhibit the characteristics of two or more classes [131].

It may seem appropriate to represent class continua by mapping the probabilities of class membership [395]. However, probabilistic image classification assumes that classes are crisply defined, that is, each class includes members and excludes non-members. Ambiguity is due to the fact that on the basis of available data we cannot be absolutely sure whether an element belongs to a class or not. Evidence needed for a definite assignment is lacking. Fuzzy classification, on the other hand, recognizes that class boundaries can be diffuse. Membership to a class is a matter of degree so that an element can be both member and non-member, to a varying degree. Ambiguity is due to the fact that an element does not exactly match the central concept of any class. Instead, it has properties that place the element somewhere in between the class centers, where membership to the classes is partial.

A commonly applied method to derive memberships to fuzzy classes is fuzzy c−means classification [39, 40]. The method computes class memberships based on some measure of the distance or dissimilarity between an element p_k and class centers:

$$\mu_{C_i}(p_k) = 1/\sum_{i=1}^{n}\left(\frac{d_{ik}}{d_{jk}}\right)^{2/(\phi-1)},$$

where $\mu_{C_i}(p_k)$ represents the membership grade of element p_k in class C_i, d_{ik} and d_{jk} are the distances between p_k and the centers of classes C_i and C_j according to some distance norm (e.g., Euclidean, Mahalanobis), ϕ is a weighting exponent, $\phi > 1$ and c denotes the number of classes.

Fuzzy c−means classification has frequently been used in soil science (e.g., [266], [59], [99]). Foody [131] demonstrated how the method can be used to represent vegetation continua from remotely sensed imagery.

4.3 Propagation of uncertainty through GIS operations

When GIS-stored data are the input to some operation, the error in the input will propagate to the output of the operation. As a result, the output may not be sufficiently reliable for correct conclusions to be drawn from it [177]. For example, a common method to detect land cover changes is to compare independently classified images by so-called post-classification comparison [330, 211]. Obviously, every error in the input classifications will also be present in the change map produced by that method. The posterior probabilities of class membership in the input classifications, $P(C_{i,t_1}|x_{t_1})$ and $P(C_{j,t_2}|x_{t_2})$ which are an intermediate result of maximum likelihood classification, can be used to assess the per-pixel probability of class successions. As the input classifications are statistically independent, the probability of a particular class succession C_{i,t_1}, C_{j,t_2} given the spectral responses x_{t_1} and x_{t_2} is simply given by

$$P(C_{i,t_1}, C_{j,t_2}|x_{t_1}, x_{t_2}) = P(C_{i,t_1}|x_{t_1}) \cdot P(C_{j,t_2}|x_{t_2})$$

[98]. By providing the latter probabilities along with the produced change map a user is informed on the level of confidence he or she can put in the map. The example demonstrates that it is crucial that geo-information derived from remotely sensed imagery be accompanied by some measure of its inherent uncertainty. Fortunately, methods like co-kriging and maximum likelihood classification can provide measures of uncertainty as a by-product of their processes [24, 132].

Usually, error propagation through GIS operations is far more complicated than the above example suggests. GIS operations are often performed by numerical models, such as crop growth models or erosion models. By definition these models are abstractions of the physical processes they represent. They may therefore significantly contribute to the error in the result of an GIS operation [177]. Also the way in which input data and model interact affect the quality of modeling results. The theory and techniques needed to analyze error propagation through numerical modeling are beyond the scope of the present chapter. The interested reader is referred to [177] and [178] for a comprehensive treatment of error propagation in spatial modeling.

4.4 Software implementation

Increasingly, image processing systems are being equipped with GIS capabilities while GIS packages are coming with image processing functionalities. Many systems support both the vector data structure and the raster data structure. Remotely sensed imagery, such as the Landsat TM image used in this book, is then modeled as a stack of raster layers that may be co-registered with vector layers representing, for example, road networks or postcode areas. Vector layers may be easily converted to the raster structure to enable integrated analyses. Yet, to our knowledge there is no off-the-shelf system that in its standard configuration is capable of performing all operations referred to in this chapter.

Many systems now in commercial use contain high level scripting languages that allow complex data processing tasks to be accomplished with relative ease. Examples of these are the Arc Macro Language of ARC/INFO and the Spatial Modeler Language of ERDAS IMAGINE. Although these languages provide suitable tools for building special purpose applications, they may imply a duplication of programming efforts if the required functionality is available in another system. For example, it is a tedious task to program statistical functions in a scripting language knowing that these functions are readily available in a package such as SPSS or S-Plus. An additional disadvantage of interpreted scripts is that they generally execute slower than compiled software modules.

Sometimes it is preferred to use different programs in complex spatial analyses. Until recently, the only way to let these programs communicate was to transfer disk stored data sets between them [180]. This often required storage of multiple copies of the same data in different formats, easily causing data integrity problems. Recent advances in software engineering such as those supporting co-operating processes, and the move to open system architectures (open GIS), provide major enhancements for integrating multiple systems [399]. By setting standards for the database parts of GIS, software from various companies can be used to display and query data that may even reside on remote computers [60]. Object-enabling system software enables separate applications to be smoothly integrated with other applications or packaged software. Eventually, these technologies will enable specially tailored GIS solutions to be built out of custom components and software modules from a variety of vendors [381].

PART II

PART II

Chapter 5

Spatial Statistics

Peter M. Atkinson

This chapter provides a review of some of the fundamentals of geostatistics [256, 257] within a remote sensing context. It is intended as a primer for some of the later chapters of this book.

5.1 Spatial variation

5.1.1 Remote sensing

Remote sensing has evolved through recent years from being a qualitative discipline based on the manual interpretation of aerial photographs to a quantitative discipline based on the computer analysis of digital multispectral images. The number of computer tools available to the investigator has increased rapidly in the last twenty years as computer hardware and software have become increasingly widely available. The basic principles underlying remote sensing and the fundamental objectives of remote sensing analysis, however, have remained the same.

The basic principle underlying the remote sensing of electromagnetic radiation is that the spectrum of radiation reflected from some property of interest defined on some 2-dimensional (2-D) surface (usually of the Earth) is determined by the way that the light interacts with that property (see chapter 3). The objective of most remote sensing is to estimate the surface property at an unobserved location from data acquired remotely. Since the radiation sensed remotely is determined by the interaction occurring at the surface, it is usually related to the surface property and this relation may be expressed by a correlation coefficient. The larger the absolute value of the correlation coefficient (the closer to 1), the greater is the likely accuracy with which the unobserved variable can be estimated.

Broadly speaking, we need to distinguish between two scenarios. In the first, the property of interest is continuous (for example, biophysical properties such as biomass and leaf area index, LAI, or biochemical properties such as lignen and nitrogen content) and can be estimated by some transformation of the spectral data to the domain of interest (for example, using simple linear regression). In the case of linear regression, a straight line is fitted to the bivariate distribution function between a sample of the property of interest and a co-located sample of the image data. This linear model is used to estimate the surface property everywhere from the remainder of the image. In the second case, the property of interest is categorical (for example, a set of land cover classes) and can be estimated from some transformation of the original image data from feature space (the spectral domain) to class space (for example, using a spectral classifier such as the maximum likelihood classifier [255]; see Chapter 9). It should be clear that in both of these scenarios the approaches are similar, both involving modeling the relation between surface and remote variables and using this relation for prediction. These approaches may be referred to as spectral approaches since the information utilized exists only in the relations between the spectral data and the property to be estimated.

Spatial statistics have not developed as rapidly in remote sensing as the spectral approaches outlined above. This is understandable since the spectral approaches exploit the fundamental principle upon which remote sensing is based. Nevertheless, spectral approaches ignore the spatial information which exists in the relations between the pixels which comprise the image [42, 267]. Thus, spatial statistics have the potential to increase both the amount of information and the accuracy of the data provided by traditional spectral remote sensing approaches. Many spatially explicit techniques for the analysis of remotely sensed images are now emerging from the research literature.

While the focus of this chapter is on the utility of certain spatial statistics in remote sensing, it is also worth drawing attention to the utility of remotely sensed images in spatial statistics. Despite the well documented explosion of data available from sources such as the world-wide-web, high quality spatial data are often difficult to obtain. Remotely sensed images are acquired synoptically and cover large areas completely. Thus, they provide excellent sources of spatial data for the development and validation of spatial statistical techniques. In the remainder of this introduction, the focus is on the effects of sensor characteristics on the remotely sensed data which result and on the inadequacy of classical statistical approaches for dealing with remotely sensed data.

5.1.2 Spatial data and sampling

It is important to distinguish between the sampling framework on the one hand and the data which we observe on the other. If we can accept the notion of an external reality then spatial data observed through measurement can be seen as a function of the underlying spatial variation in the property being measured and the sampling framework. Since all measurements obey this rule, we can never observe independently of the sampling framework (hence the ambiguity about an external reality). In general terms, we could represent this statement with the following equation:

$$Z(x) = f(Y(x), S),$$

where, $Z(x)$ is the observed variable, $Y(x)$ is the underlying property and S represents the sampling framework.

There are several components of the sampling framework. We must consider all of them because each can affect the resulting data. We can divide the sampling framework into components related to single observations and components related to the ensemble of observations. Let us consider single observations first. A single datum obtained on a continuously varying property defined as a function of 2-D space necessarily has a support. The support is defined as the area on which an observation is defined and it correspondingly has a size (positive), geometry and orientation. For example, observations of the yield of a given agricultural crop are made over a positive area (such as 1 m by 1 m orientated north-south) and expressed in units such as g/m^2. In the example, the size of the support is $1m^2$, the geometry is square and the orientation is 0°.

When more than one measurement is made, the second set of components comes into play. These components include the sample size, the sampling scheme and the sampling density: all three are necessary to define fully the sampling framework. The sample size is the number of observations, the sampling scheme is the geometry of the ensemble of observations and the sampling density relates the geometry to a unit distance on the ground. Of these, the sampling scheme requires further elaboration (see chapter 13). Examples are the random scheme, the stratified random scheme and various systematic grids such as the square and equilateral triangular grids [390]. In general, systematic sampling is more efficient than more random counterparts and geostatistics helps to show why [54, 55, 17]. The above spatial components of the sampling framework are generally mimicked by their temporal counterparts providing an even more complex sample.

5.1.3 Remotely sensed data

Remotely sensed data and, in particular, remotely sensed imagery provide a particular case in which not all of the spatial components of the sampling framework as discussed above are relevant, and others take on increased significance. Let us consider the individual observation first. In remote sensing, the size of each observation is determined by the instantaneous-field-of-view (IFOV), expressed as an angle, and the altitude and view-angle of the sensor (see chapter 1). Some sensors are fixed at non-nadir (0°) view-angles, while others are pointable. Even for sensors fixed at nadir, however, the observations at the edges of the image will have been obtained at non-nadir angles resulting in different support characteristics (in particular different sizes) to observations at the centre of the image. For the sake of simplicity, we shall assume a nadir view-angle and that all observations can be represented by the same support, referred to as a ground resolution element (GRE).

The GRE of given size, like any support, also has a geometry and orientation. The GRE is usually represented as a square pixel on the computer screen. The size of cell represented by the GRE, however, is not necessarily the same as the size of cell implied by the pixel since GREs can overlap in space. Further, the geometry of the remotely sensed GRE is not square, but rather is approximately circular in 2-D. The GRE is further complicated by the point-spread-function (PSF), a bell-shaped weighting function in which more weight is given to spatial variation at the centre of the support than that at the edges. The orientation of the support is generally determined by the direction of travel of the sensor.

When we consider the set of all observations which comprise a remotely sensed image, we see that the sampling scheme is usually fixed as a square grid with a given number of pixels in the across-track and along-track directions. So we have no control over the sampling scheme. The sample size is also usually fixed. Indeed, the only component of the sampling framework which can be varied is the spatial resolution. The spatial resolution may be defined as the size of the smallest resolvable object, but effectively is the distance between neighboring observations for a square grid. Since the distance between neighbors is fixed by the IFOV (rather than by absolute distance on the ground) such that GREs are separated by a distance in proportion to their size, it is evident that the GRE controls the entire sampling framework in remote sensing. If the sensor is flown at a low altitude then the spatial resolution will be high (fine): at higher altitudes the spatial resolution will be low (coarse). Unfortunately, for satellite sensors where sensor altitude is fixed, the investigator has no control over the sampling framework of the data to be analyzed.

5.1.4 Spatial variation and error

Generally, single observations are made with an unknown, but non-zero error associated with them. This means that remotely sensed observations $z_v(x)$ defined on a GRE v at location x can be modeled as comprising the underlying true value $y_v(x)$ and an error $e_v(x)$:

$$z_v(x) = y_v(x) + e_v(x). \tag{5.1}$$

Generally, the error is of no interest and can be considered as being random spatial variation: variation which cannot be predicted either from knowledge of location or from knowledge of other error values in the neighborhood. In contrast, $y_v(x)$ is obtained by integrating the underlying point-to-point variation over the GRE. From the viewpoint of spatial statistics it is instructive to consider the relations which exist between the pixels in a given image. The spatial variation in an image is determined by the interaction between the underlying variation and the sampling framework (which in remote sensing terms means the spatial resolution). Thus, even though in remote sensing we may not have control over the sampling framework, it is important to know about the relation between what is at the surface and how we sample it.

Since the spatial resolution is often the only component of the sampling framework which can be varied (for example, with aircraft altitude) it is sensible to question the effect of the spatial resolution on the variation which is observed in the image. In general terms, the coarser the spatial resolution, the less variation there is between the pixels in an image and the more variation that is lost through averaging and not represented in the data [396, 225, 226, 65]. Since the utility of any specific statistical technique may depend on the nature of the spatial variation in the remotely sensed data, the choice of spatial resolution in relation to the underlying frequencies of spatial variation becomes fundamentally important. Chapter 7 is devoted to the important issue of changing the spatial resolution of remotely sensed images. In the remainder of the equations of this chapter the subscript v is dropped for brevity. The important point to grasp here is that what we observe in a remotely sensed image is not the whole story: there is always something obscured from the investigator. The amount that is obscured depends on the spatial resolution.

5.1.5 Classical statistics and geostatistics

Given a remotely sensed image defined at a particular wavelength range with a given spatial resolution one can obtain for that image a frequency distribution or histogram. The histogram, like the data themselves, will be a function of both the sampling framework and the underlying property. The same is true of statistics such as the sample standard deviation s and variance s^2 obtained from the pixels of the image (chapter 2). Yet often we see statistics such as the variance reported without mention of the sampling framework upon which they depend. Clearly, for remote sensing, where the spatial resolution defines a major part of the sampling framework, this is unacceptable.

A problem with classical statistics (in the simplest sense) is that data are assumed to be independent. For pixels arranged on a regular grid it is unlikely that complete statistical independence will be achieved. In particular, it is likely that the values in neighboring pixels will be similar: if the reflectance is large in one pixel it is likely that it will also be large in its neighbor. This so called spatial dependence, defined as the tendency for proximate observations to be more similar than more distant ones, invalidates the independence assumption of classical statistics. To see the relevance of spatial dependence, consider the action of coarsening the spatial resolution of a remotely sensed image with no error in it by a linear factor of 2 (doubling the pixel size). According to classical statistics, the variance in the new image should be given by:

$$\sigma_n^2 = \frac{\sigma^2}{n}, \qquad (5.2)$$

where σ_n^2 is the grouped variance, σ^2 is the original variance and n is the number of pixels being averaged (four in the example). In practice the decrease in variance with a coarsening of spatial resolution is observed to be much less than predicted by Equation 5.2, because the data being averaged (neighboring pixels) are not independent. Thus, as the spatial resolution coarsens, less variation is 'lost' to within the support than otherwise because neighbors are statistically related. To deal with spatial dependence we turn to a set of statistical techniques known as geostatistics. Geostatistics is only a small part of the spatial statistics available, and in focusing on geostatistics it is not intended to imply exclusivity. Rather, geostatistics fits alongside many other statistical approaches (for example [83, 28]).

5.2 Geostatistical background

The most important distinction to make in geostatistics is that between model and data. A first step in most geostatistical analyses consists of computing some function such as the variogram to describe the spatial variation in a region of interest. This function must be obtained from sample data and, thus, it is a statistical function dependent on the data from which it is derived (referred to as an experimental or sample function). By contrast, if we are to use the sample function to infer further information about the region of interest then, generally, we need to adopt some formal model of the variation. Most commonly, the Random Function model is used. The variogram (or other structural function) is then defined as a parameter of this model and may itself be comprised of several parameters. To relate the experimental variogram to the variogram defining the RF model, it is necessary to fit some continuous mathematical function to the observed values. Then, the fitted function estimates the model parameter.

The style of this section emphasizes practical aspects at the expense of theory. By far the most comprehensive reference book on geostatistics available at the present time is that by Goovaerts ([158]). The reader is referred to this text for a fuller introduction to geostatistics than the brief outline given in this section.

5.2.1 Structure functions

In this section, the variogram, autocovariance function, autocorrelation function, indicator variogram and cross-variogram are defined as experimental functions.

Variogram

For continuous variables, such as reflectance in a given waveband, the experimental semivariance is defined as half the average squared difference between values separated by a given lag h, where h is a vector in both distance and direction. Thus, the experimental variogram $\gamma(h)$ (sometimes referred to as the semivariogram, but more generally abbreviated to variogram) may be obtained from $\alpha = 1, 2, ..., P(h)$ pairs of observations $\{z(x_a), z(x_a + h)\}$ at locations $\{x_\alpha, x_\alpha + h\}$ separated by a fixed lag h:

$$\gamma(h) = \frac{1}{2P(h)} \sum_{\alpha=1}^{P(h)} [z(x_\alpha) - z(x_\alpha + h)]^2.$$

Curran [88] and Woodcock and Stahler [397, 398] provide readable introductions to the variogram in a remote sensing context.

Covariance function

The covariance will be familiar from its use in describing the relation between two variables. The covariance may be used to describe the spatial relation between pairs of data $\{z(x_a), z(x_a + h)\}$ separated by a given lag h:

$$C(h) = \frac{1}{P(h)} \sum_{\alpha=1}^{P(h)} z(x_\alpha) \cdot z(x_\alpha + h) - (m_{-h} \times m_{+h}),$$

where m_{-h} and m_{+h} are the means of the corresponding head and tail values (lag means):

$$m_{-h} = \frac{1}{P(h)} \sum_{\alpha=1}^{P(h)} z(x_\alpha)$$

Chapter 5: Spatial Statistics

$$m_{+h} = \frac{1}{P(h)} \sum_{\alpha=1}^{P(h)} z(x_\alpha + h).$$

The ordered set of covariances for a set of lags $h_1, h_2...$ is called the experimental autocovariance function or simply the experimental covariance function.

Autocorrelation function
The covariance function depends on the magnitude of the data values. A standardized version of the covariance function may be obtained through the autocorrelation function $\rho(h)$:

$$\rho(h) = \frac{C_v(h)}{\sqrt{\sigma^2_{-h} \times \sigma^2_{+h}}} \in [-1, +1],$$

where σ^2_{-h} and σ^2_{+h} are the variances of the corresponding head and tail values:

$$\sigma^2_{-h} = \frac{1}{P(h)} \sum_{\alpha=1}^{P(h)} [z(x_\alpha) - m_{-h}]^2$$

$$\sigma^2_{+h} = \frac{1}{P(h)} \sum_{\alpha=1}^{P(h)} [z(x_\alpha + h) - m_{+h}]^2.$$

The ordered set of correlation coefficients is called the experimental autocorrelation function or simply the correlogram.

Indicator variogram
The available continuous data $z(x_\alpha)$ may be transformed into an indicator variable $i_v(x_\alpha; z_k)$ defined as

$$i(x_\alpha; z_k) = \begin{cases} 1 & \text{if } z(x_\alpha) \leq z_k \\ 0 & \text{otherwise} \end{cases}$$

for a given threshold (or cut-off) z_k. Then it is possible to obtain experimental indicator functions from these indicator data. Indeed, each of the three functions considered above, can be redefined for the indicator formalism. For the sake of brevity, only the indicator variogram is considered here. The experimental indicator variogram $\gamma_I(h; z_k)$ (where, $h; z_k$ is read as lag h given the threshold z_k) may be obtained from indicator data $i(x_\alpha; z_k)$ as:

$$\gamma_I(h; z_k) = \frac{1}{2P(h)} \sum_{\alpha=1}^{P(h)} [i(x_\alpha; z_k) - i(x_\alpha + h; z_k)]^2.$$

Generally, k thresholds will be defined resulting in k indicator variograms. Often, the nine deciles of the cumulative distribution function (cdf) are chosen as the indicator thresholds resulting in nine indicator variograms. The choice of number of cut-offs, however, clearly depends on the number of available data amongst other considerations. Replacing the single variogram with nine or so indicator variograms can reveal further information on the spatial variation of interest. However, this is not the main reason for expending extra effort on the indicator variogram. Primarily, the indicator formalism is used to circumvent the need for an *a priori* (for example, Gaussian) model for the cdf.

Cross-variogram
The four functions considered to this point have been univariate functions for describing the spatial continuity and pattern in single variables. Each of these functions can be extended to the bivariate case. Here the cross-variogram $\gamma_{ij}(h)$ between variable i and variable j is considered only:

$$\gamma_{ij}(h) = \frac{1}{2P(h)} \sum_{\alpha=1}^{P(h)} [z_i(x_\alpha) - z_i(x_\alpha + h)] \cdot [z_j(x_\alpha) - z_j(x_\alpha + h)].$$

Unlike the cross covariance function or the cross correlogram, the cross variogram is symmetric in i, j and $(h, -h)$; interchanging the variables or substituting $-h$ for h makes no difference. This means that the cross variogram cannot be used to detect phase shifts between two variables.

5.2.2 The Random Function model
We should first distinguish between deterministic and probabilistic models. Deterministic models allow the exact prediction of some unknown value from other data. They are usually based on the physics which generated the data in question. The problem is that such models are rare in the environmental and Earth sciences because insufficient is known about the underlying physics. In these circumstances, it is necessary to adopt a probabilistic model which allows uncertainty in the estimates. Importantly, we adopt a probabilistic model of the process which generated our data. In this section the probabilistic model known as the RF model is described.

Random Variable
To elaborate on the discussion above, let us define a Random Variable (RV) as a generating function which produces one of n possible outcomes (termed realizations) chosen randomly from a given distribution function. It is usual to distinguish between discrete (or categorical) RVs and continuous RVs. The tossing of a coin (which will generate either a head or a tail) or the rolling of a die (which will generate one value from the set {1,2,3,4,5,6}) are examples of discrete RVs. Continuous RVs $Z(x)$ are fully characterized by the cdf which gives the probability that the RV $Z(x)$ at location x is no greater than a given threshold z:

$$F(x; z) = \Pr\{Z(x) \leq z\}.$$

Similarly, for an indicator variable defined as:

$$i(x; z) = \begin{cases} 1 \text{ if } Z(x) \leq z \\ 0 \text{ otherwise} \end{cases}$$

it is possible to define a cdf:

$$F(x; z) = \Pr\{Z(x) \leq z\} = \mathrm{E}\{I(x; z)\},$$

which represents the probability that the RV $Z(x)$ is no greater than any given threshold z.

Random Function
A Random Function (RF) is a set of RVs arranged spatially such that their interdependence may be expressed as a function of separating distance. It may help to visualize a RF as a spatial arrangement (for example, in remote sensing terms, an image) of cdfs. The $\alpha = 1, 2, ..., n$ RVs $Z(x_\alpha)$ may be fully characterized by the n-variate or n-point cdf:

$$F(x_1, ...x_n; z_1, ..., z_n) = \Pr\{Z(x_1) \leq z_1, ..., Z(x_n) \leq z_n\}. \tag{5.3}$$

Equation 5.3, defined for any choice of n and any location x_α is known as the spatial law of the RF $Z(x_\alpha)$. We do not require the entire spatial law for most applications. Rather, we usually restrict our analysis to cdfs (and their moments) involving at most two locations at a time. Thus, the one- and two-point cdfs, the covariance function, autocorrelation function and variogram describe the RFs adequately for most applications. As an example, the variogram is defined in this context as:

$$2\gamma(x, x') = \mathrm{var}\{Z(x) - Z(x')\}. \tag{5.4}$$

In particular, the functions which involve two points (of the same variable $Z(x)$) describe the spatial (or auto) correlation in $Z(x)$. Where the RVs which constitute the RF $Z(x)$ are not independent it should be possible to predict one value $Z(x_0)$ from neighboring data. A realization of a RF is termed a Regionalized Variable (ReV), and in geostatistics spatial data (for example, a remotely sensed image) are modeled as a realization of a RF.

It is now possible to define a RV $Z_v(\mathbf{x}_0)$ as a function of some underlying RF and a support v:

$$Z_v(\mathbf{x}_0) = \frac{1}{v} \int_{v(\mathbf{x}_0)} Z(\mathbf{y}) d\mathbf{y}$$

where $Z(\mathbf{y})$ is the property Z defined on a punctual (point) support.

Second-order stationarity
The variogram defined in Equation 5.4 cannot be estimated unless several repetitions of the ReV $z(x_\alpha)$ are available because the function in Equation 5.4 is location-dependent. Since it simply is not possible to obtain more than one realization for a

given space-time we would appear to be stuck. This apparent impasse is circumvented in geostatistics by considering two-point functions at fixed lags h but for varying location x. In simple terms, since we cannot obtain repeated observations at fixed x_0 we restrict our analysis to fixed h and take each pair of locations $\{x_\alpha, x_\alpha + h\}$ for which data are available as further repeats of h. For this strategy to be sensible, we must choose a RF model which is stationary: that is, for which the various two-point functions, defined as parameters of the model, are invariant with x. Then the variogram can be defined as depending only on the separation vector h:

$$\gamma(h) = \frac{1}{2} \mathrm{E}\left[\{Z(x) - Z(x+h)\}^2\right].$$

When the expected value $m = \mathrm{E}\{Z(x)\}$ exists and is independent of x (is invariant within a region D) and the covariance exists and depends only on h then the RF is said to be second-order stationary (or stationary of order two). Note that stationarity is a property of the RF model and does not relate to the data. Nor is it testable from data. Rather, it is a decision taken by the investigator to allow inference [278]. Then, the covariance function $C(h)$, autocorrelation function $\rho(h)$ and variogram $\sigma(h)$ are related by:

$$\gamma(h) = C(0) - C(h) \quad \text{and} \quad \rho(h) = 1 - \frac{\gamma(h)}{C(0)}.$$

Intrinsic stationarity
The existence of the variogram does not require the second-order stationarity described above. Where only the RF increments $[Z(x) - Z(x+h)]$ are stationary (referred to as the intrinsic hypothesis) then the variogram may exist, but the covariance function and autocorrelation function cannot.

5.2.3 The sample variogram
The experimental functions defined in section 5.2.1 represent the hypothetical population: it is assumed that all of the (perfect) data are available and functions can be computed simply from the formulae. In practice, however, for continuous variables defined over a given region D one rarely has the population. In all except for one extreme case (where the support is equal to the region), the data $z(x_\alpha)$ are a sample of the possible values (modeled as realizations of $Z(x)$). In these circumstances, it is necessary to compute a sample variogram (or other structure function), choose a RF model and estimate the correlation structure of the RF model by fitting a mathematical model to the sample variogram. This section is concerned with estimating the variogram from sample data.

Preferential sampling
In ideal circumstances the sampling would be designed by the investigator for the purpose at hand and would be such as to provide (either with the whole or a part of the sample) an even coverage of the region of interest D. Usually, the sampling has already been undertaken and the observations are preferentially located over D.

This preferential sampling can lead to bias in the estimate of the variogram. Thus, prior to computing the sample variogram it is often necessary to decluster the data. Sometimes it may be possible to select an evenly distributed sample from a preferentially located one. Where this is not possible, however, some attempt should be made to decluster the resulting histogram, mean and so on. The two most common approaches are polygonal declustering and cell declustering. In each of these the objective is to weight each datum by some function of the area that it represents: large areas receiving greater weight than small ones. Of these, polygonal declustering is the most straightforward and involves weighting each value's contribution to the histogram by the area of its Dirichlet tile. Thus, the mean is computed as:

$$m = \frac{1}{|A|} \sum_{\alpha=1}^{n} \omega_\alpha \cdot z(x_\alpha),$$

where ω_α is a weight.

Computing the sample variogram
The sample variogram may be obtained using:

$$\widehat{\gamma}(h) = \frac{1}{2P(h)} \sum_{\alpha=1}^{P(h)} [z(x_\alpha) - z(x_\alpha + h)]^2.$$

Note that $\widehat{\gamma}(h)$ now has a circumflex to indicate that it is an estimate of the population parameter.

Generally, geostatistics is used to infer information about a population from a spatial sample of that population. This spatial sample may be referred to as the 'survey'. The sample variogram is usually computed from data which are independent of the actual survey, for example, pairs of values obtained from a transect (or transects) of data. Where the variation is suspected of being anisotropic (varying with orientation) then more than one transect will be necessary: at least three and generally four. Alternatively, the variogram may be computed from the survey data themselves. In either case, it is important to ensure that sufficient data are obtained to allow accurate estimation of the variogram parameter. For transect data, a minimum of 100 data for each orientation is often recommended, with the restriction that $|h| \leq 20$. The required number, however, will vary with the circumstances in hand and how much prior information is available [391]. Where the variogram is to be estimated from the survey data, the problem becomes one of estimating with sufficient accuracy the variogram at small lags since often the survey data will be arranged on a regular grid such that relevant spatial information occurs at distances less than the sampling interval.

For remote sensing, the above considerations are usually extended to the situation where two variables are being sampled. If the remotely sensed data are provided in the form of an image then the sampling for the remote variable (say reflectance in a given waveband) is a systematic square grid with complete coverage. This sample, thus, is ideal in the terms identified above. The investigator, however, will still need to design and execute a sampling program for the surface variable of interest and so the option of obtaining a sample variogram either from transect(s) or from survey data remains.

Proportional effect

Most often the sample data exhibit heteroscedasticity: a variance which varies across the region of interest. The proportional effect occurs when the local variance is related to the local mean, either directly (most common) or inversely. Thus, a proportional effect may be detected by plotting the local mean (within a moving window) against the local variance. If the proportional effect is present together with preferential sampling then problems can arise. For example, if large-valued areas are preferentially sampled then the small lags of the variogram will have an increased semivariance (direct proportionality). In such cases, it may be possible to reduce the bias by selecting an evenly distributed sample only. Where this is not possible, however, the general (or pairwise) relative variogram may reveal structure that is otherwise hidden by the traditional variogram [202].

5.2.4 Variogram models

In this section, the various models which may be fitted to the variogram are described, anisotropic models are mentioned briefly and fitting variogram models is discussed at length.

Models and their parameters

Since the sample variogram provides only discrete estimates of the semivariance γ at a set of discrete lags h_1, h_2,... it is necessary to fit a continuous mathematical model to this discrete function to make it useful in further geostatistical analyses. That is, it is necessary to obtain a complete characterization of the RF model to enable inference about the population.

Not any mathematical function will do. The model fitted to $\{Z(x), x \in A\}$ (defined as a stationary RF with covariance function $C(h)$) must be chosen such as to ensure that all finite linear combinations Y of RVs $Z(x_\alpha), x_\alpha \in A$ must have variances which are non-negative:

$$\text{Var}\{Y\} = \text{Var}\left\{\sum_{\alpha=1}^{n} \lambda_\alpha Z(x_\alpha)\right\} = \sum_{\alpha=1}^{n} \lambda_\alpha \lambda_\beta C(x_\alpha - x_\beta) \tag{5.5}$$

for any $x_\alpha \in A$ and any λ_α. This is ensured when the covariance function $C(h)$ of the RF is positive definite. A similar set of equations may be derived for the variogram (from Equation 5.5) to show that the variogram $\gamma(h)$ of the RF must be conditional negative semi definite (CNSD).

Variogram models with the CNSD property are called permissible or 'authorized'. Only a restricted set of models have this property, and in some cases this property is restricted further to a number of dimensions less than three. It is a time-consuming task to ensure that this criterion is met for all possible linear combinations of the chosen RF. For this reason, practitioners usually choose from the set of known authorized functions. The most common functions in general use are the nugget effect, spherical, exponential, Gaussian and power (including the linear as a special case) models.

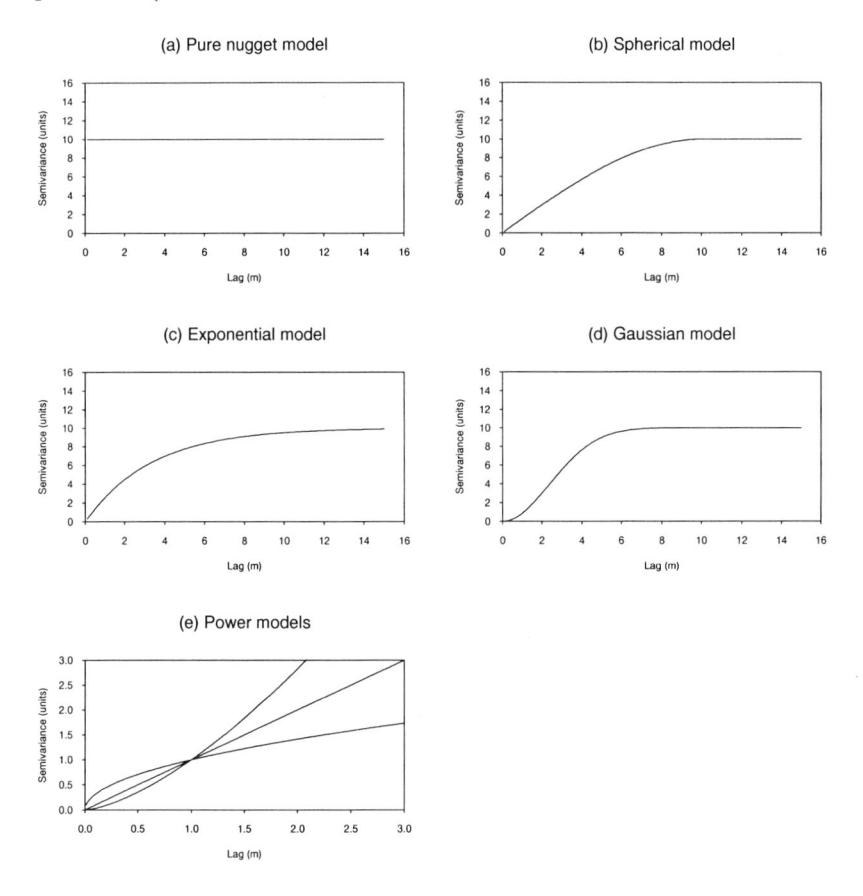

Figure 5.1: The five variogram models described in the text.

The five models are plotted in Fig. 5.1a-e. Each can be described according to its characteristic shape. First, the first four models are bounded (transitive) because they include a maximum value of semivariance known as the sill. For the spherical model the sill is reached within a finite lag known as the range a. For the exponential and Gaussian models the sill is approached asymptotically. Since the exponential and Gaussian models never reach the sill, it is necessary to define a practical range $a' = 3 \cdot r$ in place of the definite range. The nugget effect model is a 'flat' model representing a constant positive semivariance at all lags: thus, it too has a sill. Further, whereas the other models intercept the ordinate at the origin, the nugget effect model represents a discontinuity at the origin. The fifth model, the power model, is unbounded as it does not have a sill, rather increasing indefinitely with lag. The linear model (which may be familiar) is a special case of the power model with power 1.

Anisotropic models

The above five models were all written with $h = |h|$, which means that variation is the same in all directions: the variation is isotropic. Often, it is necessary to select an anisotropic model for the RF because variation is not the same in all directions. Generally, there are two types of anisotropy: geometric anisotropy and zonal anisotropy. As the name suggests, geometric anisotropy means that the spatial dimension of the variation (the range of the variogram) varies with orientation. Zonal anisotropy means that the magnitude of the variation (the sill of the variogram) varies with orientation. In both cases, the anisotropy is modeled as an ellipse (of ranges or sills) with θ representing the angle of maximum continuity (for example, maximum range) and ϕ representing the angle of minimum continuity (both measured clockwise from north by convention). The model is basically the usual isotropic model, with a range (or sill) equal to the anisotropic range a_ϕ (or sill) at angle ϕ, rescaled into an ellipse with the desired properties.

Linear model of regionalization

Most often, it will be necessary to combine (linearly) two or more basic variogram models (or structures) to provide an adequate fit to the sample variogram. This will result in a permissible variogram model if the $l = 1, .., L$ basic models $g_l(h)$ are themselves permissible and the weighted contribution b^l of each component to the sum is positive (that is, negative semivariances are not allowed):

$$\gamma(h) = \frac{1}{2}\text{E}\left[\{Z(x) - Z(x+h)\}^2\right] = \sum_{l=1}^{L} b^l g_l(h) \quad \text{with } b^l \geq 0.$$

For example, to model a discontinuity at the origin (which may be due to measurement errors or micro-scale variation that exists at a higher frequency than that of the sampling) a nugget effect is commonly combined with a structured component (such as a spherical component).

The linear model of a co-regionalization is more complex and the reader is referred to [23] and [24] for descriptions in a remote sensing context and to [158] for a straightforward complete reference. The basic rules for setting up a linear model of the co-regionalization between two RFs are (i) that every basic structure appearing on a cross variogram $\gamma_{ij}(h)$ must also appear on both auto variogram models $\gamma_{ii}(h)$ and $\gamma_{jj}(h)$ and (ii) that the reverse is not true (that is, basic structures can appear on one or more auto variograms without appearing on the cross variogram $\gamma_{ij}(h)$). In addition, the co-regionalization matrix B_l (the matrix of sill variances b'_{ll}) as defined below for the two variable case,

$$B_{l=2} = \begin{pmatrix} b_{11} & b_{12} \\ b_{21} & b_{22} \end{pmatrix}$$

must be positive definite. This is ensured when the following three inequalities hold true:

$$b_{11}^l \geq 0$$

$$b_{22}^l \geq 0$$

$$b_{11}^l b_{22}^l - b_{12}^l b_{21}^l \geq 0 \Rightarrow b_{12}^l \leq \sqrt{b_{11}^l b_{22}^l}$$

Where the cross variogram is modeled with exactly the same basic structures (a structure is used here to mean a model type with a given non-linear parameter or range) then the model is termed intrinsic.

Fitting variogram models

There are many different ways to fit models to sample variograms, including completely automatic fitting and fitting by eye. Almost inevitably the best solution is a compromise between these two (semi-automatic fitting) where the user chooses a type of variogram model (based partly on the sample function and partly on *a priori* knowledge), decides whether an anisotropic model is required, and then obtains a fit for these choices using (weighted) least sum of squares approximation given by:

$$WSS = \sum_{k=1}^{K} \omega(h_k) \cdot [\hat{\gamma}(h_k) - \gamma(h_k)]^2.$$

Alternative weighting schemes are also available [82]. Fully automatic fitting seems attractive, but given the number of variables which must be determined, the wealth of ancillary information that may be available and the need to check for non-sensical results, this option is usually insufficient and user-intervention must be added at one stage or another. Excellent discussions of model fitting may be found in [265, 388, 158].

5.3 Examples

In this section the ideas presented in section 5.2 are elaborated by applying them to the remotely sensed (Landsat TM) image which is common to the various chapters in this book.

5.3.1 Remotely sensed imagery

The details of the Landsat TM scene used for demonstrative purposes in this section is are scribed more fully in chapter 1. Only three of the available seven wavebands were selected for study in this chapter for two reasons: (i) there is much redundancy between the multiple wavebands of optical multispectral imagery with the three main dimensions of information content being the red, near-infrared and thermal-infrared wavelengths and (ii) it becomes increasingly difficult to present results for an increasing number of wavebands. The wavebands chosen were the green waveband (0.52μm-0.60μm), the red waveband (0.63μm-0.69μm) and the near-infrared waveband (0.76μm-0.90μm). It was also decided to compute from two of these wavebands a vegetation index, the normalized difference vegetation index (NDVI) since many studies involve the use of such transformations. The NDVI may be obtained as:

$$\text{NDVI} = \frac{\text{NIR} - \text{Red}}{\text{NIR} + \text{Red}}.$$

where, NIR represents the near-infrared waveband and Red represents the red waveband. The rationale behind this choice was that the green waveband might serve as a surrogate for some ground variable of interest and the NDVI might be used to predict it.

In addition to restricting the choice of wavebands, only two small subsets of the original image were considered for further analysis. First, a 128 by 128 pixel subset entirely covered by the urban area of Enschede was selected. Second, a 128 by 128 pixel area to the south-west of the centre of the image was selected. These two areas are referred to in subsequent discussions as 'urban' and 'agricultural'. Note that the areas are referred to by land use, not land cover, implying that there may be more than one land cover type within each area.

Before exploring the two subsets, it is important to state that the analysis which follows is undertaken for illustrative purposes only and may not be appropriate in practice. This is because, as outlined in the introduction, the spatial variation in spatial data is a function of the underlying variation and the sampling strategy. The underlying phenomena in the two subsets are (i) an urban area filled with sharply defined objects such as buildings, roads, gardens, parks and so on and (ii) an agricultural area with agricultural fields of homogeneous land covers (such as cereals) with abrubt boundaries between them. Neither subset, therefore, fits well the continuous RF model. To some extent the abruptness of the boundaries is smoothed by the spatial resolution of 30 m (by 30 m). Thus, in the urban area, for example, individual buildings are rarely observable and the variation appears partly smooth and partly random which is the traditional view of a regionalized variable [216]. In the agricultural area, the individual fields are visible and this makes application of the continuous RF model rather inappropriate. It would be more usual to apply geostatistics either (i) where the spatial resolution is fine enough to restrict the analysis to the continuous variation within individual fields or (ii) where the spatial resolution is coarse enough to treat the variation as continuous.

As a first step (univariate description) in exploratory data analysis, the frequency distributions of the four variables are shown for the two areas (urban and agricultural) in Fig. 5.2a-h. It is here that we encounter our first problem. It is clear from the plots that the data are in some cases skewed positively. In particular, the distribution of the near-infrared waveband is skewed positively for both areas and the distribution of the red waveband is skewed positively for the agricultural area.

Chapter 5: Spatial Statistics

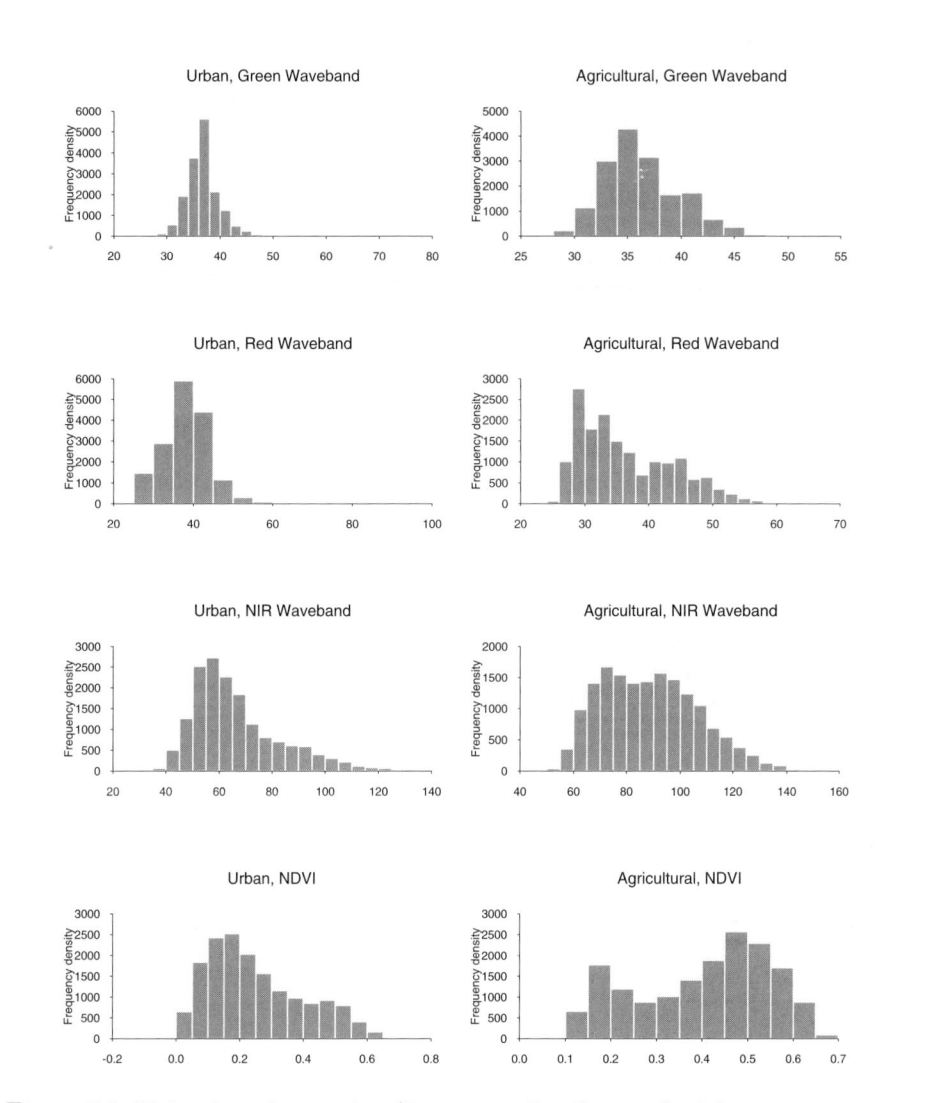

Figure 5.2: Univariate description (frequency distributions) of the various units.

Thus, we should check the data for heteroscedasticity as described in section 5.2.

Non-Gaussian distributions are a common feature of remotely sensed data, particularly where vegetation is the subject of interest since the effect on reflectance of successive layers of leaves in a canopy tends to be asymptotic such that areas with few leaves contribute to the tail of the distribution. To further complicate matters it is possible that the distributions of the red and near-infrared wavebands and the NDVI are bi-modal for the agricultural area. This is most likely to be the result of two different land covers dominating the scene.

5.3.2 Sample variograms

First, directional variograms were computed for each of the four variables and each of the two subsets. Variograms were obtained in each of four orientations (0°, 45°, 90° and 135° from north) using the gamv2m.f Fortran 77 subroutine of the GSLIB software [106]. Other software such as GSTAT and Variowin, which provide graphical user interfaces, are actually preferable for variogram (more generally, structural) analysis. The gamv2m.f subroutine requires the input of several parameters via a steering parameter file which is read by the main program. Thus, it was necessary to select (amongst other parameters) the maximum lag to which the variogram should be computed (25 pixels), the unit lag distance (1 pixel), the angular tolerance within which data will be considered as pairs of a given pixel (22.5°) and the four orientation angles given above.

Results indicate that the variation is approximately isotropic and support the decision to adopt an isotropic RF model. For this reason, it was decided to treat the variation as isotropic and the directional variograms were replaced by a single (so called omni-directional) variogram computed as the average for all orientations. Notice that this is a decision taken by the investigator and not a property of the data. To obtain the omnidirectional variograms (shown in Fig. 5.3a-h) it was necessary to re-run the gamv2m.f subroutine with a new set of steering data. Three features are immediately apparent from these plots. First, the maximum value of semivariance varies between the different variables (and most importantly, between the three wavebands). The variance in the near-infrared waveband is greater than that in the red waveband which, in turn, is greater than that in the green waveband. Second, the variograms for the agricultural area appear bounded whereas those for the urban area appear to increase indefinitely with lag. The latter phenomenon may be due to the slightly different land cover which was encountered in the south-west of Enschede resulting in a slight trend. Finally, the variograms appear fairly continuous from lag to lag and appear to approach the origin approximately linearly. This is characteristic of remotely sensed images which (i) tend to have large signal-to-noise ratios [89] and (ii) provide many data.

To characterize the salient features of the sample variograms with a continuous mathematical function we must attempt to model the regionalization. For fitting models to the variograms in Fig. 5.4a-h the S-plus software was used since this software was already used to produce the figures. Several other model fitting programs are also available including GSTAT (which allows both ordinary least squares and weighted least squares approximation using the graphical interface) and Variowin

Chapter 5: Spatial Statistics

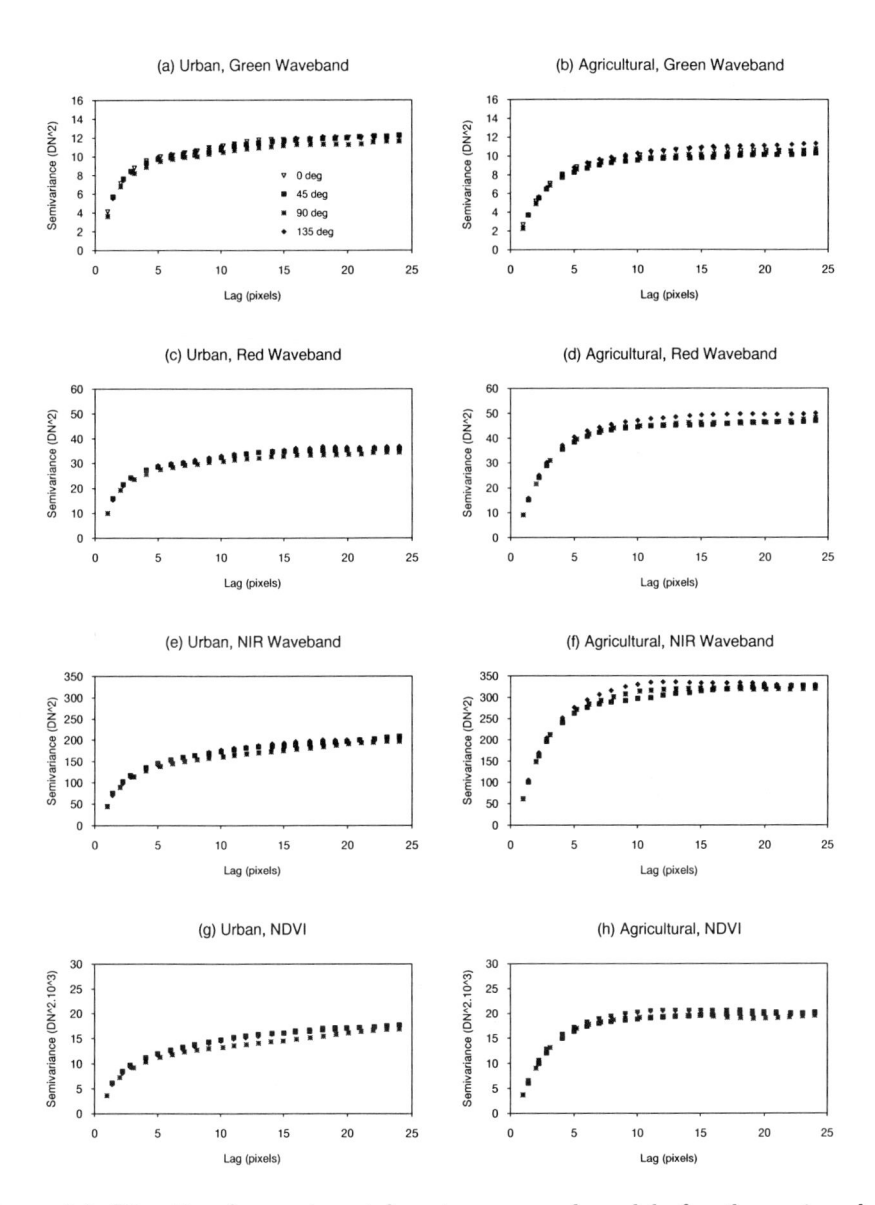

Figure 5.3: Directional experimental variograms and models for the various land cover types.

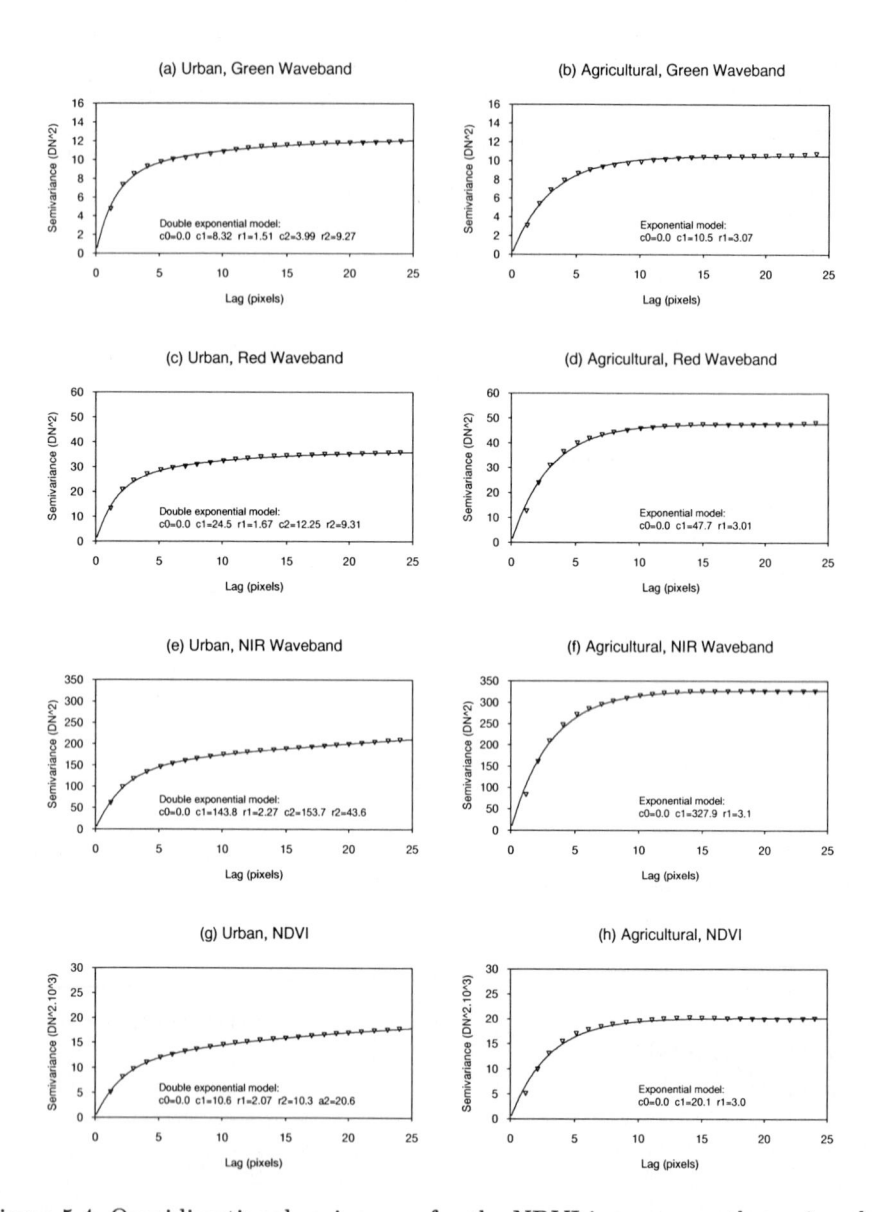

Figure 5.4: Omnidirectional variograms for the NDVI images over the various land cover types.

(which has many good graphical features, but which does not allow automatic fitting). To fit models to variograms using S-plus it was necessary to code the models in the S-plus language. These functions were then fitted to the sample variograms using the non-linear least squares fitting algorithm provided in S-plus.

Judgement was used to select a stationary isotropic RF model. Judgement was also used to narrow down the search for suitable functions for the variograms. From experience, it seemed that the exponential model was a likely candidate for the agricultural area for all variables. Thus, no other functions were fitted. It is possible that the Gaussian model may have provided a slightly better fit, and for remotely sensed images (which tend to be very continuous) this would be a reasonable choice. The Gaussian model, however, can lead to problems in kriging if used without a nugget variance and it is clear from the plots that the fitted function should pass through or very close to the origin. For the urban area, the double exponential model was selected as a likely candidate. This choice was based on experience and a feel for the likely goodness of the fit. An exponential model combined with a linear component may have provided a similar result, but without further evidence of the behavior outside of the range of the data the bounded model was considered preferable. The use of two models in combination illustrates the discussion in section 5.2 concerning modeling a regionalization as a linear combination of independent RFs. Notice that the combination amounts to adding the values of semivariance from each model at each lag: it does not mean using one model up to a finite lag and another thereafter.

Once the range of model types to be fitted was decided, these models were input to the S-plus non-linear fitting algorithm and the coefficients of the models (given on the plots in Fig. 5.5a-h) were obtained. Notice that the model was constrained to go through the origin (no nugget effect) to maintain simplicity and also to avoid negative semivariances which are not permissible. The result is that for each variable we have defined a stationary isotropic RF model with given a correlation structure as defined by the variogram model types and their estimated parameters. Since all models are bounded, the covariance function and autocorrelation function exist and the stationarity in each case can be considered second-order.

5.3.3 Indicator variograms

To illustrate the indicator formalism, indicator variograms were obtained for the NDVI variable for both subsets. To obtain a set of indicator variograms one must define a set of thresholds or cut-offs as described in section 5.2. Given that there were sufficient data, the cut-offs were selected as the nine deciles of the cdf. No declustering of the histogram was considered necessary in this case because the data were evenly spaced over the region of interest. These cut-offs along with other steering parameters were entered into the gamv2m.f subroutine to obtain the nine indicator variograms for each area as shown in Figs. 5.4a-i (urban) and 5.5a-i (agriculture).

Several features of these variograms are immediately apparent. First, one now has nine variograms in place of one so that the amount of effort involved in modeling and other analysis necessarily increases. Second, the shape of the variograms is very similar to that of the traditional variogram, but some variation between

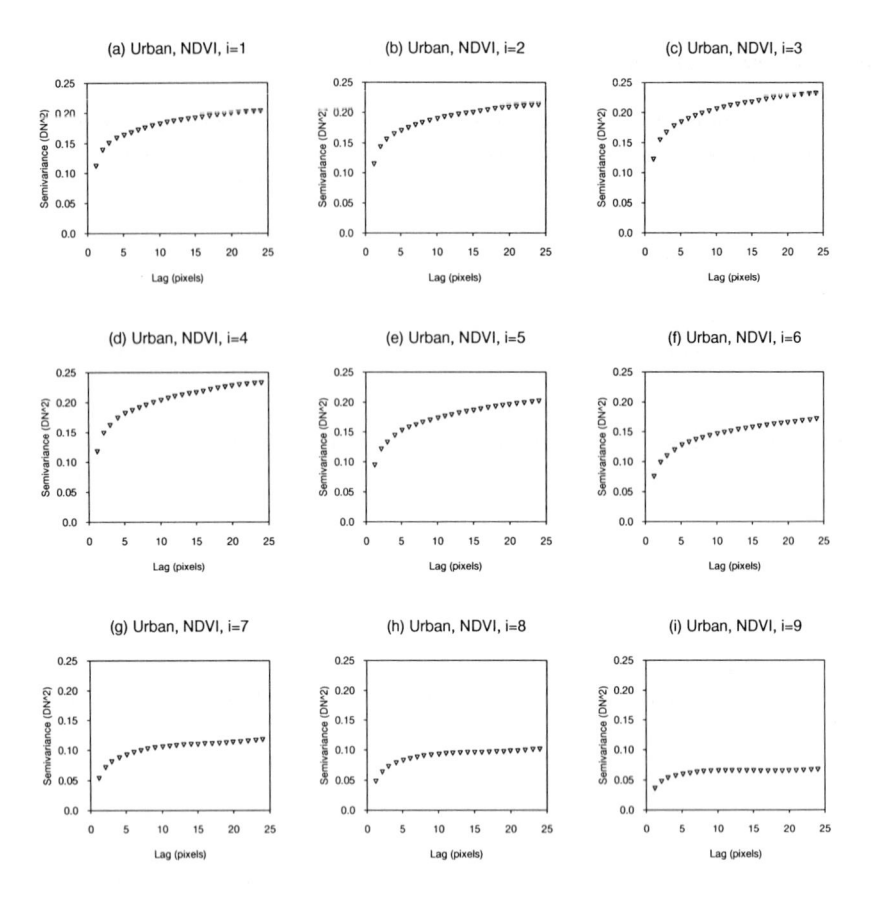

Figure 5.5: Indicator variograms for the various cutoffs for the urban units.

indicators does occur. Finally, the maximum value of semivariance for the urban area is greatest for $i = 3$ and $i = 4$ and is least for $i = 9$. For the agricultural area the semivariance is greatest for $i = 6$ and least for $i = 9$. Thus, the greatest amount of variability (not surprisingly) occurs in the middle of the distribution where the number of data is greatest.

As noted in section 5.2, the main reason for adopting the indictor approach is that it obviates the requirement for an *a priori* model of the cdf. For example, in indicator kriging the probability of the value at a given location exceeding a given threshold (cut-off) is estimated from the variogram and indictor variable for that cut-off. This is then repeated for each cut-off in question so building the entire *a posteriori* conditional cdf at that location. For remotely sensed imagery where the Gaussian model is often not a suitable choice, the indicator formalism represents a clear way forward. Further, geostatistics is increasingly about the quantification of local uncertainty as opposed to reducing it and the indicator formalism also provides neat solutions in this context through the identification of local conditional cdfs.

5.3.4 Cross-variograms

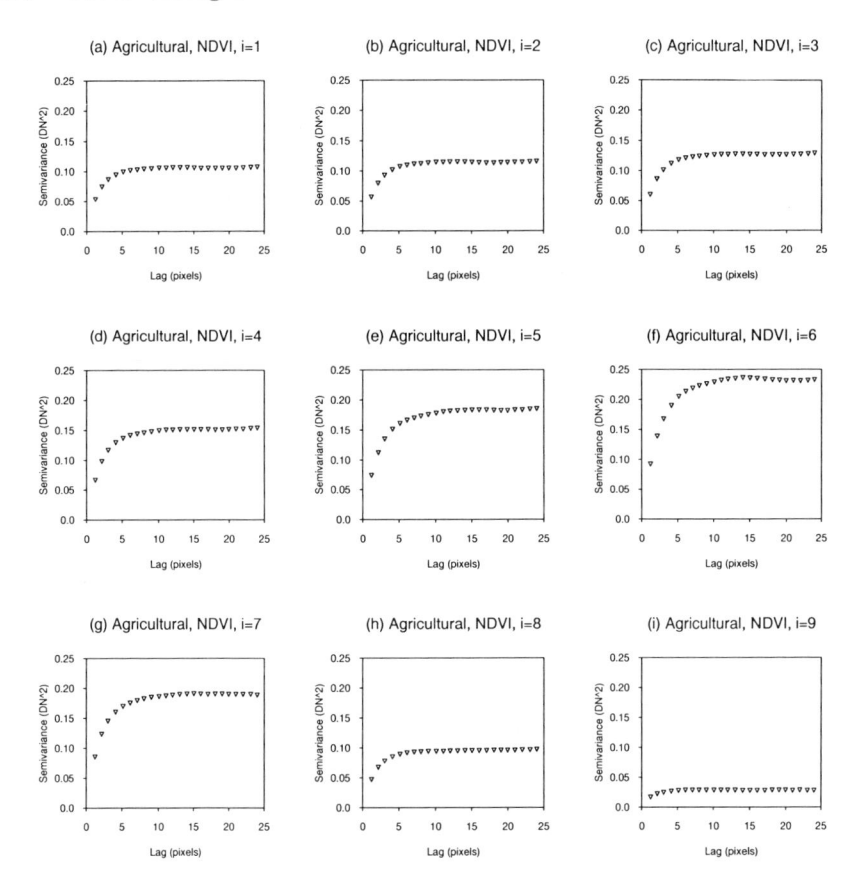

Figure 5.6: Indicator variograms for the various cutoffs for the agricultural units.

Treating the green waveband as a surrogate for some property of interest (such as LAI) at the Earth's surface, cross variograms were obtained using the gamv2m.f subroutine for the green waveband combined with each of the other variables (red, near-infrared and NDVI) for both subsets (Fig. 5.6a-f). Several features of these cross variograms are worthy of mention. First, notice that the two cross variograms for the green with red waveband interaction increase from the origin to a positive semivariance of about 20 DN^2. This means that the simple correlation coefficient between the two variables is also positive as it should be. Second, the two cross variograms for the green with near-infrared waveband interaction decrease from the origin to a negative semivariance of about 10 DN^2. This means that the simple correlation coefficient between the two variables is also negative as it should be. Notice that for cross variograms negative semivariances are allowed. Finally, the general shape of the cross variograms is similar to that of the two auto variograms.

To build a model of the co-regionalization, the first step taken was to fit models

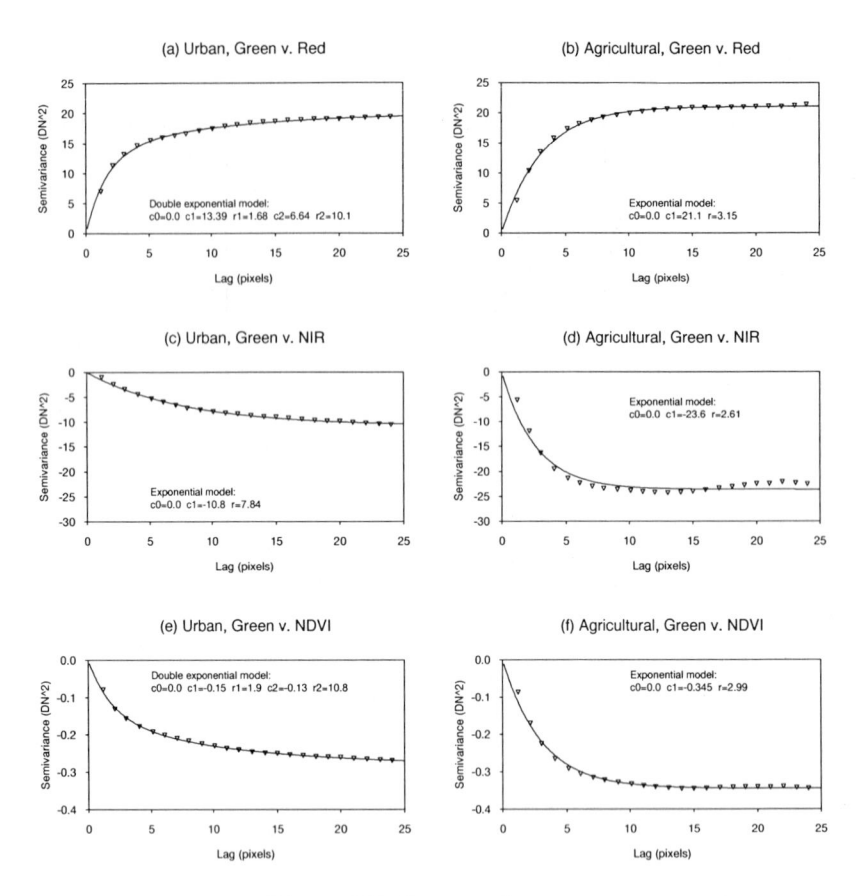

Figure 5.7: Cross variograms.

to each of the cross variograms while constraining only the model to pass through the origin (nugget effect of zero). The resulting models are shown, together with their coefficients, in Fig. 5.7a-f. For each case, the fitted models are all of the same type as for the two relevant auto variograms and the ranges of the variograms are all similar. In these circumstances, the simplest model of co-regionalization to build is that of an intrinsic co-regionalization in which the same basic structure is fitted to all three variograms (two auto and one cross) such that only the sills vary proportionally to each other. This necessitates refitting the relevant models while constraining the range to be some compromise between the ranges attained in the original unconstrained fitting. This is illustrated for the cross variogram between the green waveband and the NDVI for the agricultural case only.

The ranges of the fitted models are as follows: 3.07 pixels (green waveband), 3.0 pixels (NDVI) and 2.99 pixels (green-NDVI). It, therefore, seems reasonable to choose a range of 3 pixels as a compromise between these three. Note that certain non-linear fitting algorithms will allow one to automatically determine a single range for the three variograms simultaneously [310]. Having chosen a single range, the models can be re-fitted to the sample variograms with the constraint that the range equals 3 pixels. Since the only model involved is the exponential model the rules relating to modeling a co-regionalization are satisfied. For illustrative purposes, however, the coefficient matrix will be checked for positive definiteness as if the model was not intrinsic but rather was the more general linear model of co-regionalization. Since only one model was fitted, an exponential term, only one coefficient matrix is obtained:

$$\begin{pmatrix} \gamma_{G-G} & \gamma_{G-NDVI} \\ \gamma_{NDVI-G} & \gamma_{NDVI-NDVI} \end{pmatrix} = \begin{pmatrix} 10.5 & -0.345 \\ -0.345 & 20.1 \end{pmatrix} \cdot \exp\left(\frac{h}{3}\right),$$

where the second term on the RHS of the equation is used to mean an exponential model with $a = 3$. Here, the diagonal elements and the determinant are all positive satisfying the constraints of the linear model of co-regionalization.

co-regionalization is pertinent to remote sensing because, as outlined in the introduction, in remote sensing we are most often concerned with the relations between two or more variables. In particular, we are interested in predicting one variable from data obtained remotely which are correlated with it. Cokriging, the extension of kriging to the multivariate case, provides a viable alternative to techniques such as regression [340, 41, 155]. Cokriging may also be used to design optimal multivariable sampling strategies [263] and this has already been done in a remote sensing context [23, 24, 374]. Myers [276] gives the full matrix formulation of cokriging. Cokriging is automatic once the sample data are available and the co-regionalization has been modeled.

5.4 Acknowledgments

The author is grateful to the editors for the opportunity to write this chapter and for their patience while it materialized. The author also thanks the School of Mathematics, Cardiff University, for the study leave during which this chapter was written.

Chapter 6

Spatial prediction by linear kriging

Andreas Papritz and Alfred Stein

Kriging denotes a body of techniques to predict data in Euclidean space. By prediction[1] we mean that the target quantity is estimated at an arbitrary location, given its coordinates and some observations recorded at a set of known locations. Developed mainly by G. Matheron and co-workers in the 1960s and 1970s at the Ecole des Mines in Paris [256, 257], kriging is embedded in the framework of stochastic mean square prediction, and it is therefore related closely to earlier work by N. Wiener and A. N. Kolmogorov on the optimal linear prediction of times series.

Typically, kriging is used to answer questions of the kind *'where and how much?'*. A surveyor, as an example, might have to delineate the part of a polluted region where the concentration of some soil pollutant exceeds a threshold value, or a remote sensing specialist might need to know what kind of land use is likely for a patch of land that was obscured by clouds when the reflected radiation of the ground was recorded by some airborne sensor. In some applications, however, the part *'where?'* of the question is of lesser importance, and the quantity of interest might be the mean of an attribute over the whole study domain or the means over some sub-domains. Although kriging can deal with such tasks as well, the methods of classical sampling theory are generally better suited to tackle the so-called *global estimation* problems, which focus on the question *'how much?'*. In this chapter, we shall restrict ourselves to the problem of *local estimation*. For global estimation the interested reader should consult any of the numerous textbooks about sampling theory [69, 361, 362].

Although not immediately obvious because remotely sensed imagery normally provides full coverage data, geostatistical interpolation techniques have proven useful to tackle various problems in remote sensing. First, kriging was used to restore remotely sensed images that contained missing or faulty data [22]. A particular example is the aforementioned estimation of some characteristic of the ground surface that was hidden by clouds during the scan [311, 5]. Second, multi-variable linear

---

1. In accordance with [83] we denote both smoothing and non-smoothing, i.e., interpolating methods as spatial prediction techniques.

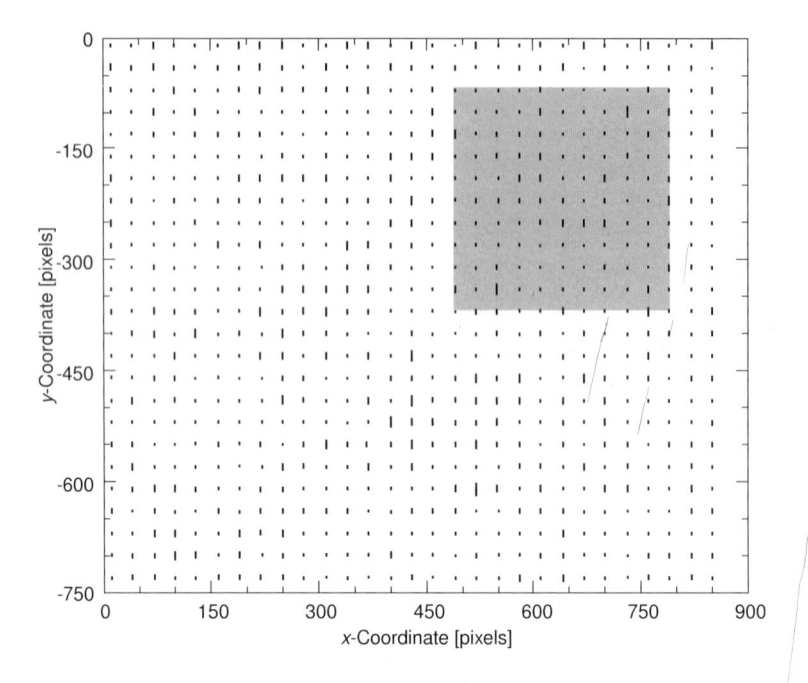

Figure 6.1: Positions of sampling locations and scaled observations of variable *BAND1*, Landsat TM scene, Twente region. The length of the symbols are proportional to $(BAND1 - \min(BAND1))/(\max(BAND1) - \min(BAND1))$. The shaded square marks the target area for which predictions are required.

prediction—termed *co-kriging* in geostatistics—was used with success to map selected characteristics of the earth's surface from sparse ground observations and abundant radiance data sensed remotely [23, 24]. Provided there is a strong relationship between the reflected radiation and the target variable then co-kriging usually improves the prediction of the ground truth compared to alternatives such a single-variable kriging of the ground data or linear regression of the target variable onto the radiance data [114, 374]. The latter author [370, 371] further showed that laboratory derived calibration curves which relate the target variable to the radiance data can sometimes substitute for the ground observations. He predicted both categorical (occurrence of two types of clay minerals [370]) and continuous target variables (degree of dolomitization [371]) from high spectral resolution images without using any ground data. A third application of geostatistical prediction is the filtering of images, including noise removal, by the so-called factorial kriging method [393, 286].

This list—far from being complete—shows that there is some scope for using spatial prediction methods in remote sensing. Therefore, this chapter presents an introductory review of the linear kriging methods. As a motivating example, which we shall analyze in detail at the end of this chapter, we consider the second problem mentioned in the previous paragraph. We assume that we have measured some

Chapter 6: Spatial prediction by linear kriging

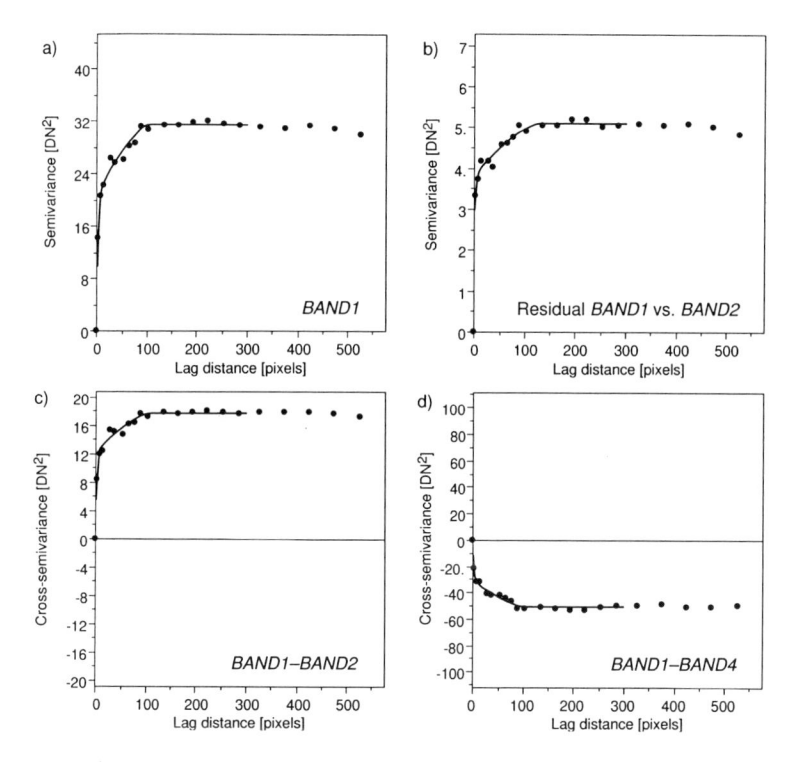

Figure 6.2: Sample (cross-)variograms and fitted model functions of data sampled from Landsat TM images, Twente region. a) Variogram *BAND1*, b) variogram of residuals, $BAND1 - 31.4 - 1.46\,BAND2$, c) cross-variogram *BAND1–BAND2*, d) cross-variogram *BAND1–BAND4*. Double-spherical models with nugget components were fitted to the sample variograms by least squares. For further details see text.

variable on the ground at a set of locations, the so-called support points, in the domain covered by the Landsat TM scenes of the Twente region (cf. chapter 1). As surrogate data, we use the digital numbers (DN) of the radiance recorded in the first waveband of the Landsat thematic mapper, but any other waveband could be used as well. In the sequel we denote this variable by *BAND1*. Fig. 6.1 shows the positions of the sampling locations along with the scaled *BAND1* values recorded at those locations. Let us assume that our task consists of mapping the ground measurements on a square grid with lower left corner in (490,-360), upper right corner in (790,-60) and unit grid spacing in both directions (all the distances and coordinates refer to the rows and columns of the original image). In Fig. 6.1 the target area is marked by the shaded square.

On examining the recorded values of *BAND1* in Fig. 6.1, we note that the data are spatially dependent: neighboring values are on average more similar to each other than observations spaced farther apart. But the data do not suggest that the unknown surface of the target quantity is continuous, let alone differentiable. There are single small or large observations (e.g. at point 730,-100), that differ quite strongly from the neighboring values. Indeed, the experimental variogram of *BAND1* (cf. Fig. 6.2a), although confirming spatial dependence up to distances of ≈ 100 pixels, shows a distinct discontinuity at the origin (nugget effect) which suggests that the surface of *BAND1* is not continuous.

Spatial data, recorded with a given minimal spacing of the observations and a given support, quite often have the character of being smooth but behaving erratically at the same time. Although the irregular part of the variation might show a well defined structure when studied with the same support at higher spatial resolution, it nevertheless cannot be predicted at the lower resolution and therefore appears irregular and chaotic to an observer.

From this we can derive two criteria that any prediction method should respect. First, we require the predicted surface to be smooth because there seems to be no use trying to reconstruct the irregular part of the variation. Second, the prediction method should be such that the support data are used locally: points that are close to a prediction location should have a larger influence on the prediction than points farther apart. If the spatial correlation decays with increasing distance then the nearest observations carry the most information about the unknown value and should be weighted accordingly.

The methods that are used in practice to predict spatial data (see [83]: sections 3.4.2 and 5.9 for an overview) do not necessarily satisfy the two requirements. Trend surface analysis, as an example, does not weight the data locally. Single observations may exert, in unexpected ways, a strong influence on remote parts of the surface. There are several ways in which the prediction surface can lack smoothness. First, the surface may or may not honour the observed data. In the former case the technique is called an *exact interpolator* (e.g. inverse-distance-squared weighted average, triangulation methods) and in the latter a *smoother* (e.g. moving average, median-polish plating). Second, the surface may not be differentiable or continuous everywhere. Delauney triangulation with linear interpolation within the triangles, as an example, results in a surface that is continuous but not differentiable at the edges of the triangles.

In section 6.2.2 we show that kriging allows a great flexibility to control the characteristics of the prediction surface. Depending on the existence and the interpretation of a nugget component in the variogram, kriging may act as a smoother or an exact interpolator. The variogram further determines whether the prediction surface is continuous or even differentiable. In addition, kriging weights the information locally, and the influence of a support point on the target value is controlled by the spatial correlation. Compared with non-stochastic methods of spatial prediction, kriging has the additional advantage that it provides the expected mean square error of prediction, i.e., a measure of its precision. In some applications this is as important as predicting the value itself.

This chapter introduces the linear model for the decomposition of the spatial variation into deterministic large-scale and stochastic small-scale variation. It addresses single-variable prediction and introduces optimal linear prediction with known mean and covariance function, discussing geometric properties of the kriging surface. It shows how to predict optimally when the mean function has to be estimated from data. Co-kriging is introduced, one of the methods to incorporate auxiliary information into the predictions. Finally, it compares the various approaches by predicting the variable *BAND1*. There are two notable omissions in the chapter. First, we do not give a detailed treatment of the so-called change of support problem, i.e. the prediction of mean values over larger areas, given observations at points. Solving this problem is straightforward as long as linear predictors are used, and the interested reader can find the material in textbooks or in numerous research papers [389, 23] (see also Chapter 7 of this volume). Second, we cannot present an introduction into non-linear spatial prediction because this would require a full chapter of its own. Papritz and Moyeed [293] give a recent review about the topic.

6.1 Linear Model

A geostatistical data set consists of pairs of observations and coordinates of sampling locations, $\{y_i = y(s_i), s_i\}$; $i = 1, 2, \ldots, n$; $s_i \in G \subset \mathbb{R}^d$, $d \leq 3$. The symbol G denotes the study area, and n is the sample size. The data are considered as a sample drawn from a single realization of a random process, $\{Y(s)\}$, with continuous index set $s \in \mathbb{R}^d$.

It is customary to decompose $\{Y(s)\}$ into a deterministic component, $\mu(s)$, the so-called *drift* or *trend*, which models the large-scale (low frequency) variation, and a stochastic component, $\{e(s)\}$, modeling the smooth small-scale fluctuations and the irregular part (high frequency components) of the variation. To make the method operational, we need assumptions about the form of $\mu(s)$ and about the covariance structure of the 'error' term $\{e(s)\}$.

Normally, there is too little understanding of the problem at hand to find a 'natural' model for $\mu(s)$. As an obvious resort, the drift is often modeled as a weighted sum of *known* functions of the coordinates, $f_p(s)$, thus leading to the well-known linear model

$$Y(s) = \mu(s) + e(s) = \sum_{p=1}^{q} \beta_p f_p(s) + e(s) = x'\beta + e(s) . \tag{6.1}$$

Here, $x = [f_1(s), f_2(s), \ldots, f_q(s)]'$ denotes a q-vector with the explanatory variables, β is the q-vector with the unknown weights, and $'$ is the transpose. Low order polynomials in the coordinates are a common choice for the explanatory variables $f_p(s)$.

To model the covariance structure of $\{e(s)\}$ some form of stationarity is required. For the time being, let us assume that the 'error' term is weakly (or second order) stationary (chapter 5), i.e.,

$$E\langle e(s) \rangle = 0, \tag{6.2}$$
$$\mathrm{Cov}\langle e(s_i), e(s_j) \rangle = C(s_i - s_j) . \tag{6.3}$$

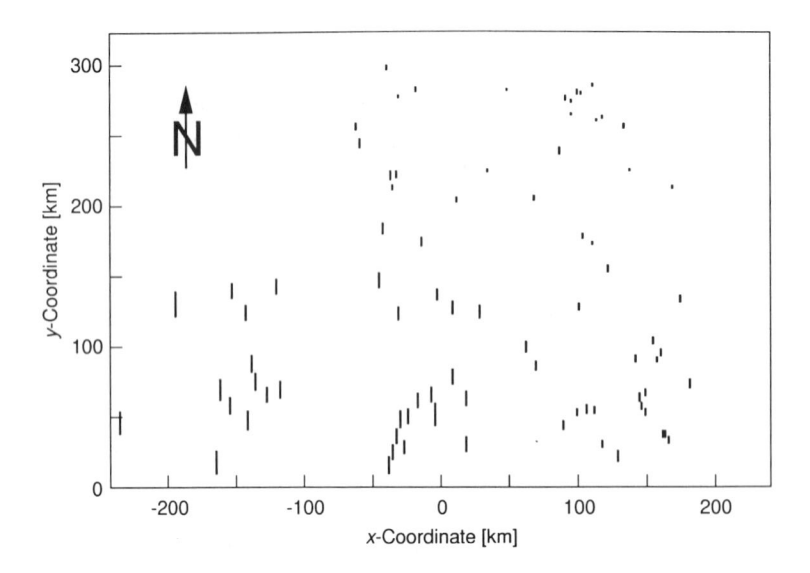

Figure 6.3: Positions of sampling locations and scaled piezometric head, H, Wolfcamp aquifer, Texas. The length of the symbols are proportional to $(H - \min(H))/(\max(H) - \min(H))$. Data drawn from [83].

The function $C(\cdot)$ is the *covariance function* of the random process. Note that $\text{Cov}\langle Y(s_i), Y(s_j)\rangle = \text{Cov}\langle e(s_i), e(s_j)\rangle$. In section 6.2.3 we shall relax the stationarity assumptions by adopting the intrinsic model, i.e., by assuming that only certain linear combinations of the data must be weakly stationary. The model (6.1) to (6.3), eventually extended by the intrinsic hypothesis (see Chapter 5), forms the basis of linear geostatistics, and it is at the heart of many statistical methods.
Before concluding this section, we add three remarks.

(1) An important special case of Eq. (6.1) is obtained on setting $q = 1$ and $f_1(s) = 1$. The random process then has a constant mean equal to $E\langle Y(s)\rangle = \beta_1$.

(2) If the mean is piecewise constant for a set of sub-domains of a tessellation then the explanatory variables are indicators which take the value one if a point lies in a particular sub-domain and zero otherwise. Evidently, $\boldsymbol{\beta}$ describes then the means of the sub-domains.

(3) In practice, the decomposition of the observed variation into drift and 'error' term is non-unique. What appears as deterministic large-scale variation to one geostatistician might be interpreted as random variation by another, especially if one allows for intrinsic stationarity. Cressie ([83]: section 4.1) presented an example which illustrates this point well. Consider Fig. 6.3 which shows the positions of the observation wells and the recorded piezometric heads of the Wolfcamp aquifer in Deaf Smith County, Texas. Clearly,

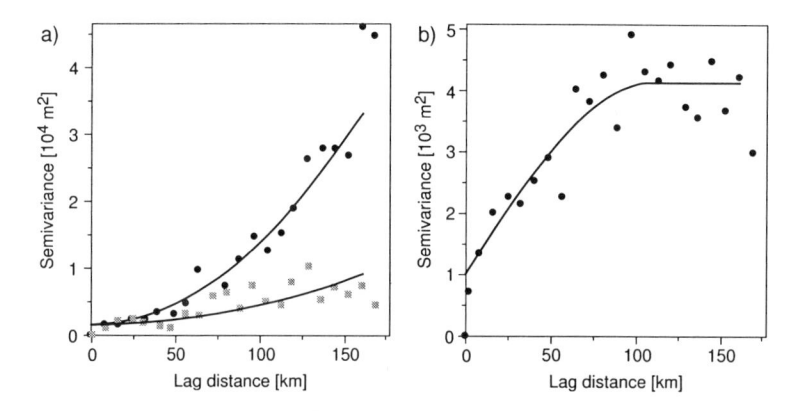

Figure 6.4: Sample variograms and fitted model functions of piezometric head, Wolfcamp aquifer, Texas. a) Geometrically anisotropic variogram of raw data: ● direction NE-SW, ■ direction NW-SE. The fitted model is a power function. b) Isotropic variogram of residuals, $H - 2595 + 6.75x + 6.01y$. An isotropic spherical variogram with a nugget component was fitted by least squares.

the piezometric head increases from north-east to south-west suggesting flow of ground water in the opposite direction.

To map the piezometric head in the region, we can adopt two models. First we can describe $\mu(s)$ by a linear function of the coordinates x and y, i.e., $\mu(s) = \mu(x,y) = \beta_1 + \beta_2 x + \beta_3 y$, and model the covariance structure of the residuals by a bounded variogram, or we can model the piezometric head by an intrinsic random process with an unbounded variogram. In the latter case, the differing slopes of the mean surface in x- and y-direction are modeled by a geometrically anisotropic variogram. Fig. 6.4 shows that both models seem compatible with the data. In sections 6.2.3 and 6.2.4 below, we shall show that the predictions, obtained from the two models, do not differ much. Thus, it appears that the decomposition of the variation into a deterministic and a stochastic part—although being non-unique—does not affect the results strongly.

(4) The linear model owes its widespread use at least partly to the fact that both optimal estimation[2] of β and optimal prediction[2] of $Y(s)$ is straightforward if we additionally assume that $\{e(s)\}$ is a Gaussian random process. The maximum likelihood estimator of β is then identical to the generalized

---

2. In the sequel, we shall distinguish between *estimation* and *prediction* as in chapter 2. By the term estimation, we refer to inference about the unknown parameters of a probabilistic model, such as the mean or variance. Estimating them by a linear function of the data as precisely as possible is a common aim, leading to the notion of the *best linear unbiased estimator* (BLUE). Unlike prediction, estimation always refers to inference about non-random quantities in classical statistical inference. This is different from kriging where we want to make a statement about a random variable, namely the value at an unvisited location. For this activity we shall use the term prediction, and we aim to find a technique that gives us an optimal predictor. Note, that this activity is also termed estimation in some geostatistical literature.

least squares estimator (cf. section 6.2.3) which can be readily obtained from the data given some estimate of the correlation structure of $\{e(s)\}$. Furthermore, under the Gaussian model, the best predictor of $Y(s)$ is a linear function of the observations.

6.2 Single Variable Linear Prediction

We turn our attention now to predicting the value of the target quantity at an unsampled location, s_0. We shall consider predictors that are *linear functions* of the recordings, y_i, $i = 1, 2, \ldots, n$; of one support variable only. The extension to several support variables will be considered in section 6.3.

6.2.1 Precision criteria

Our objective is to predict the unknown value as 'precisely' as possible. Let us denote the predicted value of $y(s_0)$ by $\widehat{y}(s_0)$. It is customary to use the squared prediction error, $\{\widehat{y}(s_0) - y(s_0)\}^2$, as a measure of how precisely a method predicts $y(s_0)$. When predicting the values for a large number, N, of points, $s_j \in G$, it is natural to choose, among several alternatives, the method that minimizes the *mean square error of prediction*, $1/N \sum_{j=1}^{N} \{\widehat{y}(s_j) - y(s_j)\}^2$. This criterion takes a small value if both the *bias* of the predictions and the *variances* of the prediction errors are small. Thus, we require a good prediction method to lack systematic error and to fluctuate as little as possible.

We use the same criteria to find an optimal predictor under the model (6.1) to (6.3). A linear predictor, $\widehat{Y}(s_0)$, is optimal if it satisfies

$$E\langle \widehat{Y}(s_0) - Y(s_0)\rangle = 0 , \qquad (6.4)$$

$$\text{minimize } E\langle \{\widehat{Y}(s_0) - Y(s_0)\}^2\rangle , \qquad (6.5)$$

and it is then called the *best linear unbiased predictor* (BLUP). Note that the validity of (6.4) and (6.5) imply that $\text{Var}\langle \widehat{Y}(s_0) - Y(s_0)\rangle$ attains a minimum because, subject to (6.4), the mean square error is equal to the variance of the prediction error.

6.2.2 Simple kriging

In this section let us suppose that both the covariance, $C(h)$, and the mean function, $\mu(s)$, are known. The latter assumption will be relaxed in section 6.2.3 where we shall only assume that the variogram is known. Our task consists of finding a *heterogeneously* linear predictor of the form

$$\widehat{Y}(s_0) = \lambda_0 + \sum_{i=1}^{n} \lambda_i Y(s_i) \qquad (6.6)$$

subject to the conditions (6.4) and (6.5). A predictor is called *homogeneously* linear if $\lambda_0 \equiv 0$. However, λ_0 does not in general vanish, and the predictor (6.6) is *heterogeneously* linear.

Bivariate case: one support datum

Let us first consider the case $n = 1$ where only one support datum, $Y(s_1)$, is available. We assume that this observation has been recorded without any measurement error. Equation (6.6) then simplifies to

$$\widehat{Y}(s_0) = \lambda_0 + \lambda_1 Y(s_1), \tag{6.7}$$

and our task consists of choosing λ_0 and λ_1 such that

$$E\langle \lambda_0 + \lambda_1 Y(s_1) - Y(s_0) \rangle = 0,$$
$$E\langle \{\lambda_0 + \lambda_1 Y(s_1) - Y(s_0)\}^2 \rangle \text{ is minimized.} \tag{6.8}$$

Now, set $Z = Y(s_0) - \lambda_1 Y(s_1)$. Minimizing the expected squared prediction error then boils down to finding the best *constant* predictor of Z. It is well-known that the expectation of a random variable has this property, and therefore $\lambda_0 = E\langle Z \rangle = \mu(s_0) - \lambda_1 \mu(s_1)$. Note that for this choice of λ_0 the predictor (6.7) is unbiased. Substituting the above expression for λ_0 in (6.8) leads to

$$E\langle \{\lambda_1[Y(s_1) - \mu(s_1)] - [Y(s_0) - \mu(s_0)]\}^2 \rangle =$$
$$\lambda_1^2 C(0) - 2\lambda_1 C(s_0 - s_1) + C(0).$$

To find the optimal weight for $Y(s_1)$, we differentiate this expression with respect to λ_1, set the resulting expression equal to zero and solve for λ_1. We then find $\lambda_1 = C(s_0 - s_1)/C(0)$, and the optimal predictor, the so-called *simple kriging predictor*, is given by

$$\widehat{Y}_{SK}(s_0) = \mu(s_0) + \frac{C(s_0 - s_1)}{C(0)}[Y(s_1) - \mu(s_1)], \tag{6.9}$$

and its mean square error, the *simple kriging variance*, equals

$$\begin{aligned} \sigma^2_{SK}\langle \widehat{Y}(s_0) \rangle &= C(0) - [C(s_0 - s_1)]^2/C(0) \\ &= C(0) - C(s_0 - s_1)[C(0)]^{-1}C(s_0 - s_1). \end{aligned} \tag{6.10}$$

Thus, the best heterogeneously linear predictor is obtained by adding the weighted residual, $Y(s_1) - \mu(s_1)$, to the expectation of $Y(s_0)$, with the weight equal to the correlation, $\rho(s_0 - s_1) = C(s_0 - s_1)/C(0)$, between $Y(s_0)$ and $Y(s_1)$.

Properties of the linear predictor

The following remarks will be helpful not only for simple kriging, but for understanding linear kriging in general.

(1) It follows immediately from the way how we constructed (6.9) that the simple kriging predictor is unbiased, and it is easy to verify this directly.

(2) It is clear from chapter 5 that we can substitute the variogram for the covariance function in the above expressions using the relation $\gamma(h) =$

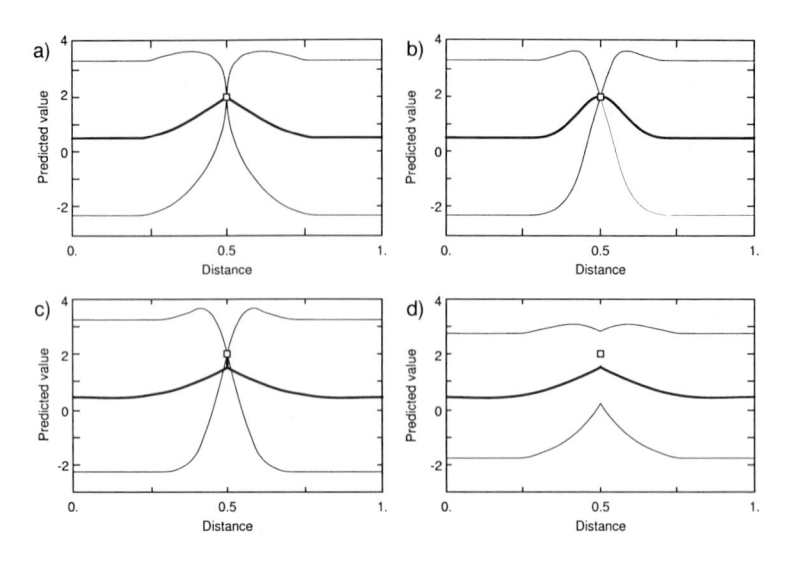

Figure 6.5: Simple kriging of $Y(s)$ on the interval $[0, 1]$ with single support datum at $s = 0.5$. The thick line denotes the predictions, and the thin lines mark a 95% prediction interval $(\widehat{Y}_{SK}(s_0) \pm 1.96\,\sigma_{SK}\langle\widehat{Y}(s_0)\rangle)$. The mean of $\{Y(s)\}$ was equal to 0.5. a) Spherical variogram, range 0.3, nugget variance 0, total variance 2. b) Gaussian variogram, scale parameter 0.3, nugget variance 0, total variance 2. c) Spherical variogram, range 0.3, nugget variance 0.67, total variance 2; prediction of 'error-contaminated' process. d) Spherical variogram, range 0.3, nugget variance 0.67, total variance 2; prediction of 'error-free' process.

$C(0) - C(h)$. The weight λ_1 then equals $1 - \gamma(s_0 - s_1)/C(0)$. This shows that the covariance function must be known to compute the simple kriging predictor: knowing the variogram is not sufficient.

(3) Instead of considering the simple kriging predictor as the sum of $\mu(s_0)$ plus the weighted residual, we can also interpret it as the sum of $\mu(s_0)$ plus the weighted correlation between $Y(s_0)$ and $Y(s_1)$, with the weight equal to $[Y(s_1) - \mu(s_1)]/C(0)$. This is the so-called *function estimate* or *dual form* of kriging ([233]: section 3.10). Apart from having computational advantages, this representation clarifies the influence of the covariance function or the variogram onto the predictions (cf. Fig. 6.5).

First, it is easy to see that simple kriging is an exact interpolator: clearly, $\widehat{Y}_{SK}(s_0) \to Y(s_1)$ as $s_0 \to s_1$ because $C(s_0 - s_1) \to C(0)$ and $\mu(s_0) \to \mu(s_1)$. Note further that $\sigma^2_{SK}\langle\widehat{Y}(s_0)\rangle \to 0$ as $s_0 \to s_1$.

Second, the behavior of the covariance function or the variogram at the origin determines the geometric properties of the prediction surface at the support points. If, for instance, the variogram has a nugget component, then the prediction surface has discontinuities at the support points. Similarly, the prediction surface will be differentiable at the support points only if the variogram has a parabolic shape close to the origin. The resulting prediction surface will be quite smooth in contrast to the surface obtained from a variogram with linear behavior near the origin which results in sharp peaks centered on the support points.

Third, Fig. 6.5 illustrates a further property of the kriging predictor. If $\rho(s_0 - s_1) \to 0$, for instance, if the distance between the target and support point increases while the range of the variogram is fixed, then the simple kriging prediction converges to the expectation $\mu(s_0)$. This behavior is sometimes referred to as *regression towards the mean*.

(4) So far, we have assumed that the support datum had been recorded without any measurement error, and the nugget component of the variogram, if present at all, was interpreted as micro-scale variation. However, we know from chapter 5 that both micro-scale variation and measurement error contribute to the nugget variance. In most instances it is then sensible to predict the *error-free* data. This can be achieved easily if the variance of the measurement error is known.

To be more specific, extend the linear model (6.1) by

$$Y(s) = \mu(s) + e_1(s) + e_2(s) + \varepsilon , \qquad (6.11)$$

where $e_1(s)$ is the stochastic component modeling smooth small-scale variation, $e_2(s)$ models the stochastic micro-scale variation with apparent pure nugget variogram, and ε is a measurement error with constant variance σ_ε^2. We assume that all stochastic components have zero mean and are mutually independent, and we want to predict the error-free datum, $Z(s_0) = Y(s_0) - \varepsilon$. By using the same approach as above and making use of the fact that all stochastic components are mutually independent, we find

$$\widehat{Z}_{\text{SK}}(s_0) = \mu(s_0) + \frac{C_1(s_0 - s_1) + C_2(s_0 - s_1)}{C_1(0) + C_2(0) + \sigma_\varepsilon^2} [Y(s_1) - \mu(s_1)] , \qquad (6.12)$$

$$\sigma_{\text{SK}}^2 \langle \widehat{Z}(s_0) \rangle = C_1(0) + C_2(0) - \frac{[C_1(s_0 - s_1) + C_2(s_0 - s_1)]^2}{C_1(0) + C_2(0) + \sigma_\varepsilon^2} , \qquad (6.13)$$

where $C_i(\cdot)$ denotes the covariance function of $\{e_i(s)\}$.

It is worth mentioning two more points. First, the comparison of Figs. 6.5c and 6.5d shows that the predicted values differ only at the support points. In contrast to $\widehat{Y}_{\text{SK}}(s_0)$, $\widehat{Z}_{\text{SK}}(s_0)$ is no longer an exact interpolator. Second, the comparison of the figures reveals that the kriging variance is smaller when the error-free process is predicted. Finally, Equation (6.12) shows that the degree by which kriging smoothes the data in the neighborhood of the support points depends on the ratio of $C_1(0)$ to $C_1(0) + C_2(0) + \sigma_\varepsilon^2$.

(5) As a last point to note, Equation (6.10) shows that the kriging variance depends only on the relative positions of the target and the support points,

but not on the observed value $Y(s_1)$. We can thus speculate that prediction intervals, computed from the kriging variance, will not necessarily be meaningful because they do not take the 'local' information of the observed values into account.

Multivariate case: many support data

In this section we shall now generalize the results for the case $n > 1$. This poses no difficulty, but we have to introduce some further notation.

Let $\mathbf{Y}' = [Y(s_1), Y(s_2), \ldots, Y(s_n)]$ denote the n-vector with the random variables modeling the observations, $\mathbf{\Sigma}$ the $(n \times n)$-matrix with the covariances of those random variables, i.e., $[\mathbf{\Sigma}]_{ij} = \text{Cov}\langle Y(s_i), Y(s_j)\rangle = C(s_i - s_j)$, and let $\boldsymbol{\sigma}$ denote an n-vector with the covariances between the values at the prediction and the observation locations, i.e., $[\boldsymbol{\sigma}]_i = \text{Cov}\langle Y(s_0), Y(s_i)\rangle = C(s_0 - s_i)$.

We predict $Y(s_0)$ by a heterogeneously linear predictor,

$$\widehat{Y}(s_0) = \lambda_0 + \sum_{i=1}^{n} \lambda_i Y(s_i) = \lambda_0 + \boldsymbol{\lambda}'\mathbf{Y}, \tag{6.14}$$

where $\boldsymbol{\lambda}' = [\lambda_1, \lambda_2, \ldots, \lambda_n]$. Our problem consists of finding λ_0 and $\boldsymbol{\lambda}$ such that the conditions (6.4) and (6.5) hold.

Using exactly the same argument as above, we find

$$\lambda_0 = \mu(s_0) - \sum_{i=1}^{n} \lambda_i \mu(s_i) = \mathbf{x}_0'\boldsymbol{\beta} - \boldsymbol{\lambda}'\mathbf{X}\boldsymbol{\beta}, \tag{6.15}$$

where $\mathbf{X}$ is the $(n \times q)$-design matrix with the explanatory variables, i.e., $[\mathbf{X}]_{ij} = f_j(s_i)$ and $[\mathbf{x}_0]_i = f_i(s_0)$.

After substituting the right-hand side of the last equation for λ_0 in Eq. (6.14), expanding $E\langle\{\widehat{Y}(s_0) - Y(s_0)\}^2\rangle$ and taking expectation, we obtain the following expression for the mean square prediction error

$$E\langle\{\widehat{Y}(s_0) - Y(s_0)\}^2\rangle = \boldsymbol{\lambda}'\mathbf{\Sigma}\boldsymbol{\lambda} - 2\boldsymbol{\lambda}'\boldsymbol{\sigma} + C(0).$$

As in the preceding section we minimize this expression by setting the partial derivatives with respect to the λ_i equal to zero. Then we obtain a system of linear equations

$$\sum_{j=1}^{n} \lambda_j C(s_i - s_j) = C(s_0 - s_i) \quad i = 1, 2, \ldots, n,$$

$$\mathbf{\Sigma}\boldsymbol{\lambda} = \boldsymbol{\sigma}.$$

If $C(\cdot)$ is a permissible covariance function (cf. chapter 5) then $\mathbf{\Sigma}$ is positive definite, and the matrix can be inverted. Thus, we obtain the solution $\boldsymbol{\lambda} = \mathbf{\Sigma}^{-1}\boldsymbol{\sigma}$, and from that the simple kriging predictor

$$\widehat{Y}_{\text{SK}}(s_0) = \mathbf{x}_0'\boldsymbol{\beta} + \boldsymbol{\sigma}'\mathbf{\Sigma}^{-1}(\mathbf{Y} - \mathbf{X}\boldsymbol{\beta}), \tag{6.16}$$

and the simple kriging variance

$$\sigma_{\text{SK}}^2 \langle \widehat{Y}(s_0) \rangle = C(0) - \boldsymbol{\sigma}' \boldsymbol{\Sigma}^{-1} \boldsymbol{\sigma} \, . \tag{6.17}$$

A comparison of the last two expressions with the corresponding equations for the bivariate case (cf. Eqs 6.9 and 6.10) reveals that the predictor and the mean square error still have the same structure. The remarks of the preceding section are therefore also valid for the case $n > 1$.

6.2.3 Universal kriging

In practice, simple kriging is not very useful because it assumes that the expectation of $Y(s)$ is known for any location $s \in G$. This is hardly ever the case, and $\mu(s)$ must be estimated from the data. If we model the drift as a weighted sum of known functions then it is natural to estimate the vector of coefficients, $\boldsymbol{\beta}$, by least squares. The standard result from ordinary least squares theory gives us

$$\widehat{\boldsymbol{\beta}}_{\text{OLS}} = (\boldsymbol{X}'\boldsymbol{X})^{-1}\boldsymbol{X}'\boldsymbol{Y}. \tag{6.18}$$

However, this estimator is not optimal if the $\{e(s)\}$ are spatially dependent. Then there exists another linear estimator, the so-called *generalized least squares estimator*, $\widehat{\boldsymbol{\beta}}$, with the property $\text{Var}\langle \widehat{\beta}_p \rangle \leq \text{Var}\langle \widehat{\beta}_{p,\text{OLS}} \rangle$, $p = 1, 2, \ldots, q$. One can show that $\widehat{\boldsymbol{\beta}}$ has the smallest variance among all the unbiased linear estimators of $\boldsymbol{\beta}$. It is given by

$$\widehat{\boldsymbol{\beta}} = (\boldsymbol{X}'\boldsymbol{\Sigma}^{-1}\boldsymbol{X})^{-1}\boldsymbol{X}'\boldsymbol{\Sigma}^{-1}\boldsymbol{Y}, \tag{6.19}$$

and its covariance matrix is equal to

$$\text{Cov}\langle \widehat{\boldsymbol{\beta}}, \widehat{\boldsymbol{\beta}}' \rangle = (\boldsymbol{X}'\boldsymbol{\Sigma}^{-1}\boldsymbol{X})^{-1} \, . \tag{6.20}$$

If we substitute $\widehat{\boldsymbol{\beta}}$ for $\boldsymbol{\beta}$ in Eq. (6.16) then we obtain the so-called *universal kriging predictor*

$$\begin{aligned} \widehat{Y}_{\text{UK}}(s_0) &= \boldsymbol{x}_0'\widehat{\boldsymbol{\beta}} + \boldsymbol{\sigma}'\boldsymbol{\Sigma}^{-1}(\boldsymbol{Y} - \boldsymbol{X}\widehat{\boldsymbol{\beta}}) \\ &= \boldsymbol{\kappa}'\boldsymbol{Y}, \end{aligned} \tag{6.21}$$

where the vector with the weights, $\boldsymbol{\kappa}$, is given by the expression

$$\boldsymbol{\kappa}' = \{\boldsymbol{\sigma} + \boldsymbol{X}(\boldsymbol{X}'\boldsymbol{\Sigma}^{-1}\boldsymbol{X})^{-1}(\boldsymbol{x}_0 - \boldsymbol{X}'\boldsymbol{\Sigma}^{-1}\boldsymbol{\sigma})\}'\boldsymbol{\Sigma}^{-1} \, . \tag{6.22}$$

Evidently, by replacing $\boldsymbol{\beta}$ with $\widehat{\boldsymbol{\beta}}$, we obtain a *homogeneously* linear predictor. But this raises the question whether this predictor is optimal among all the homogeneously linear predictors. Item 1 of the following remarks shows that this is indeed the case, and it derives the expression for the universal kriging variance, which is equal to

$$\sigma^2_{\text{UK}} \langle \widehat{Y}(s_0) \rangle = C(0) - \boldsymbol{\sigma}' \boldsymbol{\Sigma}^{-1} \boldsymbol{\sigma}$$
$$+ (\boldsymbol{x}_0 - \boldsymbol{X}' \boldsymbol{\Sigma}^{-1} \boldsymbol{\sigma})' (\boldsymbol{X}' \boldsymbol{\Sigma}^{-1} \boldsymbol{X})^{-1} (\boldsymbol{x}_0 - \boldsymbol{X}' \boldsymbol{\Sigma}^{-1} \boldsymbol{\sigma}). \quad (6.23)$$

In summary, we maintain:
- The simple kriging predictor is the *heterogeneous* BLUP. Its evaluation requires that both the mean and the covariance function are known.
- The universal kriging predictor is the *homogeneous* BLUP. As we shall show below, its evaluation requires only that the variogram is known. The unknown drift parameters are estimated from the data. The error resulting from estimating $\boldsymbol{\beta}$ is responsible for the fact that in general $\sigma^2_{\text{UK}} \langle \widehat{Y}(s_0) \rangle \geq \sigma^2_{\text{SK}} \langle \widehat{Y}(s_0) \rangle$.

A few remarks will clarify the matter further.

(1) We shall now show that the universal kriging predictor is the homogeneous BLUP. Readers disinterested in this rather theoretical question may wish to skip this and may continue with item 2.

We have to check whether there exists another predictor from the class of unbiased homogeneously linear predictors that has a smaller mean square error than (6.21). It is easy to see that $E\langle \widehat{Y}_{\text{UK}}(s_0) \rangle = \mu(s_0)$, on noting that $\widehat{\boldsymbol{\beta}}$ is an unbiased estimator of $\boldsymbol{\beta}$. To see that the universal kriging predictor is indeed the homogeneous BLUP, we rearrange Eq. (6.21)

$$\begin{aligned}
\widehat{Y}_{\text{UK}}(s_0) &= \boldsymbol{\sigma}' \boldsymbol{\Sigma}^{-1} \boldsymbol{Y} + (\boldsymbol{x}_0 - \boldsymbol{X}' \boldsymbol{\Sigma}^{-1} \boldsymbol{\sigma})' \widehat{\boldsymbol{\beta}} \\
&= \boldsymbol{\sigma}' \boldsymbol{\Sigma}^{-1} \boldsymbol{Y} + (\boldsymbol{x}_0 - \boldsymbol{X}' \boldsymbol{\Sigma}^{-1} \boldsymbol{\sigma})' \widehat{\boldsymbol{\beta}} \\
&\quad + (\boldsymbol{x}_0 - \boldsymbol{X}' \boldsymbol{\Sigma}^{-1} \boldsymbol{\sigma})' \boldsymbol{\beta} - (\boldsymbol{x}_0 - \boldsymbol{X}' \boldsymbol{\Sigma}^{-1} \boldsymbol{\sigma})' \boldsymbol{\beta} \\
&= \boldsymbol{x}'_0 \boldsymbol{\beta} + \boldsymbol{\sigma}' \boldsymbol{\Sigma}^{-1} (\boldsymbol{Y} - \boldsymbol{X} \boldsymbol{\beta}) + (\boldsymbol{x}_0 - \boldsymbol{X}' \boldsymbol{\Sigma}^{-1} \boldsymbol{\sigma})' (\widehat{\boldsymbol{\beta}} - \boldsymbol{\beta}) \\
&= \widehat{Y}_{\text{SK}}(s_0) + (\boldsymbol{x}_0 - \boldsymbol{X}' \boldsymbol{\Sigma}^{-1} \boldsymbol{\sigma})' (\widehat{\boldsymbol{\beta}} - \boldsymbol{\beta}) .
\end{aligned}$$

It turns out that the universal kriging predictor can be written as the sum of the simple kriging predictor plus the weighted estimation error of $\boldsymbol{\beta}$. Now, it can be shown that for *any* linear estimator, $\widetilde{\boldsymbol{\beta}} = \boldsymbol{SY}$, of $\boldsymbol{\beta}$—and therefore also for $\widehat{\boldsymbol{\beta}}$—the following relation holds

$$\text{Cov}\langle [\widehat{Y}_{\text{SK}}(s_0) - Y(s_0)], [\widetilde{\boldsymbol{\beta}} - \boldsymbol{\beta}] \rangle = 0 .$$

This result enables us to workout an explicit expression for the *universal kriging variance* and to check the optimality of (6.21)

$$\begin{aligned}E\langle[\widehat{Y}_{\text{UK}}(s_0) - Y(s_0)]^2\rangle \\ = E\langle[\widehat{Y}_{\text{SK}}(s_0) - Y(s_0) + (x_0 - X'\Sigma^{-1}\sigma)'(\widehat{\beta} - \beta)]^2\rangle \\ = E\langle[\widehat{Y}_{\text{SK}}(s_0) - Y(s_0)]^2\rangle \\ + (x_0 - X'\Sigma^{-1}\sigma)'\text{Cov}\langle\widehat{\beta}, \widehat{\beta}'\rangle(x_0 - X'\Sigma^{-1}\sigma).\end{aligned}$$

Both terms of the right-hand side of this expression attain a minimum: we cannot find any other unbiased linear predictor of $Y(s_0)$ and any other unbiased linear estimator of β that have smaller variances. Therefore, $\widehat{Y}_{\text{UK}}(s_0)$ is indeed the best unbiased homogeneously linear predictor of $Y(s_0)$, and its mean square error is given by Eq. (6.23). To derive this result, we made use of Eqs (6.17) and (6.20).

(2) One can show that all the remarks except item 2, section 6.2.2, are also valid for universal kriging. Note in particular, that equation (6.21) still has a function estimate representation because it can be represented as a weighted sum of covariances with the weights depending on the generalized least squares residuals.

(3) To compute κ we have implicitly assumed that the design matrix X has full column rank equal to q. In practice, the explanatory variables have to be selected with some care to ensure this.

(4) Most textbooks introduce universal kriging as the result of minimizing the mean square error of the homogeneously linear predictor under some constraints to ensure its unbiasedness. In the sequel, we summarize the argument shortly.

Our objective is to find a homogeneously linear predictor, $\widehat{Y}_{\text{UK}}(s_0) = \sum_{i=1}^{n} \kappa_i Y(s_i)$, such that it is unbiased and that it has minimum mean square error. From the first condition we can derive the following constraints

$$\begin{aligned}E\langle\widehat{Y}_{\text{UK}}(s_0) - Y(s_0)\rangle &= \sum_{i=1}^{n} \kappa_i E\langle Y(s_i)\rangle - E\langle Y(s_0)\rangle \\ &= \sum_{i=1}^{n} \kappa_i \sum_{p=1}^{q} f_p(s_i)\beta_p - \sum_{p=1}^{q} f_p(s_0)\beta_p = 0.\end{aligned}$$

Evidently, the last equality holds if

$$\sum_{i=1}^{n} \kappa_i f_p(s_i) = f_p(s_0) \tag{6.24}$$

for all $p = 1, 2, \ldots, q$. This defines the q constraints, which must be taken into account when minimizing $E\langle[\widehat{Y}_{\text{UK}}(s_0) - Y(s_0)]^2\rangle$. The weights κ_i can then be found using standard Lagrangian techniques. Of course, this approach leads also to the results (6.21) and (6.23).

For the case where the explanatory variables are polynomials in s, the above approach has some advantages, because the universal kriging predictor exists for a class of non-stationary random process, the so-called *intrinsic processes of order k* (cf. section 6.2.5, item 3, below, and the textbook by Kitanidis [233], which are more general than the intrinsic processes considered so far. The unbiasedness conditions (6.24) then guarantee that the universal kriging error, $\widehat{Y}_{\mathrm{UK}}(s_0) - Y(s)$, is a so-called *kth order increment*, i.e., it has zero expectation and its variance can be expressed as a function of the *generalized covariance function* only. Thus, the knowledge of the generalized covariances is sufficient to derive the universal kriging predictor. In the next section, we shall demonstrate this for ordinary kriging, a special case of universal kriging with $\mu(s) = \beta_1$.

For the sake of completeness we give here the expression for the universal kriging weights and the universal kriging variance in terms of the variogram matrix, $\mathbf{\Gamma}$, and variogram vector, $\boldsymbol{\gamma}$,

$$\boldsymbol{\kappa}' = \{\boldsymbol{\gamma} + \mathbf{X}(\mathbf{X}'\mathbf{\Gamma}^{-1}\mathbf{X})^{-1}(\boldsymbol{x}_0 - \mathbf{X}'\mathbf{\Gamma}^{-1}\boldsymbol{\gamma})\}'\mathbf{\Gamma}^{-1}, \quad (6.25)$$

$$\sigma^2_{\mathrm{UK}}\langle \widehat{Y}(s_0) \rangle = \boldsymbol{\gamma}'\mathbf{\Gamma}^{-1}\boldsymbol{\gamma} \\ - (\boldsymbol{x}_0 - \mathbf{X}'\mathbf{\Gamma}^{-1}\boldsymbol{\gamma})'(\mathbf{X}'\mathbf{\Gamma}^{-1}\mathbf{X})^{-1}(\boldsymbol{x}_0 - \mathbf{X}'\mathbf{\Gamma}^{-1}\boldsymbol{\gamma}) \quad (6.26)$$

(5) Some authors [106, 385] reserve the term 'universal kriging' for applications where the explanatory variables depend directly on the coordinates (e.g., polynomials in s) and use the expression *external drift kriging* for the more general case when the $f_p(s)$ are functionals of other regionalized variables. The explanatory variables then depend only indirectly on s. However, similar to [83], we use 'universal kriging' to denote all applications where a spatially varying mean function is estimated from data.

Any auxiliary information that is correlated with the target variable, say $Y_1(s)$, can serve as an explanatory variable, too, provided it is available for any prediction point. In remote sensing this is the case when the target variable is predicted on a grid with the same spacing as the image of the auxiliary variable(s). The dependence of the target on the auxiliary information is then taken into account by modeling the conditional mean of $Y_1(s)$, given the auxiliary variables, $Y_i(s)$, $i = 2, 3, \ldots, r$; as a linear function of the latter, i.e., $E\langle Y_1(s)|Y_2(s), Y_3(s), \ldots, Y_r(s)\rangle = \beta_1 + \sum_{i=2}^{r} \beta_i Y_i(s)$. The generalized least squares estimate (cf. Eq. 6.19) then provides the optimal estimate of $\boldsymbol{\beta}$ conditional on the observed $Y_i(s)$.

If the auxiliary variables are not available for any prediction point then universal kriging can no longer be used. Co-kriging (cf. section 6.3) or *regression kriging* [8] provide then ways to incorporate the auxiliary information into the predictions.

6.2.4 Ordinary Kriging

Ordinary kriging is a special case of universal kriging which is obtained on setting $q = 1$ and $f_1(s) = 1$ so that $\mu(s) = \beta_1 = \text{const}$. The design matrix X then reduces to a vector with all elements equal to 1. Of course, the solution vector κ is still given by Eq. (6.21) and the *ordinary kriging variance* by Eq. (6.23). In the sequel we show that the same result is obtained from a minimization of the mean square error subject to the unbiasedness constraint $\sum_{i=1}^{n} \kappa_i = 1$.

We have to choose the coefficients κ_i such that they minimize

$$E\langle [\widehat{Y}_{OK}(s_0) - Y(s_0)]^2 \rangle - 2\psi (\sum_{i=1}^{n} \kappa_i - 1),$$

where ψ is a Lagrange multiplier. Minimization with respect to κ_i and ψ leads to the following system of linear equations

$$\sum_{j=1}^{n} \kappa_j C(s_i - s_j) - \psi = C(s_0 - s_i) \quad i = 1, 2, \ldots, n,$$

$$\sum_{j=1}^{n} \kappa_j = 1, \qquad (6.27)$$

and the ordinary kriging variance is equal to

$$\sigma_{OK}^2 \langle \widehat{Y}(s_0) \rangle = C(0) - \sum_{i=1}^{n} \kappa_i C(s_0 - s_i) + \psi. \qquad (6.28)$$

By writing the expressions (6.27) in matrix notation and solving for κ we obtain the result of Eqs (6.21) and (6.23).

Although we have worked so far with the stationary covariance function we shall now explicitly show that the knowledge of the variogram is sufficient to derive the ordinary kriging predictor. As mentioned above, an analogous generalization exists for the intrinsic processes of order k. We recall from chapter 5 that a valid variogram function must be conditionally negative semi-definite, i.e., for any linear combination of $Y(s_i)$ with weights, a_i, summing to zero, the expectation of the squared weighted sum, $\{\sum_i a_i Y(s_i)\}^2$, depends only on the variogram. The ordinary kriging error, $\widehat{Y}_{OK}(s_0) - Y(s_0)$, is a linear combination with the required property because $\sum_{i=0}^{n} a_i = 0$, with $a_0 = -1$ and $a_i = \kappa_i$, $i = 1, 2, \ldots, n$. Thus, we can express the expected squared prediction error by

$$E\langle[\widehat{Y}_{\text{OK}}(s_0) - Y(s_0)]^2\rangle = -\sum_{i=0}^{n}\sum_{j=0}^{n} \kappa_i \kappa_j \, \gamma(s_i - s_j)$$

$$= -\sum_{i=0}^{n} \kappa_i \left(\kappa_0 \, \gamma(s_i - s_0) + \sum_{j=1}^{n} \kappa_j \, \gamma(s_i - s_j) \right)$$

$$= 2\sum_{i=1}^{n} \kappa_i \, \gamma(s_0 - s_i) - \sum_{i=1}^{n}\sum_{j=1}^{n} \kappa_i \kappa_j \, \gamma(s_i - s_j) \, .$$

using $\kappa_0 = -1$ and $\gamma(s_0 - s_0) = 0$. If we minimize this expression subject to the constraint $\sum_{i=1}^{n} \kappa_i = 1$, we obtain a system of linear equations

$$-\sum_{j=1}^{n} \kappa_j \, \gamma(s_i - s_j) - \psi = -\gamma(s_0 - s_i) \quad i = 1, 2, \ldots, n \, ,$$

$$\sum_{j=1}^{n} \kappa_j = 1, \tag{6.29}$$

and the ordinary kriging variance is equal to

$$\sigma^2_{\text{OK}}\langle\widehat{Y}(s_0)\rangle = \sum_{i=1}^{n} \kappa_i \, \gamma(s_0 - s_i) + \psi \, . \tag{6.30}$$

In the course of this derivation we have at no point assumed that a stationary covariance function exists. The existence of the variogram is therefore sufficient for the existence of the ordinary kriging predictor.

As an illustration, we computed the ordinary kriging predictions for the piezometric head of the Wolfcamp aquifer data (Figs. 6.6a and 6.6b). We used the geometrically anisotropic power function variogram shown in Fig. 6.4a, and we predicted the error-free process on a 1.6×1.6 km grid and contoured the predictions and the kriging standard errors by the MATHEMATICA function `ListContourPlot`. We used all the support points to predict any target value. The predicted surface is very smooth and close to a plane, rising from north-east to south-west. The function estimate representation of (6.21) provides a heuristic explanation for the smoothness. It tells us that any predicted value is obtained by adding up 'basis functions' with approximately parabolic shape. The prediction surface is smooth because the slope of the variogram is close to zero near its origin. The kriging standard errors are smallest in that part of the domain where the support points are densest (cf. Fig. 6.3), and they increase towards the borders of the domain, where the prediction surface extrapolates the data.

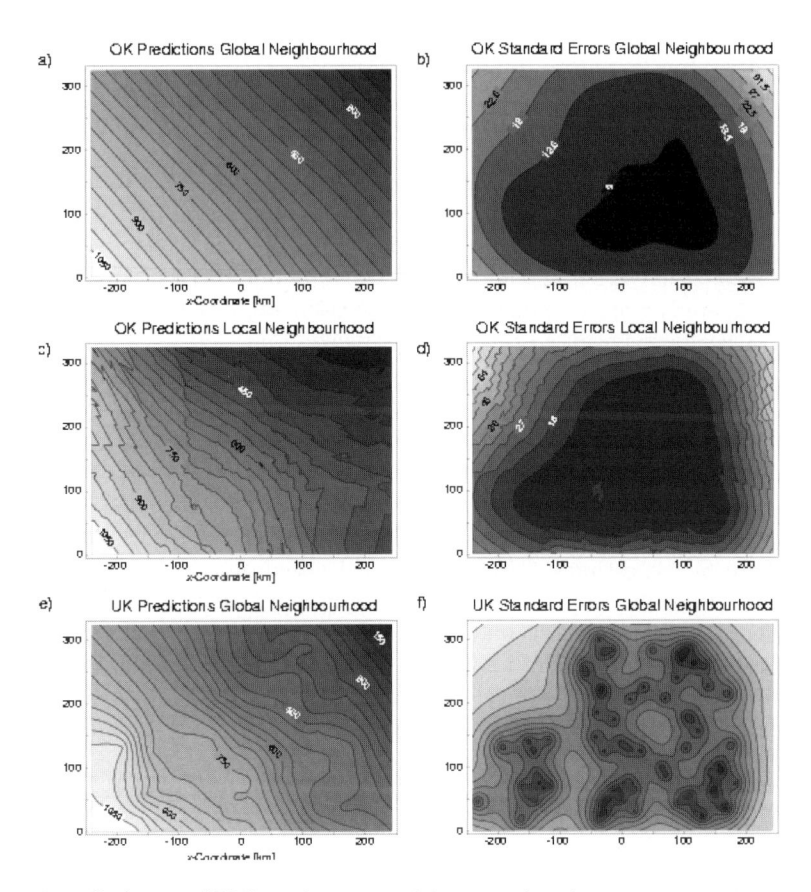

Chapter 6: Spatial prediction by linear kriging   101

Figure 6.6: Ordinary (OK) and universal kriging (UK) predictions of piezometric head [m], Wolfcamp aquifer, Texas. The predictions were computed with a geometrically anisotropic power function variogram (OK) or an isotropic spherical variogram (UK) (cf. Fig. 6.4). The 'pure nugget component' was treated as measurement error, and the 'error-free' process was predicted. The ordinary kriging predictions were computed from a global (a, b) as well as a local kriging neighborhood (c, d), consisting of the 20 nearest support points. For universal kriging the results are only reported for the global neighborhood.

6.2.5 The practice of kriging

In this section we want to discuss some issues which arise when ordinary and universal kriging are used in practice.

Estimating the variogram of the error process

When deriving the universal kriging predictor, we have assumed that the variogram of $\{e(s)\}$ was known. However, this is hardly ever the case, and the variogram must be estimated from data. Several approaches have been suggested to tackle this problem.

(1) When the drift is modeled as a linear function in the coordinates then it is sometimes possible to find the direction along which $\mu(s) \approx$ const. The sample variogram is then estimated for this direction only. As an obvious drawback, only a small number of pairs of observations is usually considered, and there is reason to suspect that the estimate, although being unbiased, is rather poor because of its large variance.

(2) In the second approach, the variogram is estimated from the residuals, $R(s_i) = Y(s_i) - \widehat{\mu}(s_i)$. Usually, $\mu(s)$ is estimated by ordinary least squares, thus $\widehat{\mu}(s_i) = x_i' \widehat{\beta}_{\text{OLS}}$. However, the variogram of the residuals is not equal to the variogram of the error process, and we obtain a biased estimate when we work with the residuals. Some authors have therefore suggested to estimate the variogram iteratively from the residuals $R(s_i)^{(k+1)} = Y(s_i) - x_i' \widehat{\beta}^{(k)}$, where $\widehat{\beta}^{(k)}$ is estimated from Eq. (6.19) based on a covariance matrix inferred from the residuals of the kth iteration. Although such an approach may approximate the generalized least squares estimate of β, it will nevertheless not provide an unbiased estimate of the variogram. It is not even guaranteed that the procedure will converge. Cressie ([83]: section 3.4.2) further points out that the estimate of the variogram will be biased even if $\widehat{\beta}$ is computed with the true covariance matrix.

There is no general agreement about the relevance of the bias. While Wackernagel ([385]: p. 182) warns that the bias is generally considerable, Cressie [83] expresses the view that the bias is not likely to affect the predictions strongly because they mainly depend on the behavior of the variogram near the origin which is not strongly impaired by the bias. But he also warns that the kriging variance may be significantly affected by the bias.

We have smoothed the piezometric head of the Wolfcamp aquifer by universal kriging, using a planar mean function and the bounded variogram shown in Fig. 6.4b. The variogram was estimated form the ordinary least squares residuals. All the support points were used to predict the values. Fig. 6.6 shows the prediction surface (e) along with the mapped standard errors (f). Compared to the results obtained by ordinary kriging, both predictions and modeled error variances differ somewhat. The contours of the predictions show more detail than those of the intrinsic model. This can be attributed to the fact that close to the support points the prediction surface is built by adding up functions with an approximately linear slope. Further away from the support points (e.g. in the upper left corner) the prediction surface is planar, approaching thus the generalized least squares estimate of $\mu(s)$. The standard errors of universal kriging are clearly larger than those obtained by ordinary kriging. This is in accordance with theory, and it shows that there is a prize to pay for the more general parameterization of the mean function.

(3) The difficulty of inferring the variogram of the error process eventually lead to the development of the theory of intrinsic random processes of higher order [258]. The idea is no longer to estimate the covariance function of the error process, but to infer the so-called generalized covariance function. Generalized covariances can be considered as stationary covariances of higher order increments, i.e., of linear combinations of the $Y(s_i)$ which satisfy the unbiasedness constraints of Eq. (6.24). For a kth order increment they are unique up to a polynomial of degree $2k$ in h ([233]: section 6.5). Various methods have been suggested to estimate the parameters of the generalized covariance functions. The most straightforward approach is given by Kitanidis ([233]: section 6.12) who suggests to use the ordinary least squares residuals as increments and to compute a sample estimate of the generalized covariance function from them by the method-of-moments estimator. At first sight this seems to lead again to a biased estimate. Kitanidis [233], however, points out, that although the intrinsic process and the residual process have not the same variogram, they nevertheless have the same generalized covariance function, and this is the quantity which is inferred from the residuals.

As an illustration, consider the case of estimating the covariance structure of a weakly stationary process $\{Y(s)\}$. The covariance function might be estimated from the centered observations, $Y(s_i) - \bar{Y}$, but this estimate is biased (note that the sample mean $\bar{Y}$ is equal to $\widehat{\beta}_{1,\text{OLS}}$). But the variogram—which is equal to minus the generalized covariances in this case—can be estimated without any bias from either the differences of the observations, $Y(s_i)$, or the OLS-residuals, $Y(s_i) - \bar{Y}$. In both cases we are dealing with zero order increments which filter a constant form the data. For any difference $Y(s_i) - Y(s_j) = R(s_i) - R(s_j)$ the weights obviously satisfy $\sum_k a_k = 0$ because $a_i = 1$, $a_j = -1$, $a_k = 0$ for $k \neq (i,j)$. Estimating the variogram for a given lag distance then amounts to inferring the (stationary) variances of the increments for this distance. Since by definition $E\langle Y(s_i) - Y(s_j)\rangle = 0$, the variances can be unbiasedly estimated by averaging the squared differences.

(4) Parametric or semi-parametric estimation methods such as maximum likelihood or restricted maximum likelihood methods can be used for inference about the covariance structure of $\{e(s)\}$. However, these topics are clearly beyond the scope of this chapter and the interested reader is referred to the literature (see references in [408, 233]).

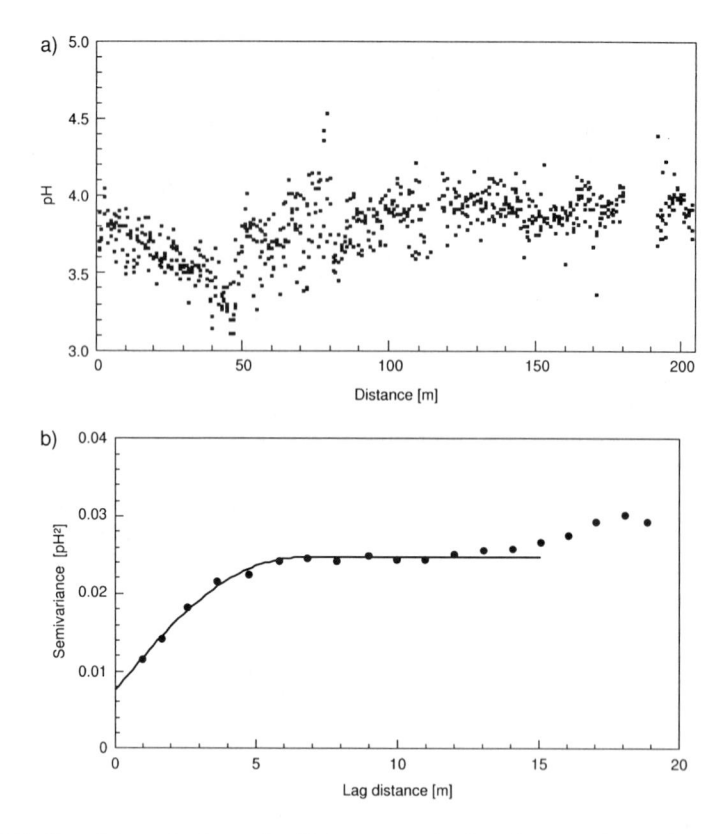

Figure 6.7: Spatial variation of pH in topsoil of a forest, Kestenberg, Wildegg, Switzerland. a) Distribution of pH along four parallel transects, which were separated by a distance of 1 m. b) Sample variogram and fitted pentaspherical model function.

Kriging with a local neighborhood
So far, we have assumed that all the support data are used to predict any target value. In geostatistical terminology, this is called kriging with a *global neighborhood*. Practitioners, however, commonly use a *local moving neighborhood*, consisting only of the points nearest to the target point. The heuristic basis for this is the observation that very often the kriging weights of the more distant points are negligible compared to those of the closer points. In addition, some authors argue that the computational burden to invert the $n \times n$ variogram matrix repeatedly is prohibitive. However, most data sets contain only a few hundred observations and with modern computers this poses no serious obstacles. Furthermore, the function estimate form of the kriging predictor (cf. item 3, section 6.2.2) demonstrates that the covariance matrix must be inverted only once. Thus, the computational argument is in most instances not very meaningful.

But there is one argument occasionally favoring a local moving neighborhood.

Consider the data set presented in Fig. 6.7. The scatter plot shows the topsoil pH recorded on four narrowly spaced parallel lines in a forest. The plot reveals that the mean function varies spatially: the mean decreases between $x = 0$ m and $x \approx 45$ m, then increases again, and remains approximately constant beyond $x \approx 70$ m. Although a constant mean function appears not appropriate, the presence of a distinct sill in the variogram demonstrates that locally the mean can be reasonably well modeled by a constant. Geostatistical terminology uses the term *quasi-stationary* to denote this behavior. For such data, ordinary kriging with a local neighborhood offers an alternative to fully-fledged universal kriging: the mean is allowed to vary smoothly at the larger scale, and at each point it is estimated locally from the data lying within the neighborhood.

To demonstrate the effect of using a local neighborhood we have recomputed the ordinary kriging predictions for the Wolfcamp data, using at each target location only the nearest 20 support points. Figs. 6.6c and 6.6d show the prediction surface along with a map of the standard errors. Clearly, the contours 'wiggle' more strongly than in the case of a global neighborhood. With the local neighborhood the intrinsic model results in a surface which follows the observations more closely, and therefore, it is more similar to the surface obtained by universal kriging. As expected, the standard errors are larger when a local neighborhood is used, especially in those parts of the domain where the surface extrapolates the data. However, the general pattern of the standard errors still is quite similar.

The 'wiggles' of the prediction contours result from the fact that the prediction surface is no longer continuous when a local kriging neighborhood is used. This can be seen well around the point (-120,100) in Fig. 6.6c. It is therefore questionable whether the seemingly more detailed structure of the surface has much meaning. This is often neglected by practitioners who are not aware of the problem.

Interpretation of the kriging standard errors

We mentioned shortly in section 6.2.2, item 5, that the kriging variance depends only on the relative positions of the target to the support points and on the variogram. The comparison of the mapped kriging standard errors of Fig. 6.6 with the locations of the support points (Fig. 6.3) illustrates this well. In particular, the funnel-shaped zones with smaller kriging error around the support points, obtained by universal kriging, reflect only the shape of the spherical covariance function (see also Fig. 6.5), and they do not describe the local roughness of the data.

Nevertheless, the kriging standard errors are occasionally used to construct prediction intervals, $\widehat{Y}(s_0) \pm 2\sigma \langle \widehat{Y}(s_0) \rangle$, to quantify the prediction uncertainty. In addition, several authors [389, 23, 24] recommended to select optimal sampling design based on the kriging variance. But both applications will give meaningful results only if the empirical distribution of the data can be reasonably well approximated by a normal distribution. If this is not the case, then one should resort to non-linear kriging techniques (for an overview see [293]).

If the modeling of prediction uncertainty is the main objective then there is another point to be considered. The so-called parameter uncertainty, which arises from the estimation of the covariance structure of $\{Y(s)\}$, is neglected by the 'plug-in' approach of traditional geostatistics. Recent studies [169, 108] show that the prediction uncertainty can be noticeably larger if parameter uncertainty is properly taken into account.

6.3 Multi-Variable Linear Prediction

So far we have assumed that the recordings of the target variable at the support points were the only information available. In this section, we shall now consider the more general case where several variables have been observed simultaneously. Including the additional variables as auxiliary information will not always reduce the kriging variance, but it will never make it larger than single-variable prediction. The precision of prediction can be considerably improved, provided the auxiliary variables were observed at other locations than the target variable and provided the target and the auxiliary variables are spatially cross-correlated. An interesting example is the prediction of values of a remotely sensed image, hidden by clouds on a particular day, but available from the image of the previous clear day. The image of the previous day can then be used as auxiliary data [5].

Although the notation becomes rather cumbersome, linear multi-variable prediction or *co-kriging* is straightforward. Let us assume that we have observed the target quantity, say $Y_1(s)$, at a set of locations, $s_{1,i}$, $i = 1, 2, \ldots, n_1$. In addition, data about $m - 1$ auxiliary variables, $Y_j(s)$, are available for the locations $s_{j,i}$, $i = 1, 2, \ldots, n_j$, $j = 2, 3 \ldots, m$. Note that the locations $s_{j,i}$ and $s_{l,k}$ need not match for $j \neq l$ and any i and k. We assume that either the auto- and cross-covariance functions or the auto- and cross-variograms are known. Furthermore, let us suppose that the mean function of the jth variable can be modeled by $\mu_j(s) = \sum_{p=1}^{q_j} \beta_{j,p} f_{j,p}(s) = x_j' \beta_j$, but we do not know the mean parameters, β_j. Note, that in general the functions $f_{j,r}(s)$ and $f_{l,s}(s)$ will differ for $j \neq l$ and any r and s.

6.3.1 Co-kriging weakly stationary processes

Our task consists of predicting $Y_1(s_0)$ at location s_0 as a homogeneously linear function of the $Y_j(s_{j,i})$, $i = 1, 2, \ldots, n_j$, $j = 1, 2, \ldots, m$. Thus, we are looking for a predictor,

$$\widehat{Y_{1,\text{cUK}}}(s_0) = \sum_{j=1}^{m} \sum_{i=1}^{n_j} \kappa_{ji} Y_j(s_{j,i}),$$

which satisfies the conditions of Eqs (6.4) and (6.5). Even under this more general setting, the solution for the optimal predictor is given by Eqs (6.21), (6.22) and (6.23) [348]. But now, the vectors β, x_0, Y, κ, σ and the matrices Σ and X have an extended structure.

Let Y denote the $(\sum_{j=1}^{m} n_j)$-vector with the observations of all variables in order, i.e., $Y' = [Y_1', Y_2', \ldots, Y_m']$ with $[Y_j]_i = Y_j(s_{j,i})$, and let κ be the vector with the kriging weights, such that the universal co-kriging predictor can be written as

$$\widehat{Y_{1,\text{cUK}}}(s_0) = \kappa' Y. \tag{6.31}$$

The covariance matrix is now an $(m \times m)$-block matrix with the ith row and jth column block equal to the matrix $\Sigma_{ij} = \text{Cov}\langle Y_i, Y_j{'}\rangle$, thus $[\Sigma_{ij}]_{kl} = \text{Cov}\langle Y_i(s_{i,k}), Y_j(s_{j,l})\rangle$. Thus, the dimensions of Σ are $\sum_{j=1}^{m} n_j \times \sum_{j=1}^{m} n_j$. The vector with the covariances between the target variable at the prediction location and all m variables at the observation locations has the same length as Y, and it is equal to $\sigma' = [\sigma_1{'}, \sigma_2{'}, \ldots, \sigma_m{'}]$ with $[\sigma_j]_i = \text{Cov}\langle Y_1(s_0), Y_j(s_{j,i})\rangle$.

The $(\sum_{j=1} q_j)$-vector β is equal to $\beta' = [\beta_1{'}, \beta_2{'}, \ldots, \beta_m{'}]$ with $[\beta_j]_i = \beta_{j,i}$. The vector x_0 has the same length as β, but all its entries are equal to zero, expect the first q_1 elements which are equal to $[x_0]_i = f_{1,i}(s_0)$, $i \leq q_1$. Finally, the design matrix X is a $(m \times m)$−block matrix with the ith diagonal entry, X_{ii}, equal to the design matrix of Y_i and all the off-diagonal entries, X_{ij}, equal to 0. Of course, $[X_{ii}]_{kl} = f_{i,l}(s_{i,k})$. Note that the dimensions of X_{ij} are $n_i \times q_j$, and therefore, X has the dimensions $\sum_{j=1}^{m} n_j \times \sum_{j=1}^{m} q_j$.

If universal co-kriging is considered as a constrained minimization problem, finding κ then amounts to solving the following system of linear equations

$$\begin{bmatrix} \Sigma & X \\ X' & 0 \end{bmatrix} \begin{bmatrix} \kappa \\ \psi \end{bmatrix} = \begin{bmatrix} \sigma \\ x_0 \end{bmatrix}.$$

The vector ψ contains the $\sum_{j=1}^{m} q_j$ Lagrange multipliers. Note that the unbiasedness constraints (Eq. 6.24) can be expressed by

$$X'\kappa = x_0.$$

6.3.2 Co-kriging intrinsic processes

So far we have assumed that the random processes $\{Y_j(s)\}$ jointly are weakly stationary. Since the universal co-kriging error is an increment of a given order, we can hypothesize that the co-kriging predictor can be expressed in terms of the auto- and cross-variograms, or more generally, in terms of the generalized (cross-)covariances [236]. The implementation, however, requires some care. To express the variance of the ordinary co-kriging error in terms of the auto- and the cross-variograms, we additionally require the 'symmetry condition' [276]

$$E\langle[Y_i(s_r) - Y_i(s_s)][Y_j(s_t) - Y_j(s_r)]\rangle = E\langle[Y_j(s_r) - Y_j(s_s)][Y_i(s_t) - Y_i(s_r)]\rangle.$$

This restriction arises from the fact that the cross-variogram is an even function, and therefore requires the cross-correlations patterns to be 'symmetric'. The use of the pseudo cross-variogram [292] avoids this restriction, but this function exists only if the differences, $Y_i(s) - Y_j(s)$, are weakly stationary and this may not hold for processes that are jointly intrinsic.

But leaving this rather theoretical issue aside, the universal co-kriging system can be implemented either with the traditional cross-variogram or with the pseudo cross-variogram [382], and the solution is found from Eqs (6.25) and (6.26). The definition of γ and Γ is obvious from the previous section, and we can do without a formal definition. Basically, the variogram matrix contains all the auto- and cross-semivariances between the support data, and the variogram vector contains the (cross-) semivariances between the target quantity and the support data.

6.3.3 Benefits of co-kriging

Notwithstanding the rather simple principles, the implementation of co-kriging may appear cumbersome, and it is therefore legitimate to ask what will be the advantages of the procedure. As stated above, co-kriging provides a consistent framework to incorporate auxiliary information into the predictions. Unlike universal kriging, the auxiliary information need not be available at each target point, and it may be gathered by an arbitrary sampling design.

In a typical application of co-kriging we have to map a target variable that is costly to observe. Therefore, only a few measurements are available. At the same time, we can record cheaply other variables at a large number of points, and these auxiliary variables are somehow related to our target quantity. The relation between the target and the auxiliary variables need not be linear. By using linear methods, we approximate the dependence, however, by a linear function. As we shall see in the next section, co-kriging can greatly improve the predictions in such an *under-sampling case*.

If all the variables are recorded at all the support points then co-kriging does not usually result in more precise predictions than univariate kriging. The predictions are even exactly the same if all the auto- and cross-variograms are just a multiple of some basic elementary variogram, i.e., if $\gamma_{kl}(\boldsymbol{h}) = b_{kl}\gamma(\boldsymbol{h})$ for any k and l. Helterbrand and Cressie [175] discuss the so-called intrinsic coregionalization in detail.

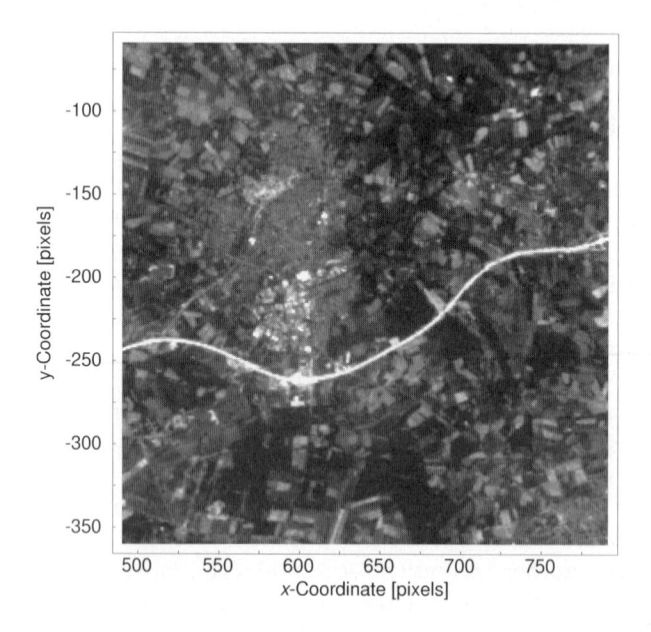

Figure 6.8: Grey scale image of *BAND1*, Landsat TM scene, Twente region. The grey levels vary from 72 DN (black) to 110 DN (white).

Co-kriging has another advantage over univariate kriging ([385]: page 148). Its predictions are coherent in the sense that co-kriging a linear combination of some variables is equivalent to co-kriging the variables first and building the linear combination of the predicted values afterwards. Univariate kriging does not in general have this property.

6.4 Mapping the Variable *BAND1* of the Twente Images

As set out in the introduction of this chapter, we now want to map the spatial distribution of the surrogate ground observations, *BAND1*, on a square with lower left corner in (490,-360) and upper right corner in (790,-60). We shall compare the results obtained by various kriging techniques with the true distribution of *BAND1* in the target area. Fig. 6.8 shows a grey scale image of *BAND1*. The pixels values varied between 72 and 175 in the target area. To increase the spread of the grey levels in Fig. 6.8 we truncated *BAND1* at 110 and interpolated the grey levels linearly on [72, 110]. Pixels with larger values mostly occurred along the motorway, crossing the target area from west to east, and in a few isolated spots scattered around (600,-225). For prediction we sampled *BAND1* on a 30 × 30 grid from the whole image (cf. Fig. 6.1), and we added clusters of points close to randomly selected grid points to estimate the variograms at short lag distances (not shown in Fig. 6.1). The shortest distance between any grid and cluster point was 4 pixels. Both auto- and cross-variograms were estimated from the same design.

The variograms were shown in Fig. 6.2. They all had a distinct discontinuity at the origin. The joint spatial dependence was modeled by fitting a linear model of Co-regionalization to the set of sample auto- and cross-variograms. The model consisted of a nugget component and two spherical variograms, one with a range equal to 10 pixels and the other with range $\approx$ 100 pixels. This confirms the visual impression that there were two distinct scales of spatial dependence: the field and the regional scale.

We predicted the target variable first by univariate ordinary kriging, using the variogram shown in Fig. 6.2a. The nugget variance was treated as micro-scale variation, and kriging was therefore an exact interpolator, honoring the support data. We used a local neighborhood consisting of 121 points at most, and all the points had to be less than 150 grid units apart from the prediction location. Fig. 6.9 shows the ordinary kriging predictions along with the kriging standard errors. Apart from the discontinuities at the support points, the prediction surface was very smooth. Locally, the surface followed the support data, forming circular peaks and funnel-shaped depressions, but these local deflections did not adequately describe the small-scale structure of *BAND1*. This could hardly be expected because the minimal distance between any two support points was 30 pixels whereas the small-scale dependence extended only up to 10 pixels. The large-scale structure of *BAND1* was better represented: the darker parts represent zones where the true values were on average smaller than the overall mean, and the brighter regions had indeed larger values. Once again, the map of the kriging standard deviations merely revealed the positions of the support points. Of course, the standard errors vanished at the support points, and they reached their maximum value at the centers of the sampling grid

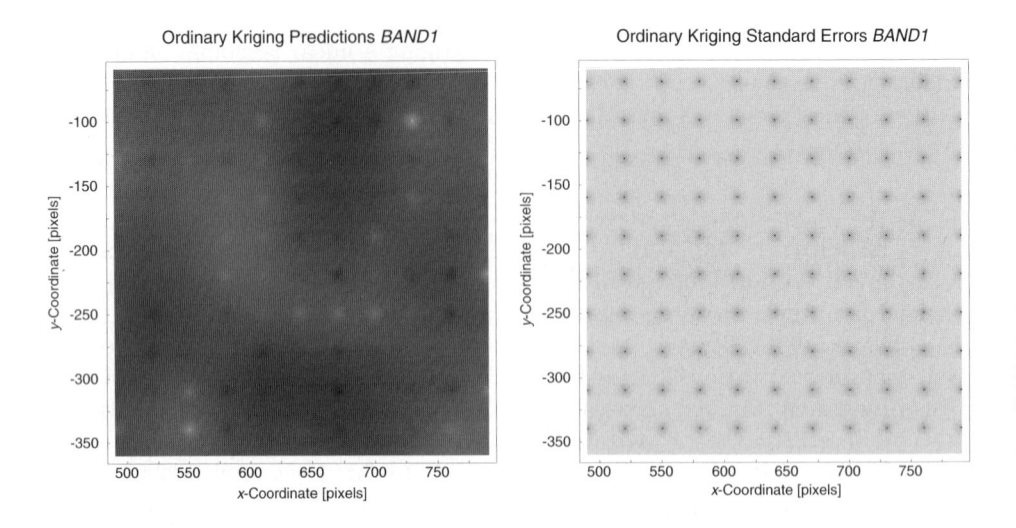

Figure 6.9: Grey scale image of ordinary kriging predictions and kriging standard errors, *BAND1*, Landsat TM scene, Twente region. The grey levels vary from 72 DN (black) to 110 DN (white) for the predictions and from 0 DN (black) to 6 DN (white) for the standard errors. See text for further explanations.

cells. Since the support points extended outside the target area, no increase of the kriging variance was visible at the edges of the target area.

Second, we predicted *BAND1* by co-kriging, using the radiance in either the second (*BAND2*) or the fourth waveband (*BAND4*) of Landsat TM as the auxiliary variable. The comparison of the cross-variograms (cf. Fig. 6.2c and 6.2d) shows that *BAND1* was strongly and positively correlated with *BAND2*, but only weakly and negatively correlated with *BAND4*. The coefficients of correlation, estimated from the fitted variogram parameters, were 0.92 for *BAND2* and -0.48 for *BAND4*, respectively. For prediction, the auxiliary variables were sampled on a 3×3 grid, and we used any auxiliary support point that was less than 21 pixels apart from the target point. The moving neighborhood of *BAND1* was the same as above.

Fig. 6.10 shows the predictions along with the kriging standard errors. Using *BAND2* as an auxiliary variable greatly increased the precision of the predictions: the small-scale structure of *BAND1* was much more accurately reproduced from the support data: the motorway and the larger fields, as examples, were captured quite well. There were discontinuities in the prediction surface both at the support points of the target and the auxiliary variable.

The map of the kriging standard errors shows basically three magnitudes of the prediction errors. The errors vanished completely at the support points of *BAND1*, they were small at the support points of *BAND2*, and they were largest but approximately constant at the remaining points. The funnel-shaped depressions around the support points of *BAND1* were no longer present. Evidently, the auxiliary variable

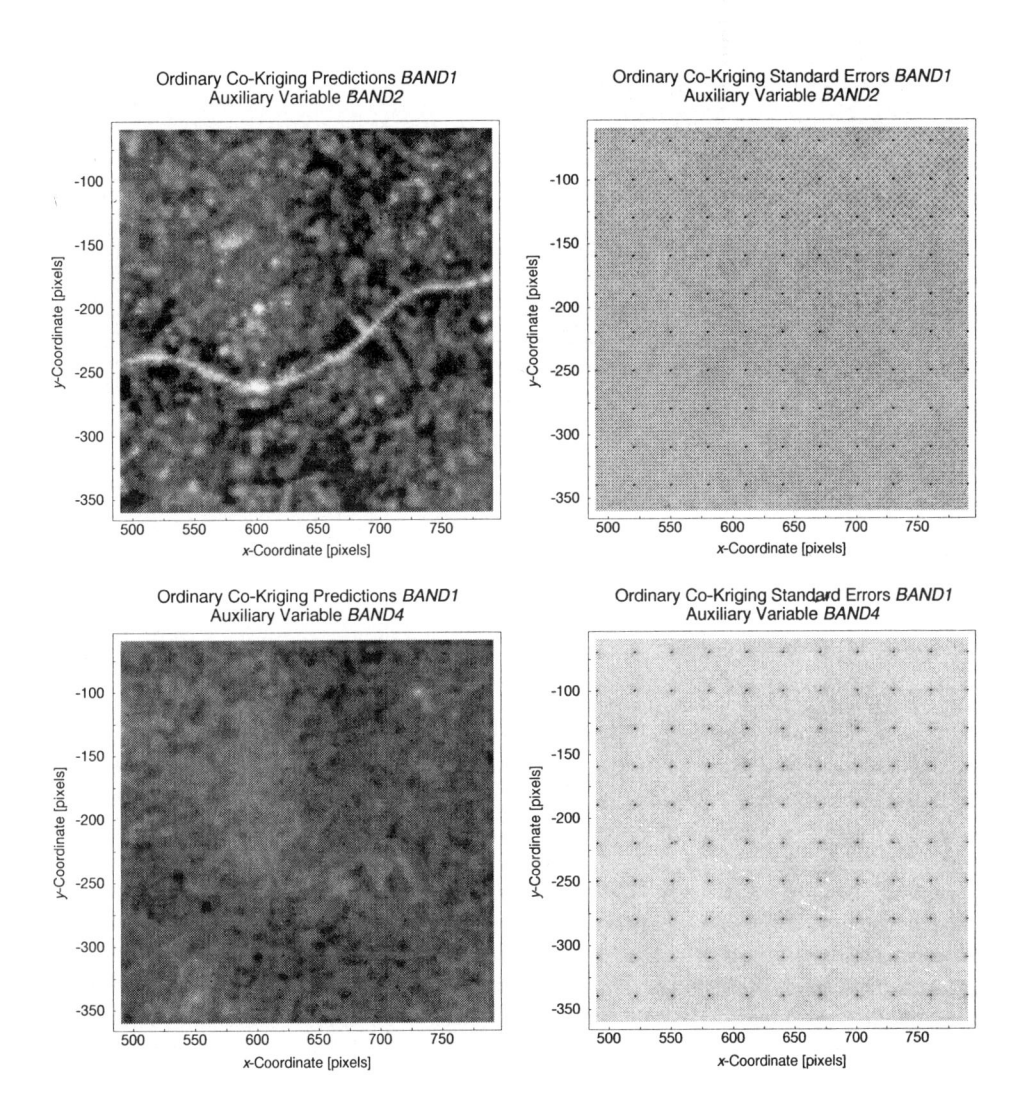

Chapter 6: Spatial prediction by linear kriging 111

Figure 6.10: Grey scale image of ordinary co-kriging predictions and kriging standard errors, *BAND1*, Landsat TM scene, Twente region. The grey levels vary from 72 DN (black) to 110 DN (white) for the predictions and from 0 DN (black) to 6 DN (white) for the standard errors. See text for further explanations.

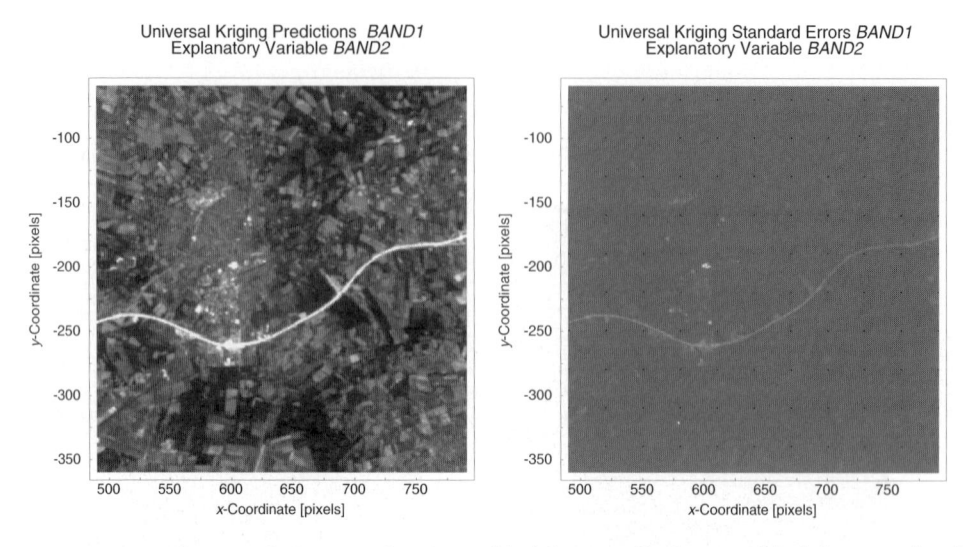

Figure 6.11: Grey scale image of universal kriging predictions and kriging standard errors, *BAND1*, Landsat TM scene, Twente region. The grey levels vary from 72 DN (black) to 110 DN (white) for the predictions and from 0 DN (black) to 6 DN (white) for the standard errors. See text for further explanations.

Table 6.1: Mean error *MEP* and mean square error *MSEP* for predicting *BAND1*, Landsat TM scene, Twente region.

	Ordinary Kriging	Co-kriging		Universal Kriging	
Aux. Variable	—	BAND4	BAND2	BAND4	BAND2
MEP [DN]	0.715	0.679	0.559	0.587	0.024
MSEP [DN]	40.0	37.1	15.9	36.5	4.42

provided most of the information to predict *BAND1*. This is not at all surprising, in view of the strong correlation and the large density of its support points.

The co-kriging predictions using *BAND4* as the auxiliary variable showed more detail than the univariate predictions but much less than the co-kriging predictions with *BAND2*. The motorway, for instance, was hardly visible. Discontinuities of the predictions were visible only at the support points of *BAND1*. Similarly, there was hardly any reduction of the kriging standard errors at the support points of *BAND4*, and the funnel-shaped depressions in the error surface were again discernible. Thus, although available at the same number of support points, *BAND4* contributed much less to the predictions than *BAND4* due to its weaker correlation with the target variable.

Third, Fig. 6.11 shows the results of universal kriging using *BAND2* as the explanatory variable for the mean function. To this end, we assumed that the auxiliary information was available at any point in the target region, and the variogram of the ordinary least-squares residuals (cf. Fig. 6.2b) was used for prediction. The moving local neighborhood was the same as for univariate ordinary kriging.

The prediction surface described the spatial distribution of *BAND1* almost perfectly. The small-scale structures were accurately reproduced, and it is quite difficult to spot some discrepancies between the predictions and the true values. The map of the standard errors differed strongly from those obtained by the other methods. The largest error occurred at the brightest pixels of *BAND1*, i.e., at the positions where also *BAND2* was largest. It is well-known from linear regression theory, that the prediction standard error depends on the value of the explanatory variable. Thus, at some points, the universal kriging variance was dominated by the uncertainty caused by estimating the mean function (cf. Eq. 6.23). Apart from the discontinuities at the support points of *BAND1*, the contribution of the simple kriging error to the total mean square error did not vary strongly.

To summarize the results we compared the predictions obtained by the various approaches with the true values of *BAND1* and computed the mean errors, *MEP*, and mean square errors, *MSEP*, of prediction. Table 6.1 reports these statistics and shows that the use of auxiliary information can considerably improve the precision of the predictions. Universal kriging with *BAND2* as explanatory variable had an excellent precision because *BAND1* was strongly correlated with the *BAND2* and the latter was available at each prediction location. Co-kriging with *BAND2* as auxiliary variable was still more precise than the remaining methods, but it was less precise than universal kriging because the auxiliary information was available only on a 3×3 grid. Using *BAND4* as auxiliary variable did not improve the precision strongly, and the density of the points at which *BAND4* was available had only a small influence on the precision. The weak correlation of *BAND1* and *BAND4* was responsible for these findings. Thus, the benefits of using auxiliary variables depends both on the density of the auxiliary information and on the correlation with the target quantity. The larger the number of auxiliary support points and the stronger the correlation with the target quantity, the more precise will be the predictions.

Chapter 7

Issues of scale and optimal pixel size

Paul J. Curran and Peter M. Atkinson

"there is now ample theoretical and empirical evidence for recognizing this resolution dependency as a basic aspect of remote sensing. Its implications must be fully pursued" ([246]: p.103).

7.1 Introduction

In this chapter we will address issues of scale and the choice of an optimum spatial resolution (herein termed pixel size) for the study of our environment using remotely sensed data. We will introduce notions of scale and scaling-up and ask why pixel size is important. We will end by outlining and illustrating the spatial statistics that can be used to select an appropriate pixel size for a given remote sensing application.

7.1.1 Scale and scaling-up

Small plates on the side of a fish, graduated series of musical notes, flaky deposit, weighing instrument, climbing over a wall and removing tartar from teeth are some of the over thirty meanings of the word scale [50]. However, in a discussion of space the word scale has just two, although unfortunately opposite, meanings. First, cartographic scale, which is the ratio of the size of a representation of an object to its actual size and colloquial scale, which is a synonym for the word size. For example, a map of a field and a few trees would be at a large cartographic scale (e.g., 1:100) but a small colloquial scale (e.g., 'it's a small scale phenomenon'). In contrast, a map of the world would be at a small cartographic scale (e.g., 1:1 million) but a large colloquial scale (e.g., 'it's a large scale phenomenon') [91, 19, 350]. To avoid confusion and for the purposes of this chapter, descriptive terms such as 'local', 'regional' or 'global' scale will be used.

Over the past few hundred years we have developed a reasonably sound understanding of how our world 'works' at the local scale (i.e., a scale somewhere between cells/leaves and fields/ trees). However, many pressing environmental problems, such as carbon cycling or climate change, are regional to global in scale [133]. It would be reasonable to question if our local scale understanding could be scaled-up in order to resolve such regional to global scale problems. The answer is intuitively yes. For example, if we were to measure (i) the NO_2 flux from a typical paddy field and (ii) the area of all of the World's paddy fields we could then (iii) estimate the contribution that paddy fields make to global NO_2 flux. Unfortunately, in practice the answer to our scaling-up question is usually no. This is because the component parts of the environment usually interact [164] and so, for example, a region containing many paddy fields will usually have high levels of evaporation and rainfall and consequently a depressed NO_2 flux. Even when the degree of interaction is known we are never sure that local scale processes are operating at the regional to global scale or that any global scale predictions made on the basis of local scale processes are not wildly in error. Climatologists and more recently geomorphologists [234] and ecologists [357] have tried to identify phenomena that operate over a wide range of scales, for example, the absorption of photosynthetically - active radiation by a leaf and a forest [95]. Such fractal phenomena, in addition to non-fractal phenomena occurring on spatially autocorrelated landscapes, possess spatial dependence such that areas near to each other tend to be more similar than those further apart [92]. As a result there is usually, a relationship between the characteristics of an environmental phenomenon and the area over which it is observed. In remote sensing the observed area is defined, to a first approximation, by the pixel size of the resultant image [20].

7.1.2 Why is pixel size important?

Consider the following 'thumbnail sketch' of the now famous [249] paper.

"An observer trying to estimate the length of England's coastline from a satellite will make a smaller guess than an observer trying to walk its coves and beaches, who will make a smaller guess in turn than a snail negotiating every pebble" ([151]: p.96).

The insight provided by this observation has passed into our common understanding and we know that the length of a phenomenon depends on the spatial resolution of our measurements [246]. Paradoxically (in the context of section 7.1.1), the analogous conclusion that a remotely sensed observation of a phenomenon depends on its area (*i.e.* its pixel size) is rarely drawn.

If this point is unclear imagine measuring the perimeter of your hand with a 1 cm long ruler and a 1 mm long ruler. The 1 mm long ruler would capture more of your hands variability and the perimeter of your hand would be longer than if the 1 cm ruler had been used. Likewise, if we asked you to estimate the brightness of your hand over 1 cm x 1 cm squares and 1 mm x 1 mm squares then although the average brightness would be the same the variability would not. The hundred 1 mm x 1 mm squares within each 1 cm x 1 cm square would not all possess the same value of brightness and in capturing more of your hands variability would have a larger range. In other words, the variability in the brightness of a surface is dependent upon pixel size and it follows that the degree of this dependency is related to the spatial properties of that surface.

The implications of this observation can be visualized for the remote sensing of categorical variables [392, 366]. For example, in agricultural regions where the dominant spatial frequency of the scene is determined by the dimensions of a typical field, the remote sensing of land cover will be most accurate when using approximately field-sized pixels. If pixels much smaller than a typical field are used then the variability in remotely sensed observations for each land cover will increase (as within-field variations will be recorded) and consequently classification accuracy will decrease [365]. If pixels much larger than a typical field are used then the variability in the remotely sensed observations for each land cover will increase (as between-field variations will be recorded) and consequently classification accuracy will decrease [397, 398]. Remote sensing practitioners have a feel for this interaction between pixel size, spatial frequency of the scene and classification accuracy. For example, there is a sizeable literature on the use of per-field classification to remove within field variability [296, 14] and spectral unmixing to reduce between field variability [136, 93]. The influence of pixel size on the remote sensing of continuous variables has received relatively little attention in the remote sensing literature. Some notable exceptions being [65, 318, 20, 21, 19, 92] and [116]. However, the literature is replete with examples where the influence of pixel size on remotely sensed observations is not considered (discussion in [129]). Perhaps some of the most striking examples are when canopy reflectance models are used to derive a-spatial or small area radiation/biophysical variable relationships on the assumption that these relationships will be the same at the pixel sizes associated with airborne or satellite sensors [153, 163, 152]. The folly of this assumption is discussed in [51].

As is implied by the hand example (earlier in this section) the variability and range of remotely sensed observations will, in most cases, decrease with an increase in pixel size [103]. Not surprisingly, therefore, researchers have observed that

"relationships between spectra and biophysical variables at one scale were not the same at another scale" ([90]: p.893).

The case study provides an illustration of these relationships between pixel size, remotely sensed data and a biophysical variable. The social science community refer to similar triangular relationships as the 'scaling effect'. In addition, the case study demonstrates that the location of observations is also related to the relationship between remotely sensed data and a biophysical variable. The social science community refer to similar triangular relationships as the 'zoning effect'. Discussion of the influence of spatial resolution on the relationship between two variables has a long history in the social sciences [149, 279]. For example, in the statistical textbook by Yule [405] the importance of space and place were illustrated by correlating wheat and potato yields for 48 agricultural counties in England. A correlation coefficient (r) of 0.22 was increased to 0.58 and then 0.99 as counties were aggregated into first 12 and then 3 spatial units (i.e., scaling effect). In addition there were many possible r values for each configuration of counties (i.e., zoning effect). In the thirty or so years that followed there developed a considerable body of literature on what became known as the 'modifiable areal unit problem' or MAUP [305, 360, 94, 68] culminating in the reviews by [290] and [289]. At first sight the MAUP of the social sciences appears to be very similar to the effect of pixel size on the relationship between remotely sensed observations and biophysical variables. However, there are important differences, as in the social sciences the sizes and shapes of the areal unit (e.g., voting area) varies and the area of the spatial unit is often not related directly to what is being observed. For example, census data are based on areas that vary in size and shape and have a different number of people within them [253]. The MAUP has tantalized social scientists [148, 402, 142], many of whom have searched for a 'solution' [139, 165] using techniques drawn from geographical information systems, geo-computation and operational research [15, 141, 183, 347]. Such research has emphasized the complexity of the MAUP and determined that the magnitude of 'scaling' and 'zoning' effects were related directly to the spatial autocorrelation of the area being observed. In remote sensing we usually have a constant size, shape and meaning for our pixels. As a result, geostatistical techniques (Webster and Oliver, 1990) that utilize this spatial autocorrelation can be used to model, understand and predict the influence of pixel size on remotely sensed observations and the environment they represent [407, 202, 83, 312].

The use of geostatistics within remote sensing is relatively new [350]. If we look back only two decades we see that commentators at the time had identified, as a weakness, our inability to define what pixel size we wanted.

"Surprisingly, earth scientists have as yet largely failed to provide comprehensive quantitative data in a form which is comparable with deriving objective estimates of resolution requirements" ([365]: p. 50).

The productive synergy between geostatistics and remote sensing is the result of a series of papers in the late 1980s (see [92] for a review) that discussed images in terms of the spatial autocorrelation between pixels. For example, homogenous areas were seen as having high spatial autocorrelation (neighboring pixels were similar) such that a coarse spatial resolution (i.e., large pixel size) would be appropriate and heterogeneous areas were seen as having low spatial autocorrelation (neighboring pixels were dissimilar) and a fine spatial resolution (i.e., small pixel size) would be needed. This conclusion was formalized by Woodcock and Strahler [396] who showed how between-pixel variance varied according to the spatial variability of the scene and the spatial resolution of the sensor and Curran [88] who used geostatistics to provide a methodological link between pixel size requirements for different land covers. The basis of this link was a plot of variability against ground distance, otherwise known as the variogram [18, 20, 102]. In the following section we will explore the use of the variogram to estimate the optimum pixel size.

7.2 Geostatistical theory and method

The geostatistical background to this chapter has already been given in Chapter 5. In this section, the geostatistical operation of regularization is discussed and some statistics which are useful for selecting a pixel size for remote sensing investigations are described.

7.2.1 Regularization

Observations are usually obtained over a positive finite support (i.e., pixel) v, and this is certainly the case in remote sensing. As a consequence, the sample variogram

Chapter 7: Issues of scale and optimal pixel size

of some variable such as reflectance in a given waveband is itself defined for a support of given positive size. If reflectance were measured on a different support it would have a different form. Therefore, it is important to know the support (or pixel) for which the variogram is defined.

Figs. 7.1a and b show two sets of variograms that have been computed from airborne multispectral scanner system (MSS) data in red (0.63-0.69 μm) (Fig. 7.1a) and near-infrared (0.76-0.90 μm) (Fig. 7.1b) wavelengths, and at two pixel sizes of 1.5 m (upper curves) and 2 m (lower curves) [18]. The variograms have been fitted using the exponential model with a nugget component (chapter 5). The effect of coarsening the pixel size is to remove short-range variation from the variograms so that generally the semivariance decreases. The variograms representing a pixel size of 2 m then describe the longer-range variation that remains. In the remainder of this section a model of regularization is described which allows one to change the support of the variogram, that is, change the pixel size and observe its effect on the variogram.

We start by defining a RV $Z_v(x_o)$ (which we use as a model of the generating process for an observed datum $z_v(x_o)$ as a function of some underlying RF $Z(y)$ defined on a punctual (or point) support and a support of size v:

$$Z_v(x_o) = \frac{1}{v} \int_{v(x_o)} Z(y) dy$$

The relation between the punctual (or point) semivariance and the regularized (defined on a support of positive size) semivariance at a lag h is given [66, 216, 225, 226] by:

$$\gamma_v(h) = \overline{\gamma}(v, v_k) - \overline{\gamma}(v, v), \tag{7.1}$$

where, $\overline{\gamma}(v, v_k)$ is the integral of the punctual semivariance between two pixels of size v whose centroids are separated by h, given formally by:

$$\overline{\gamma}(v, v_k) = \frac{1}{v^2} \int_v \int_{v(h)} \gamma(y, y') dy dy',$$

where, y sweeps a pixel of size v and y' sweeps independently another pixel of equal size and shape at a lag h away. The quantity $\overline{\gamma}(v, v)$ is the average punctual semivariance within a pixel of size v and is written formally as:

$$\overline{\gamma}(v, v) = \frac{1}{v^2} \int_v \int_v \gamma(y, y') dy dy',$$

where y and y' now sweep the same pixel independently.

Given Equation 7.1 and a model for the punctual variogram it is possible to estimate the regularized variogram for any pixel of size v. The regularized function is estimated through numerical approximation, by discretizing the new coarser pixel into a given number of points arranged on a regular grid (in the examples used within

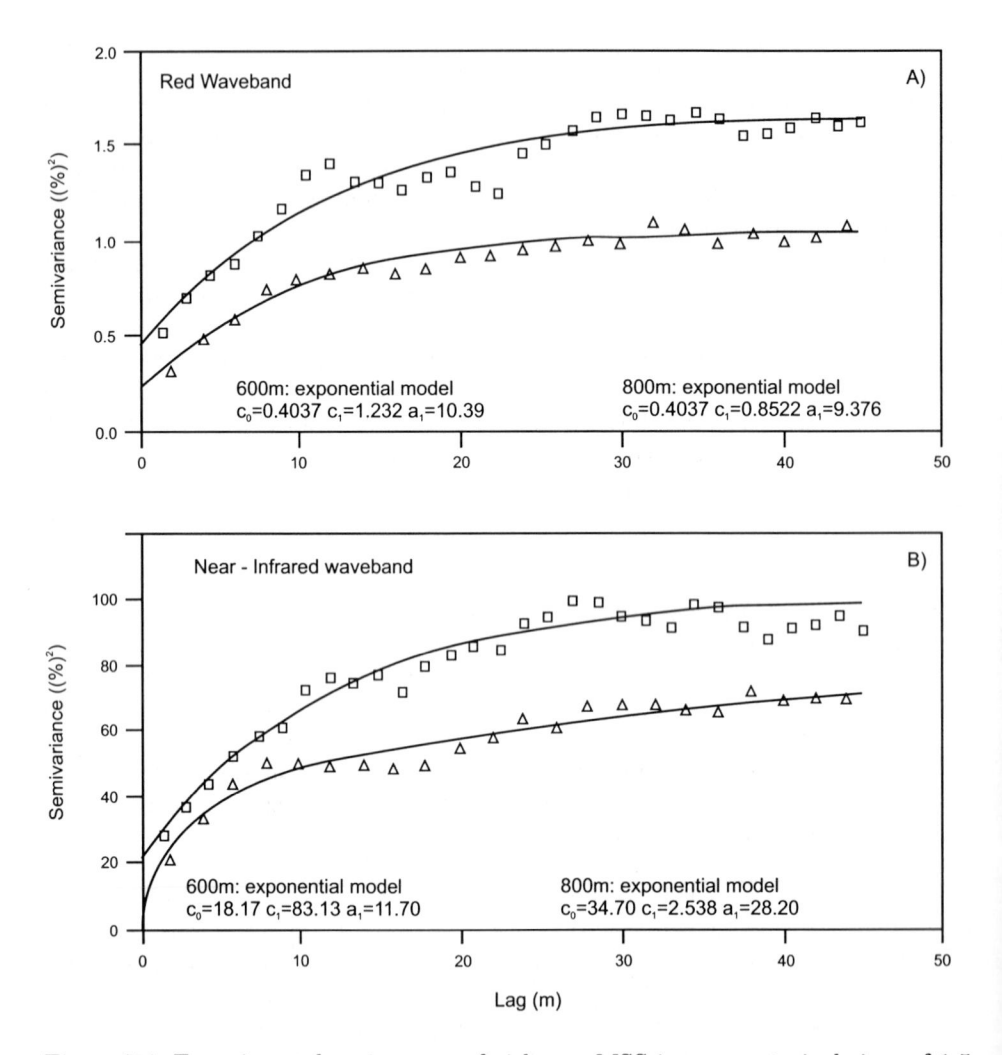

Figure 7.1: Experimental variograms of airborne MSS imagery at pixel sizes of 1.5 m. and 2 m. in the (a) red and (b) near-infrared wavebands (after [18]).

this chapter, 8 by 8 points were used). Thus, expressing Equation 7.1 in numerical terms we obtain:

$$\hat{\gamma}_v(h) = \frac{1}{M_x M_y N_x N_y} \sum_{i=1}^{M_x} \sum_{j=1}^{M_y} \sum_{k=1}^{N_x} \sum_{l=1}^{N_y} \gamma(x_{ij} - (x_{kl} + h))$$

$$- \frac{1}{M_x M_y N_x N_y} \sum_{i=1}^{M_x} \sum_{j=1}^{M_y} \sum_{k=1}^{N_x} \sum_{l=1}^{N_y} \gamma(x_{ij} - x_{kl}), \qquad (7.2)$$

where M_x and M_y are the number of points in the x and y directions respectively for pixel v and N_x and N_y are the number of points in the x and y directions respectively for pixel v_h. The estimated function, since it is obtained analytically, is necessarily discrete and the investigator may choose to fit a continuous mathematical model to the discrete values prior to further analysis.

Equations 7.1 and 7.2 provide the investigator with a useful tool for exploring the relation between the observed variation and the support of the sampling frame. In particular, it provides information on which to base a choice of pixel size for a given remote sensing investigation [396]. Since no measurement is required except on the original support it amounts to scaling the model rather than the data. The importance of Equation 7.1 has to do not only with understanding the effect of the support and regularizing the variogram. Summary statistics such as the dispersion variance $D^2(v, V)$ (for data on a support v within a region V) and a priori variance $D^2(v, \infty)$, and simulations may be produced from the regularized variogram for the new support without actually measuring on that support [407].

7.2.2 Selecting an optimum pixel size

In this section, we turn our attention more specifically to the problem of choosing a pixel size for a given remote sensing investigation. In particular, we define four statistics, which may be expressed as functions of pixel size, which can be used to aid this choice.

We have already seen how the variogram changes with spatial resolution. However, it is not clear how the variogram might be useful in selecting a spatial resolution. It is useful, therefore, to consider what information the variogram conveys about the measured spatial variation it represents. To do this, consider the parameters of an exponential model fitted to the sample variogram. The sill c of the exponential model relates information on the amount of variation present in V. In fact, the sill is equal to the a priori variance $D^2(v, \infty)$ of Z. The non-linear parameter r of the model relates information on the scale of spatial variation. Since the exponential model is asymptotic a series of different, superimposed scales of spatial variation is implied, with no clear maximum range. The spherical model, on the other hand, does have a clear maximum range a. In fact, for 3-D the exponential model lies somewhere between the spherical model (one maximum scale of variation) and the power model (fractal variation). A nugget model component is often fitted together with a structured component such as an exponential or spherical model (see Chapter 5). The nugget variance (the sill of the nugget component) is related to both variation which exists at a micro-scale (a shorter interval than that of the sampling frame) and measurement error.

Given the above interpretations of the parameters of the fitted model, it is possible to see how the variogram might aid in the selection of an appropriate combination of pixel size and estimation technique for a given investigation [396]. For example, suppose that the investigator wishes to map land cover within V and the range a of a fitted spherical model is 300 m. In the simplest case, a range of 300 m implies that the largest land cover 'objects' (say agricultural fields) in the scene are of this 'diameter'. Thus, the investigator might select imagery with a pixel size of 30 m (to capture the variation of interest) and use maximum likelihood classification to estimate hard land cover classes for each pixel. However, if the only imagery available had a pixel size of 260 m, then the investigator might apply some fuzzy classifier instead to estimate fuzzy memberships to each land cover class. A problem with using the variogram in this way is that it may be difficult to interpret a variogram model which does not have a definite range (such as the power model). Thus, four statistics are defined below which provide more tractable information on the relation between the observed variation and pixel size.

The average local variance has been used previously to help select a suitable pixel size [396, 397, 398]. The average local variance σ_{vw}^2 may be estimated from a moving (3 by 3) window w applied to an image of L rows by M columns of pixels with support v using:

$$\bar{\sigma}_{vw}^2 = \frac{1}{L \cdot M} \sum_{l=1}^{L} \sum_{m=1}^{M} \frac{1}{9} \sum_{j=-1}^{+1} \sum_{k=-1}^{+1} [\bar{z}_v(l+j, m+k) - z_v(l+j, m+k)]^2, \tag{7.3}$$

where it is assumed that there is a buffer of one pixel surrounding the image to be analyzed. The local variance σ_{vw}^2 is calculated for a range of integer multiples of the original pixel size $|v|$ and expressed as a function of pixel size. The plot usually rises to a peak and thereafter decreases with increasing pixel size. The pixel size at which the peak occurs may help to identify the predominant scale of spatial variation in the image. Thus, the pixel size at which the plot of $\bar{\sigma}_{vw}^2$ against pixel size is maximum is related to the range of the variogram model and conveys similar information [396].

The semivariance $\gamma_v(p)$ at a lag of one pixel interval p (where $|p| = \sqrt{v}$) may be obtained from $\alpha = 1, 2, \ldots, P(p)$ pairs of pixel values $\{z_v(x_\alpha), z_v(x_\alpha + p)\}$ defined on a pixel of size v at locations $\{x_\alpha, x_\alpha + p\}$:

$$\widehat{\gamma}_v(p) = \frac{1}{2P(p)} \sum_{\alpha=1}^{P(p)} [z_v(x_\alpha) - z_v(x_\alpha + p)]^2.$$

Thus, $\gamma_v(p)$ can be estimated for several different pixel sizes by coarsening the pixel size of (degrading) the imagery and computing $\widehat{\gamma}_v(p)$ each time. Then, $\widehat{\gamma}_v(p)$ plotted against spatial resolution can be used in a way similar to the local variance. The quantity $\widehat{\gamma}_v(p)$ could equally be defined for a local moving window (as for the local variance) in which case the central pixel $\{x_i, i = 1, 2, \ldots, N\}$ would be compared with its eight neighbors. In this chapter we use $\gamma_v(p)$ where either this or σ_{vw}^2 could be used to relate similar information.

One advantage of using $\hat{\gamma}_v(p)$ is that it can be estimated readily by regularizing the punctual variogram as an alternative to degrading the imagery. This means that $\hat{\gamma}_v(p)$ can be obtained for any pixel size, not just integer multiples of the original pixel size, that image data are not a prerequisite for estimation of $\gamma_v(p)$ (for example, transect data would do) and that it is possible to deal with measurement error and the point-spread-function more appropriately [21].

The familiar dispersion (or sample) variance $D^2(v, V)$ of a variable defined on a support v within a region V may also be used to relate information pertinent to the choice of pixel size (see also [370]):

$$D^2(v, V) = \frac{1}{N} \sum_{i=1}^{N} [\bar{z}_v(x_i) - z_v(x_i)]^2,$$

where, $z_v(x_i)$ is the mean of all $\{i = 1, 2, \ldots, N\}$ values $z_v(x_i)$ within the image. The image variance $I^2(v, V)$ may be obtained from $D^2(v, V)$ as follows:

$$I^2(v \cdot 2^k, V) = D^2(v \cdot 2^k, V) - D^2(v \cdot 2^{k+1}, V),$$

where, 2^k varies between 0 and $\sqrt{N}$. Clearly, this implies that the image is comprised of $\sqrt{N}$ rows by $\sqrt{N}$ columns and that $\sqrt{N} = 2^k$ for some value of k. Thus, a prerequisite for calculation of the image variance is that the image must be of 2^k by 2^k pixels. However, the image variance can be estimated from the punctual variogram using Equation 7.2 (see similar argument above for the quantity $\gamma_v(p)$). Formally, the dispersion variance $D^2(v, V)$ may be obtained from the regularized variogram using:

$$D^2(v, V) = \frac{1}{V^2} \int_V \int_V \gamma_v(y, y') dy dy',$$

that is, the integral of the regularized variogram over the region V. The image variance may then be obtained using Equation 7.3.

7.3 Remotely sensed imagery

As in chapter 5 only three of the available seven wavebands were selected for study in this chapter. The wavebands chosen were the green waveband (0.52 μm-0.60 μm), the red waveband (0.63 μm-0.69 μm) and the near-infrared waveband (0.76 μm-0.90 μm). Further, the normalized difference vegetation index (NDVI) was computed as NDVI = (NIR-Red)/(NIR+Red), where, NIR represents radiance in the near-infrared waveband and Red represents radiance in the red waveband.

In addition to restricting the choice of wavebands, only two small subsets of the original image were considered for further analysis. First, a 128 by 128 pixel subset covered entirely by the urban area of Enschede was selected. Second, a 128 by 128 pixel area to the south-west of the center of the image was selected. These two areas are referred to in subsequent discussions as 'urban' and 'agricultural'.

As stated in chapter 5, the subsets are chosen from the imagery for illustrative purposes only. The underlying phenomena in the two subsets are (i) an urban area filled with sharply defined objects such as buildings, roads, gardens, parks and so on and (ii) an agricultural area with agricultural fields having relatively homogeneous crop covers (such as cereals) with abrupt boundaries between them. Neither subset, therefore, fits well the continuous RF model. However, to some extent the variation is smoothed by the pixel size of 30 m (by 30 m). Thus, in the urban area, individual buildings are rarely observable and the variation appears partly smooth and partly random. In the agricultural area, individual fields are visible and this makes application of the continuous RF model less than ideal. It would be more usual to apply geostatistics either (i) where the pixel size was fine enough to restrict the analysis to the continuous variation within individual fields or (ii) where the spatial resolution was coarse enough to treat the variation as continuous.

7.4 Examples

The two subsets were analyzed in chapter 5 providing histograms and omnidirectional modeled variograms for each of the green, red and near-infrared wavebands and the NDVI. The variograms (obtained using the GSLIB software, [106] and the fitted models are shown in Fig. 7.2.

As an initial step, the variogram models of Fig. 7.3 were regularized to a support of 30 m by 30 m as if the fitted models represented punctual (point-to-point) variation. The regularized variograms thus obtained are shown in Fig. 7.4 as the lower dashed curves. The regularization was undertaken using code written in the Splus^{TM} software following Equation 7.2 . Fig. 7.3 provides some guidance on the order of magnitude of the variogram parameters required for a punctual model.

Punctual models were fitted to the variograms defined for a pixel size of 30 m by 30 m by trial and error. The procedure was as follows. First, we decided to use an exponential model for the punctual variogram. Next, a value was chosen for the non-linear parameter. With this parameter fixed, the Splus code was used to fit iteratively the sill of the exponential model over a range of lags chosen to be $\{h = 300, 330, \ldots, 750m\}$ using Equation 7.2 . Thus, an initial sill value was provided by the user, the punctual model was regularized to obtain estimates for lags between 300 m and 750 m and the average difference for these fifteen lags was used to estimate a new sill as $\{c_{new} = c_{old} + avg(\gamma_{v(obs)}(h) - \gamma_{v(est)}(h))\}$. The process was repeated (about five iterations was usually sufficient) with the most recently estimated sill as input for the next iteration until a satisfactory fit was obtained. Visual inspection of the fitted models allowed the whole process to be repeated with new estimates for the non-linear parameters of the models. The resulting punctual models and their coefficients are shown in Fig. 7.4, together with the original sample functions (discrete symbols) and the regularized estimates obtained

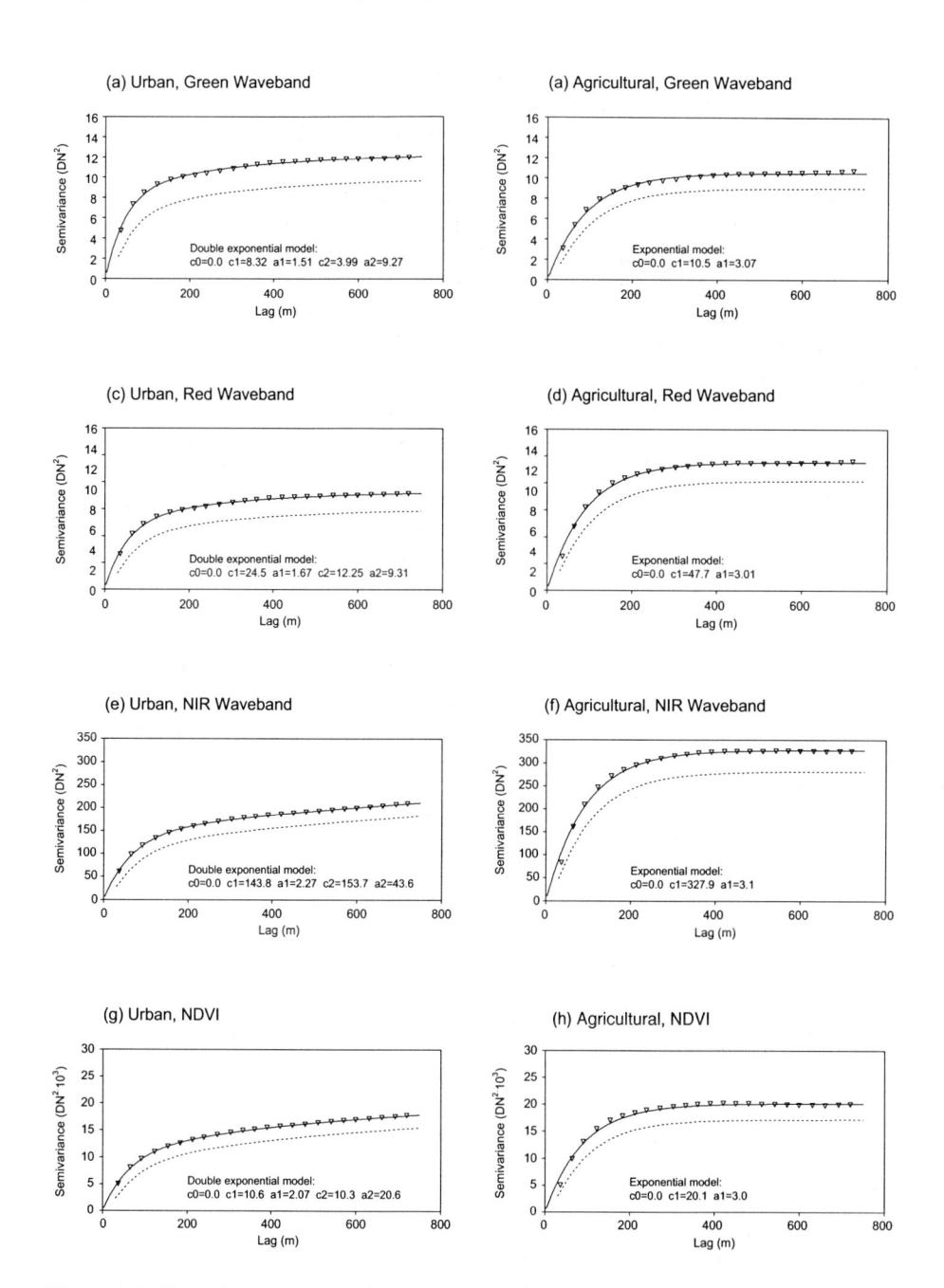

Chapter 7: Issues of scale and optimal pixel size 125

Figure 7.2: Sample variograms (discrete points) fitted with the exponential or double exponential model (solid curves). The fitted models have been further regularized over a support of 30 m by 30 m (dashed curve) to help determine the order of magnitude of the starting values for the coefficients of a punctual model.

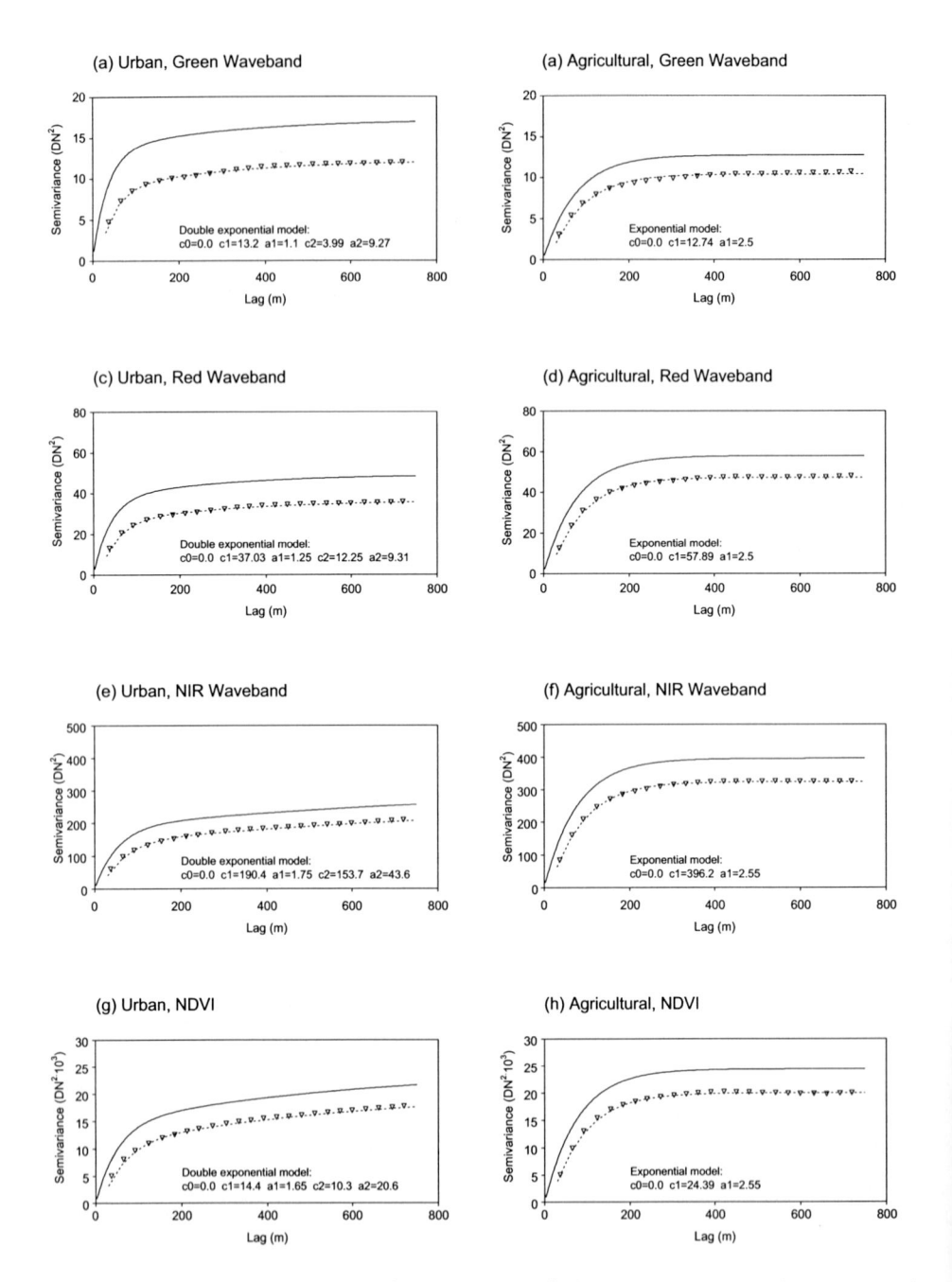

Figure 7.3: Sample variograms (discrete points) fitted with punctual exponential or double exponential models (solid curves). The punctual models have been regularized over a support of 30 m by 30 m (dashed curves) for comparison with the sample variograms.

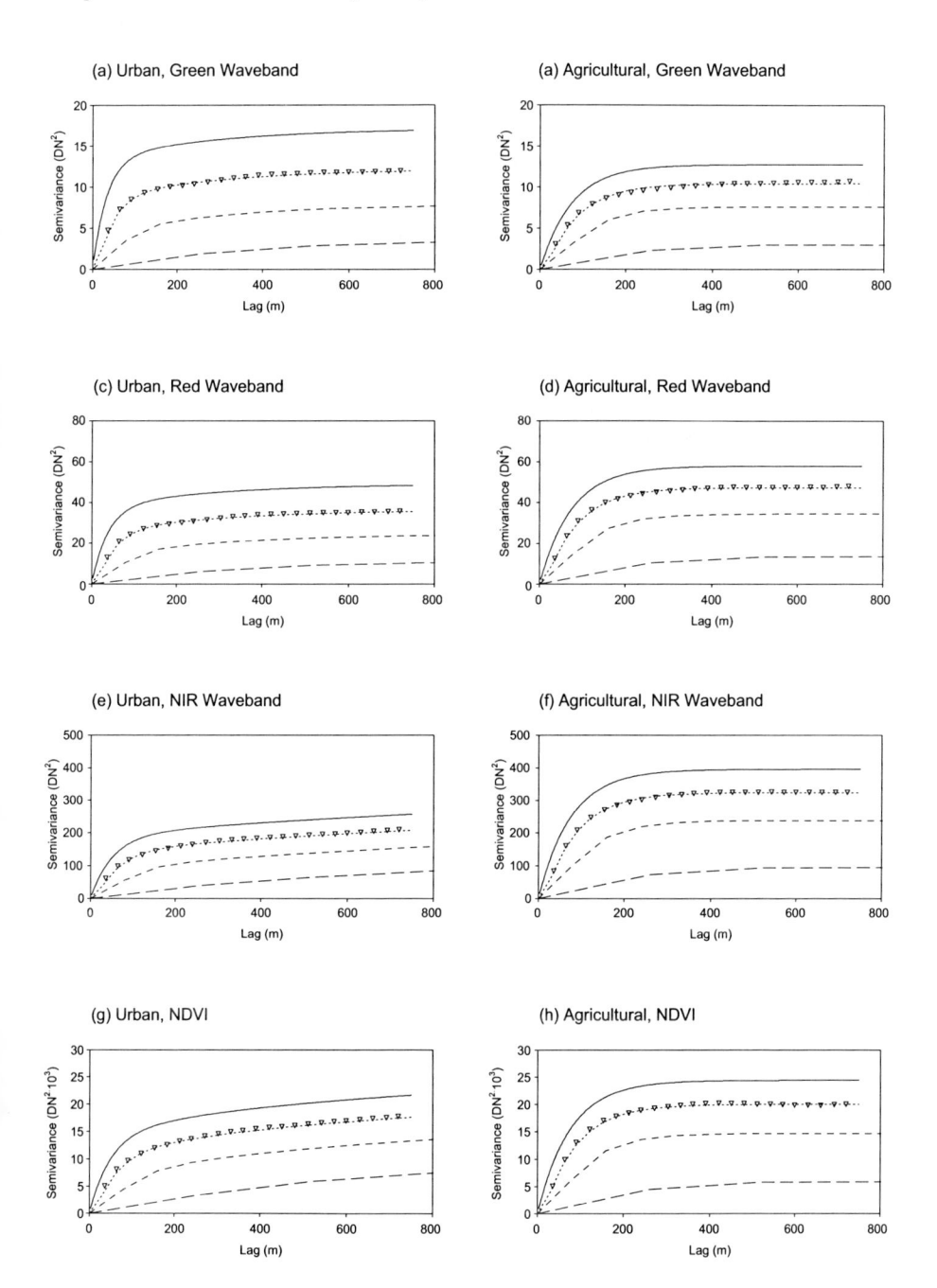

Chapter 7: Issues of scale and optimal pixel size   127

(a) Urban, Green Waveband

(a) Agricultural, Green Waveband

(c) Urban, Red Waveband

(d) Agricultural, Red Waveband

(e) Urban, NIR Waveband

(f) Agricultural, NIR Waveband

(g) Urban, NDVI

(h) Agricultural, NDVI

Figure 7.4: Sample variograms (discrete points) fitted with punctual exponential or double exponential models (solid curves) regularized over supports of 30 m by 30 m (upper dashed curves), 80 m by 80 m (middle dashed curves) and 260 m by 260 m (lower dashed curves).

from the punctual models (lower dashed curves). Clearly, greater automation would be helpful, but the procedure had the advantage of allowing user intervention.

Once a satisfactory punctual model had been fitted for each of the eight variograms shown in. Fig. 7.4, it was possible to regularize the variograms to any desired pixel size using Equation 7.2 . We chose to regularize the variograms to pixel sizes of 80 m by 80 m and 260 m by 260 m: 80 m represents approximately the spatial resolution of Landsat multispectral scanner (MSS) imagery and 260 m represents that of the forthcoming Medium Resolution Imaging Spectrometer (MERIS) imagery [383]. These regularized variograms are shown along with the punctual models from which they were obtained as the two lower dashed curves in Fig. 7.4. Clearly, the variation one would observe in MERIS imagery is very different to that observable in Landsat TM imagery. This has implications for the kinds of analysis and the set of techniques which might be applied.

As an example, the punctual variograms obtained in Fig.7.4 were regularized to new pixel sizes $\{v_k, k = 1, 2, \ldots\}$ to estimate the quantities $\gamma_{vk}(p)$. The resulting plots of $\gamma_{vk}(p)$ against v_k are shown in Fig.7.5.

The semivariance increases to a peak at a pixel size of about 60 m for the urban examples and at about 120 m for the agriculture examples. If the objective were to map the variation in the urban subset then the pixel size chosen should be smaller than the 30 m of the Landsat TM imagery, say about 10 m at the most. This makes sense when one considers that the urban scene is comprised of individual buildings and blocks of buildings interspersed with gardens and parks. On the other hand, one might consider using a fuzzy classifier if Landsat TM imagery were all that were available. For the agriculture subset and for most wavebands the peak occurs at a larger pixel size indicating that the Landsat TM pixels may be just fine enough to map the agricultural fields, but that one would be better (depending on the objective) using, for example, SPOT HRV, XS or Panchromatic imagery with pixel sizes of 20 m by 20 m and 10 m by 10 m respectively.

As an alternative to the quantity $\gamma_{vk}(p)$ the data for the two subsets were degraded to successively coarser pixel sizes to obtain plots of dispersion variance $D^2(v, V)$ and image variance $I^2(v, V)$ against pixel size. Both functions are plotted in Fig.7.6, where the monotonically decreasing plot is $D^2(v, V)$. The image variance is interesting because it effectively partitions $D^2(v, V)$ into different scales of operation, such that the sum of all k values of $I^2(v, V)$ is equal to $D^2(v, V)$. Further, whereas for the local variance σ_{vw}^2 and the quantity $\gamma_{vk}(p)$ there is likely to be only one peak, for the image variance there is quite often more than one peak. For example, in Fig.7.6 there is clear evidence for a second peak for the urban area plots (near-infrared and NDVI), and this is most likely to be related to the small trend component discussed (see Chapters 5 and 6). Thus, the plots of $I^2(v, V)$ against v suggest that there is a peak at around 60 m for the urban subset and 120 m for the agriculture subset, but also that there is evidence for a longer-range structure in the near-infrared and NDVI for the urban subset. Much of the information which is provided by these plots can also be obtained from the variogram by an experienced interpreter.

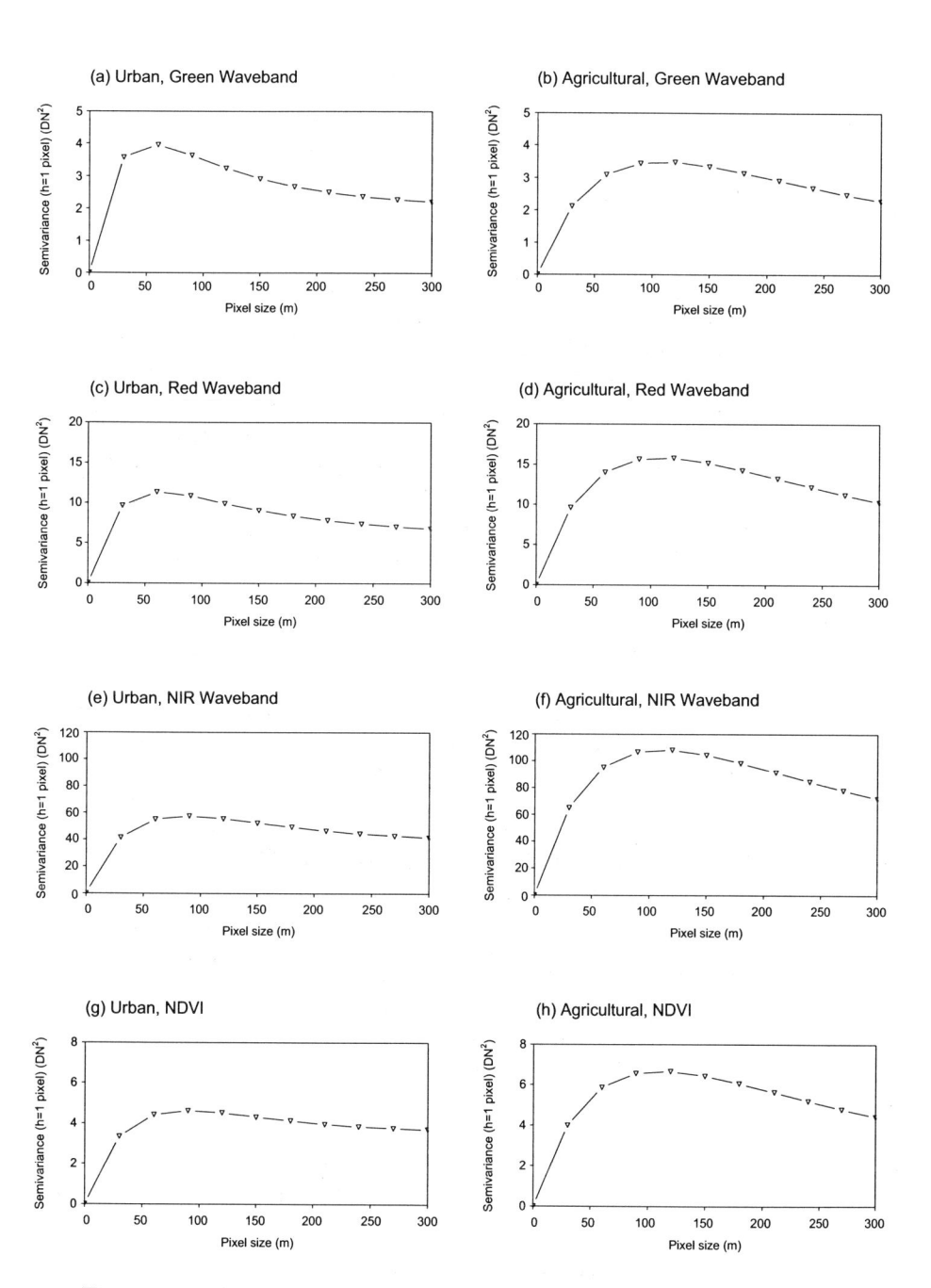

Figure 7.5: Plots of semivariance at a lag of one pixel against pixel size.

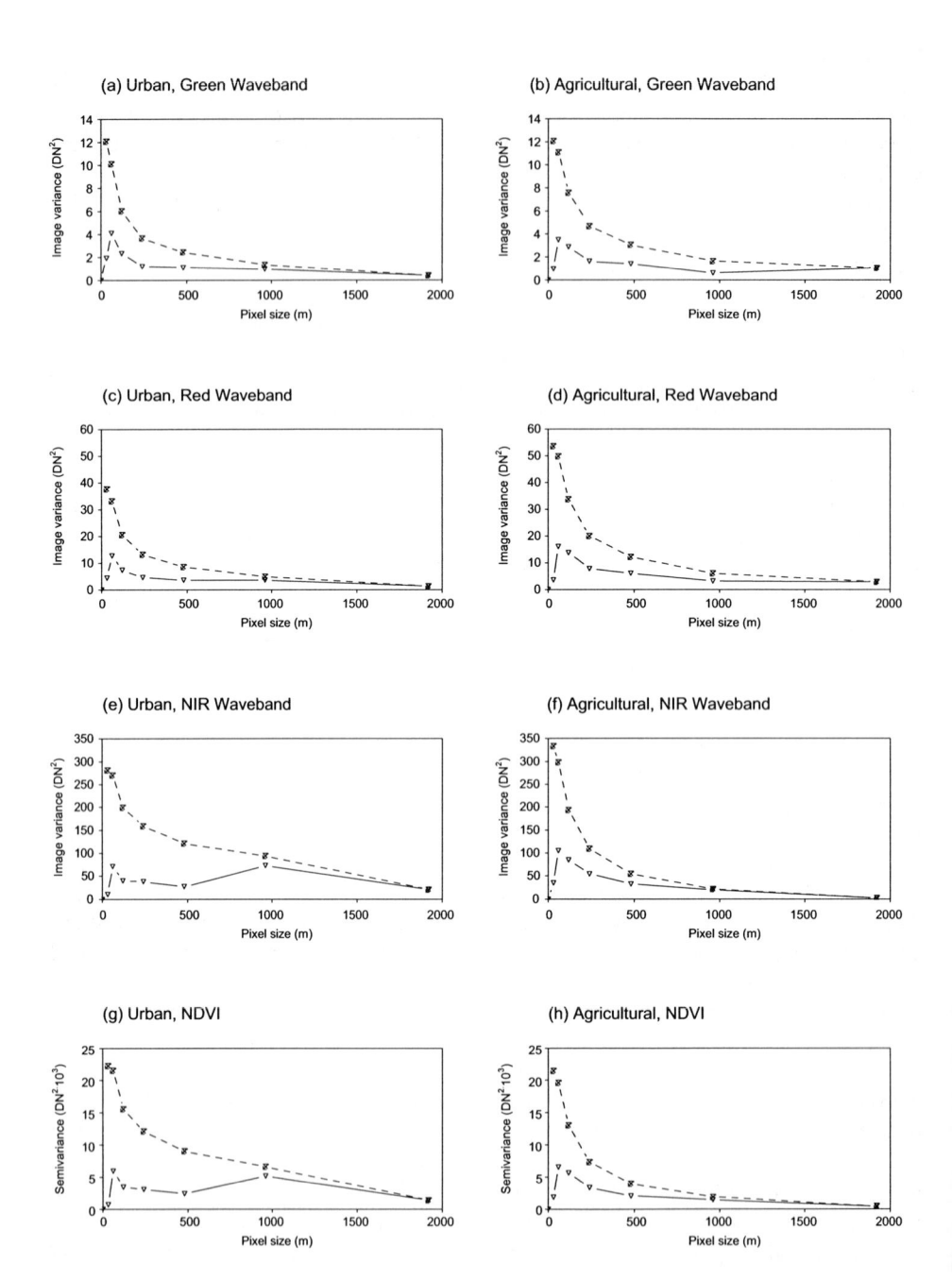

Figure 7.6: Plots of dispersion variance (upper dashed curves) and image variance (lower dashed curves) against pixel size.

Case study
The aim of this case study was to determine how much of the variability in the chlorophyll content of wheat canopy could be accounted for using the remotely sensed position of the 'red edge' (the boundary between chlorophyll absorption and within-leaf scattering).

The experiment took place at Rothamsted Experimental Station at Harpenden in Southern England. The site chosen was Claycroft, a short-term wheat (var. Marcia) fertilization experiment that ran from 1991 to 1994. The site was divided into 21 plots each 21 m x 23 m in size. An unfertilized control plot was selected as it had a mature (mean 87 cm) and complete canopy but the plants within it exhibited a wide range of vigor. Using dark colored string the central portion of this plot was divided into twenty 1m by twenty 1 m sub-plots leaving a border of between 0.5 to 1.5 m. These subplots represented 1m 'pixels' and within them remotely sensed measurements and a biophysical measurement were made.

On 15-18 June 1993 and within 2-3 hours of solar noon a radiance spectrum was recorded simultaneously for each sub-plot and associated reference target using an IRIS MKIV spectroradiometer [270]. The sensor head was mounted on average 1.4 m above the top of the canopy to give a constant field-of-view of 1 m. The radiance spectrum was converted to reflectance using the radiance spectrum from the reference target. The reflectance spectrum was then smoothed with a five point moving average and converted to a first derivative spectrum on which the point of maximum slope on the red edge (red edge position, REP) was located. The REP range was between 703-728 nm.

On 19-20 June three replicate measurements of relative chlorophyll content were recorded at each plot using a SPAD-502 chlorophyll meter [404]. This hand-held meter was clamped onto the flag leaf and measured transmission. The three 'SPAD values' were averaged by sub-plot, and ranged from 21-38. In the previous year a destructive experiment on other plots on this site had shown that the SPAD values were correlated strongly with chlorophyll content ($r= 0.95, n = 100$) [299]. Therefore, the SPAD values were treated as measures of relative chlorophyll content. The correlation between the REP and SPAD values was typical of that recorded in the earlier experiment above [299] at which samples had also been collected for subplots of 1m x 1m. The following relation was found:

SPAD value $= -492.2 + 0.64 \cdot \text{REP} (r = 0.66, n = 400)$

To illustrate the magnitude of the scaling effect and zoning effect the SPAD values and REP for the sub-plots were aggregated spatially. The table illustrates both the 'scaling effect' and the 'zoning effect'. There was a positive exponential relationship between the pixel size and correlation coefficient and the correlation coefficient also varies with the pixel location. Therefore, the correlation between REP and SPAD value was dependent upon not only the biophysical/biochemical interactions within the canopy but also on the size and location of the 'pixels' over which these interactions were measured.

7.5 Summary

The environment interacts differentially with electromagnetic radiation according to its essential physical, chemical and biological properties. The electromagnetic radiation recorded by an imaging sensor has value because (i) it can represent these essential physical, chemical and biological properties spatially and (ii) can be used to provide information on our environment; for example, its biomass, cover, temperature. However, points in the environment that are near to each other are more alike than those that are further away and the degree of dissimilarity depends upon spatial autocorrelation in the environment and the pixel size of our remotely sensed observations. In other words, space isn't a parameter it is a variable and relationships between remotely sensed data and ground data derived at one scale will not be the same at another. The spatial dependence in remotely sensed data is both a burden and a challenge. A burden in the sense that we need to account for the influence of space (other chapters in this book) but a challenge in that in choosing the spatial dimensions of our measurements we can do so to our advantage. This chapter has discussed one such example where spatial statistics were employed to select the optimum pixel size for a given remote sensing investigation.

The realization that we need to choose a pixel size is the first stage in exercising that choice. The message of this chapter is that as remote sensing is inherently spatial

> "it is a geographical fact of life that the results of spatial study will always depend on the areal units that are being studied" ([289]: p.37).

7.6 Acknowledgment

This chapter was written while Dr Peter Atkinson was on study leave at the School of Mathematics, University of Wales, Cardiff.

Table 7.1: The correlation between remotely sensed red edge and chlorophyll content for different pixel sizes and locations. The locations of the pixels was also an influencing factor.

Number of Pixels	Pixel size (m)	Correlation coefficient (r)
400	1x1	0.66
300	2x1	0.66
300	1x2	0.67
100	1x4	0.72
100	4x1	0.71
100	2x2	0.74
80	1x5	0.76
80	5x1	0.74
50	2x4	0.85
50	4x2	0.82
40	1x10	0.77
40	2x5	0.84
40	5x2	0.81
40	10x1	0.80
25	4x4	0.87
20	1x20	0.85
20	2x10	0.87
20	4x5	0.90
20	5x4	0.89
20	10x2	0.84
20	20x1	0.86
16	5x5	0.93
10	2x20	0.90
10	4x10	0.92
10	10x4	0.95
10	20x2	0.92
8	5x10	0.96
8	10x5	0.95
5	4x20	0.94
5	20x4	0.92
4	5x20	0.96
4	10x10	0.98
4	20x5	0.96

Chapter 8

Conditional Simulation: An alternative to estimation for achieving mapping objectives

Jennifer L. Dungan

An important objective of many environmental studies is to create maps, wall-to-wall descriptions of a quantitative variable, such as sediment load (g l^{-1}), land or water surface temperature (°C), soil organic matter content (mg cm^{-1}), precipitation (mm) or timber stand volume (m^3 m^{-2}). These are specialized, usually two dimensional, types of maps and are most often represented as contour plots or rasters. They are inherently different from maps for navigation and thematic or choropleth maps. A value is represented at every lattice point or grid cell. Examples of these maps are included in [240, 245, 379]. They are used to summarize the state of the environment and describe spatial trends. Increasingly, they constitute intermediate stages in an environmental analysis and are used as layers in geographical information systems. Remote sensing is often used to provide data to construct these maps.

In this chapter, we consider the purposes for creating maps representing spatial fields of one variable, and how these purposes can be met by a geostatistical method called conditional simulation. We will call the biophysical variable to be mapped across a region the primary variable. We will assume that ground-based measurements of the primary variable at a number of locations throughout the study region are available, whether they are used to create the map itself, to model the correspondence between ground and image data, or to check accuracy of the map. Other variables used in the mapping of the primary variable, in this case the spectral variables from remotely sensed measurements, will be called ancillary variables.

One way to map a primary variable using ground measurements is to interpolate, using ad hoc, kriging or spline methods [239] (see also Chapter 6 of this volume). Geostatistical interpolation, referred to as "estimation," is comprised of a variety of kriging methods. Matheron [259] shows that kriging and splines are equivalent. Assume for a moment that our objective is to map radiance (mW·cm^{-2}·st^{-1}·m^{-1}) in a particular waveband. Readings from a ground-based field radiometer and images from a space-borne radiometric sensor are the two data sources for the primary variable. Further assume that the ground area represented by a single field radiometer reading is equal to the area represented by one pixel. To illustrate this hypothetical situation, we will borrow a sub-scene of the Thematic Mapper data introduced in chapter 1, and use band 5 to represent the true radiance. If only the field radiometer data is used, and 150 samples are taken at locations well-distributed throughout the area to be mapped (Fig. 8.1a), estimates could be obtained using an interpolation method (Fig. 8.1b). Alternatively, if the imaging sensor is used, no interpolation would be needed, as long as the sensor was calibrated accurately for radiance. The digital numbers from the image would simply be rescaled to the appropriate radiometric units. The map from the imaging sensor (Fig. 8.1c) is a raster defined with a given pixel size (chapter 7) and, if the calibration was perfect, the map would be without error.

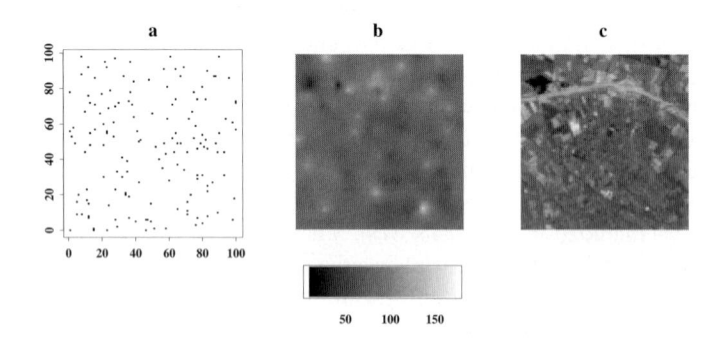

Figure 8.1: a) Locations of 150 samples representing ground measurements of the primary variable (from band 5 in the subset TM image). Note that symbols are not to scale. b) Estimates of radiance obtained using only 150 samples and a kriging method. c) Estimates obtained using a hypothetical calibrated image sensor (actually the "true" map of the primary variable).

The values of radiance at the ground-sampled locations are equal in Figs. 8.1b and 8.1c. This is where the similarity between these two maps ends. The sampled locations represent only 1.5% of the entire map, which is a 100 x 100 grid. The interpolated map (8.1b) has a generally smooth character while the remotely sensed map (8.1c) is much more spatially variable or non-smooth. Real measurements that are collected exhaustively over the landscape generally do not have the smooth character exhibited by interpolation. Therefore the smooth map, while founded on local optimality criteria, miss a feature of reality that is often important to capture—spatial variability. Spatial variability can be considered a global feature of map, since it only has meaning in the context of the relationship of all values to one another in space.

Naturally, mere radiance is rarely the ultimate objective of an environmental mapping study. Radiometric variables are only indirectly related to many of the biophysical, ecological or other environmental processes of interest to researchers. Direct measurements usually can be made at only a small number of locations, and remote sensing is used to fill in or extrapolate from these locations. This could be done with a multivariate interpolation method such as cokriging (Papritz and Stein this volume), but cokriging is no different than kriging in its failure to represent spatial variability. This chapter describes conditional simulation, an alternative to the kriging-type estimation methods that could be fruitfully employed in the spatial description of environmental variables to capture spatial variability.

Conditional simulation, also called stochastic interpolation, stochastic simulation or stochastic imaging, is a way of describing the variability in spatial fields [215, 223]. Conditional simulation is the generation of synthetic realizations of a random function that possess the same spatial statistics (moments up to the second order) as the data that have been collected about a variable. It is therefore an especially appropriate algorithm for mapping, because it emphasizes the global attributes of a spatial field, its histogram and spatial pattern. Conditional simulation algorithms yield not one, but several, maps, each of which is an equally likely outcome from the algorithm. The "conditional" part of simulation is the property that simulated fields have the same values at sampled locations as the measurements. This property, called "exactness," is a feature of geostatistical methods in general. Conditional simulation is a geostatistical method because it starts out with the assumption that spatial autocorrelation exists and exploits that autocorrelation. This is different from many statistics that assume independence (and identical distributions) for the random variables under consideration [83].

Besides spatial variability, the second major contribution of conditional simulation is a new spatial description of uncertainty from geostatistical regionalized variable theory. Both spatial variability and the characterization of uncertainty are relevant objectives for mapping biophysical variables. Using conditional simulations to represent spatial fields of a variable emerged relatively recently, in the early 1970's. Like other modern developments in resampling [119], it is made possible by the capability of computers to generate many millions of calculations. The purpose of this chapter is to familiarize the reader with conditional simulation, describe a few useful algorithms and discuss its unique role in exploring uncertainty about spatial fields. Though conditional simulation is well accepted in other geophysical disciplines, its place in remote sensing data analysis and applications is as yet very small—perhaps this chapter will spur further investigations of its utility.

8.1   Contrasting simulation with estimation

A map is but a model of reality. Every map is created with certain objectives, and certain properties of the reality it is meant to represent are emphasized while others are ignored. Therefore, the objectives in making maps from ground and image data must be clearly understood prior to choosing a method. Possible objectives are to predict a regional mean [17, 341], to locate regions of high or low values [26], to calculate areas meeting a criterion [338], or to accurately describe a primary variable's correspondence with another environmental variable [242, 80, 284]. Estimation and simulation methods are suited to different objectives.

If the objective for mapping requires spatial variability to be represented, we have already seen an illustration of a shortfall of estimation. That is, maps created via estimation are overly smooth compared to reality. Though the cartographic qualities of such smooth interpolated maps may be appealing, and in fact may be preferred by map users, it is clear that reduced spatial variability may pose problems for analysis. It is because of this overly smooth representation of reality that Journel ([221]: p. 31) cautions against the actual mapping of kriged results: "In all rigor, estimates based on a local accuracy criterion such as kriging should only be tabulated; to map them is to entice the user to read on the map patterns of spatial continuity that may be artifacts of the data configuration and the kriging smoothing effect."

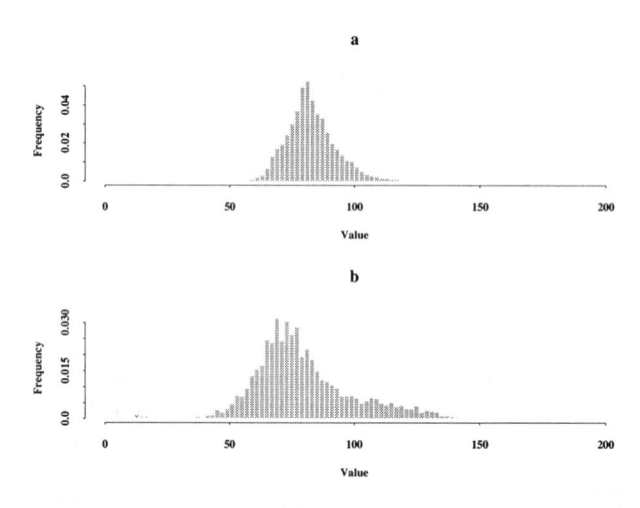

Figure 8.2: a) Histogram of data from 1b (interpolated map) b) Histogram of data from 1c ("true" map).

The spatial smoothness of interpolated maps has an aspatial counterpart in that the set of the kriged values will have less variance than the actual values, with fewer values in the tails. This effect is illustrated using a histogram (Fig. 8.2) taken from our hypothetical example of mapping radiance values. The center of that histogram, measured by the mean of the kriged map, will usually be a good estimate of the actual regional mean. In the radiance example, the mean of the kriged map is 82.66; that of the map from image data is 80.13. In general, this similarity is not true with higher statistical moments (e.g. the variance of the kriged map is 114.08; that of the exhaustive data is 412.88). Journel and Huijbregts ([216]: p. 451, equation VI.1) give a formula for predicting the reduction in variance of the kriged map relative to that of the data. Other second order moments, such as the variogram, will also be underestimated by the kriged map. A variogram calculated from a kriged map will not be similar to the variogram model that was used to implement the kriging (and which was modeled from data).

In chapter 6 the local accuracy (unbiasedness) and precision (minimum error) criteria are stated for kriging, which define their optimality. Conditional simulation is not as locally accurate as kriging, but is more globally representative of an entire spatial field of a quantitative variable. Simulated maps achieve mean, variance and variogram statistics that are representative of the data set used to construct them. Estimation and simulation are treated separately in the research literature because of these different objectives.

There are at least four general objectives for which the properties of conditional simulation may be more suitable than estimation. One objective is the requirement for information about the probability that a group of pixels is equal to a value or range of values, rather than probabilities for each one independently. This is important in the detection of probable groups of contiguous pixels, which have been called "flow paths" [218]. Flow paths are relevant to, for example, environmental remediation and wildlife habitat analysis. Another objective occurs in situations when the correlation of the primary variable with other environmental variables is important in the analysis. Multiple layers in a GIS may be inputs to deterministic models or overlay analysis, so the accurate correspondence among layers is an important criterion [38, 31]. If all layers are from kriged maps, their correlation may be significantly altered from the true relationship between variables. Reduced numbers of extreme-valued pairs could have a disproportionate influence on results, for example if the deterministic model is non-linear [30].

A third objective is to test consequences of different segmentations of the spatial field of primary variable values. It is well known in geography that the way spatial data are segmented (zoned or aggregated) can alter the statistical relationships found and conclusions drawn [138]. This is known as the modifiable areal unit problem [289] because for some geographical phenomena, spatial units are modifiable— their support can be changed arbitrarily. Most of the primary variables that might be mapped using geostatistical techniques are inherently modifiable. A common need is to aggregate units of smaller spatial support, called up-scaling ([19], see also Chapter 7 of this volume). Aggregation usually decreases the variance and the sill of the variogram, among other characteristics. If these statistics were already underestimated through kriging, false conclusions could be drawn. Simulation offers a promising alternative for up-scaling [224, 367].

Finally, a fourth objective in mapping is a requirement for a map of uncertainty. Along with the spatial model of the primary variable described by a map, ideally there exists an associated spatial model of the error or uncertainty. With ad hoc and spline point-based interpolation methods, there is no error model. Webster and Oliver [390] and Switzer [355] have stated that error models from regionalized variable theory comprise the main contribution of geostatistics.

Values of kriging variance (frequently called "estimation variance") have often been displayed in map form. In the same way that maps of kriged values are inadequate spatial representations of a variable, maps of kriging variances are an unrealistic spatial representation of uncertainty. A kriging variance map does not provide a useful spatial description of error, and only in ideal, multivariate Gaussian circumstances is a useful description of local precision [355, 220, 311]. Kriging variance does not depend on the magnitude of variable values, only their location and geometry. High variance (low precision) is calculated far away from sampled locations and low variance (high precision) is calculated near samples. Conditional simulation, via multiple realizations, can be used to express uncertainty [223]. More on this will be discussed in the last section of this chapter.

There is an apparent paradox in the idea that the optimality criteria (minimum bias and maximum precision) used in kriging results in estimates that do not achieve the global statistics of the spatial field. Conversely, conditional simulation mimics the field properties but is locally less accurate—it has a variance twice that of kriging [216, 83]. In fact, there is some suggestion that local optimality and global accuracy cannot be achieved simultaneously. Some researchers have proposed compromise solutions intermediate between estimation and simulation [285, 159]. Csillag [85] states "There is a contradiction between the requirements of constant attribute accuracy and constant spatial resolution." Generally speaking, this dichotomy may be a geographer's analogue to the Heisenberg uncertainty principle. Whether or not this paradox represents something fundamental, the current state of the art in mapping quantitative variables makes it clear that objectives must be defined clearly, preferably a priori because the choice of a method affects the possibility of achieving those objectives and no one method allows the simultaneous achievement of all possible objectives.

8.2 Geostatistical basis of conditional simulation

At the heart of both geostatistical estimation and simulation is the random function model (chapter 5). A random function is a collection of random variables defined for a number of dimensions D, $\{Z(x, s) : x \in D, s \in \Omega\}$. One and only one value is defined for each s in the set of all possible outcomes (Ω) and each value has a probability. When D, the index set, is two or three dimensions (i.e. a plane or volume) the random function lends itself to modeling natural phenomena existing in space and x becomes a two or three element vector defining a location in the region being studied. Christakos [64] calls these random functions, usually abbreviated as $Z(x)$, "Spatial Random Fields" (SRFs).

Sets of values that arise from random functions are called realizations. The simple one-dimensional example often given is to regard throws of a die as a one-dimensional random process, the index set being defined by the number of throws being made. For one throw, the number that appears on the die can be regarded as a realization. Generalizing this to a two dimensional random function, a single realization could be described as a vector $[z(x_1), z(x_2), \ldots, z(x_n)]$ where $z(x_i)$ are values at the coordinates x_i. Any variable distributed in space and described in two or three dimensions is called a regionalized variable, and regionalized variables are often modeled as realizations from a random function. Multiple realizations are possible from a random function that meet criteria derived from measurement data (such as a given histogram and variogram), therefore multiple conditional simulations are possible.

Special types of spatial random functions are used for geostatistical methods of estimation and simulation. These SRFs feature autocorrelation and thereby incorporate a fundamental property of geographical phenomena. They also have certain forms of stationarity, which are required in order to accomplish any kind of prediction, inference, estimation or simulation. For example, second-order stationarity (a form of weak stationarity) is the stationarity of second moments, that is, the covariance of every pair of random variables, $C[Z(x), Z(x+h)]$, depends only on the distance, h, between them. Neither the mean nor the variance of $Z(x)$ varies with x. Second-order stationarity of the increment $[Z(x) - Z(x+h)]$ is called intrinsic stationarity. In an intrinsically stationary random function, the spatial covariance need not exist, only the variogram $\gamma(h)$ must exist and is dependent on h. If the spatial covariance does exist, it is related to the variogram in both second-order stationary and intrinsically stationary SRFs by

$$\gamma(h) = C(0) - C(h) \tag{8.1}$$

So, in many cases the variogram is assumed equivalent to a constant (the variance) minus the spatial covariance. This relation is behind the fact that many geostatistical algorithms flexibly interchange $C(h)$ and $\gamma(h)$.

Another feature of regionalized variable theory is that each variable is measured on a constant support. Support, the shape, size and orientation of the area or volume that has been measured (chapter 7), can be as small as a point or as large as the entire region being studied. A change in any characteristic of the support defines a new regionalized variable. Two sets of measurements of a variable made on different supports (say 10 cm^2 and 1 m^2) would have different second-order statistics, such as the variance and variogram. Therefore, these two sets of measurements would be considered as two different variables in a simulation model, just as in an estimation model.

A random variable within an SRF at location x and with a support v can be described, not only by its summary statistics such as expected value, but by its full cumulative distribution function (cdf). This is used in geostatistical models as a prior distribution in a Bayesian framework. A complete lack of knowledge (no data) about a variable of interest can be described using uniform prior cdfs (any value, or any value within a range of physically possible values, are equally likely) from a stationary random function. If data are available, the prior cdf at x, $F(x)$, can be modeled from all the data, weighted equally without regard to their distance to the location x:$F(z) = \Pr\{Z(x) \leq z\}$. Data at locations near x provide information about the variable, and this information is used to update the prior cdf, making some values more likely, others less likely. The random variable is then described using a conditional cumulative distribution function (ccdf), the probability that is $Z(x)$ less than z given the data, expressed as $F(z|(n)) = \Pr\{Z(x) \leq z|(n)\}$. Using stationary random functions, conditioning by nearby data is accomplished through a variogram, or by equation 8.1, spatial covariances. This ccdf can be used to specify the expected value, which is called the estimate $E\{Z(x)|(n)\}$. The fact that the random function $Z(x)$ and the prior cdf are stationary does not mean this conditional expected value is constant; in fact it is non-stationary. Going beyond the estimate to explore the uncertainty described by these non-stationary ccdfs involves drawing many realizations from them; this is called simulation.

Both parametric and non-parametric approaches to constructing ccdfs exist. The parametric approach uses the simple or ordinary kriging estimate and its variance to construct Gaussian ccdfs at every location. The non-parametric approach uses indicator kriging (chapter 6) which discretizes the ccdf by defining several specific z values, or thresholds. These ccdfs must be interpolated in between thresholds and extrapolated in the tails. The ccdfs constructed using either of the parametric or non-parametric approaches, using functions of $C(h)$ or $\gamma(h)$, completely define the random function from which multiple realizations are drawn. There are a variety of ways this has been done, as we will see in the next section.

8.3 Algorithms for conditional simulation

Algorithms for conditional simulation have been constructed according to three general strategies. The first strategy is to add the missing spatial variability "back in" to an estimate made by kriging. The second strategy is to construct the simulation from Bayesian first principles. The third is to use optimization techniques to converge on solutions that meet statistical criteria. Each of these general approaches has been implemented in several forms and the number of algorithms is still growing today. At present, no single text or software package covers all published algorithms, but conditional simulation is discussed in [216, 83, 64, 158, 233] and [106]. It is implemented in software packages: GSLIB (Stanford University, Stanford, CA), Gstat (University of Amsterdam) and Isatis (Geovariances, Avon, France), among others.

8.3.1 Adding variability back in

Several algorithms regard the problem of simulation as the model [215],

$$Z_c(x) = Z^*(x) + [Z_u(x) - Z_u^*(x)] \tag{8.2}$$

where $Z_c(x)$ is the conditionally simulated primary variable at location x, $Z_u(x)$ is an unconditionally simulated value, $Z^*(x)$ is the value of the primary variable estimated from actual z data using a smooth interpolator and $Z_u^*(x)$ is the estimate provided by the interpolator but using the unconditionally simulated values at the same locations as $z(x_i)$. Though $Z^*(x)$ and $Z_u^*(x)$ must be obtained separately, they are calculated using the same covariance model and data locations. The unconditional simulation is not conditional to the actual z data values, but is conditional to the variogram modeled from them. The difference term in square brackets on the right hand side of this equation can be considered a simulation of the error caused by smoothing. In general, any algorithm that generates unconditional realizations, $Z_u(x)$, can be made conditional by adding a simulated kriging error as in (2). While the exact relation 2 is only correct when $Z^*(x)$ and $Z_u(x)$ are obtained through simple kriging, some allowance has been made for substituting other varieties of kriging.

The first algorithm that was developed for univariate conditional simulation is known as turning bands simulation [258, 215]. It utilizes equation 8.2 for conditioning. The unconditional simulation portion is accomplished by projecting realizations of a one dimensional stationary process simulated along independent lines onto the 2D or 3D simulation grid. This single dimension (a line) is simulated using a moving average process, the covariance simulated along the lines is such that after projection, the resulting 2D or 3D covariance matches the target model. The more lines, the more exact the process. Turning bands is more tractable in three dimensions than in two, though Brooker [49] and Mantoglou [250] described solutions for the latter. It is a fast algorithm because of the requirement to do one dimensional simulations only, then sum for two and three dimensional results. Realizations from turning band algorithms often contain artifacts due to the practical limit on the number of lines and rarely include the capability to reproduce anisotropy (variograms that change with direction) not aligned with the simulation grid axes. These problems have reduced its use in favor of alternative methods [112].

Another algorithm that adds the spatial variability back into the kriged estimate is LU decomposition [97, 9]. This is a fast, analytical solution to equation 8.2, framed as

$$z_c(x) = L_{21} \cdot L_{11}^{-1} \cdot z_0(x) + L_{22} \cdot w \tag{8.3}$$

where w is a vector of values randomly drawn from a standard normal distribution. The length of w is s, the number of cells to be simulated. L_{11}, L_{21} and L_{22} come from the LU decomposition of the covariance matrix, C:

$$C = L \cdot U = \begin{bmatrix} L_{11} & \\ L_{21} & L_{22} \end{bmatrix} \cdot \begin{bmatrix} U_{11} & U_{12} \\ & U_{22} \end{bmatrix} \tag{8.4}$$

where C is

$$C = \begin{bmatrix} C_{11} & C_{12} \\ C_{21} & C_{22} \end{bmatrix},$$

an $(n+s) \times (n+s)$ matrix: C_{11} is the $(n \times n)$ covariance matrix among the data, C_{12} (equal to C_{21}^T) is the $(n \times s)$ covariance matrix between the data and cells to be simulated and C_{22} is the $(s \times s)$ covariance among the cells to be simulated. $L_{21} \cdot L_{11}^{-1} \cdot z_0(x)$ are the simple kriging estimates at x where $z_0(x)$ are the data. Additional realizations are created for every new vector w; easily done using pseudo random number generators. To date, this approach has found common application only when the simulation grids are small, so as to keep the matrix decomposition problem tractable. Therefore, for the large grids sometimes encountered in remote sensing applications it may have limited application. Its speed makes it attractive for situations when many hundreds of realizations, each of small size, must be generated.

8.3.2 Using Bayes Theorem

Many of the commonly used algorithms developed after turning bands are variants of sequential simulation [219]. The sequential approach exploits Bayes Theorem, which describes how the prior (marginal) probabilities for a random variable can be revised (or updated) to reflect additional information contributed from nearby (and therefore dependent) data to create a posterior, joint probability. According to the theorem (chapter 9), the joint probability of A_1 and A_2, $\Pr(A_1 \cap A_2)$, is equal to the probability that event A_1 has occurred multiplied by the probability that A_1 has occurred given A_2 has occurred:

$$\Pr(A_1 \cap A_2) = \Pr(A_2) \cdot \Pr(A_1|A_2) \tag{8.5}$$

This can be extended to m events as follows:

$$\begin{aligned}
\Pr(A_1 \cap A_2 \cap A_3 \cap \ldots \cap A_m) &= \Pr(A_m|A_1 \cap A_2 \cap \ldots \cap A_{m-1}) \cdot \\
& \quad \Pr(A_1 \cap A_2 \cap A_3 \cap \ldots \cap A_{m-1}) \\
&= \Pr(A_m|A_1 \cap A_2 \cap \ldots \cap A_{m-1}) \cdot \\
& \quad \Pr(A_{m-1}|A_1 \cap A_2 \cap \ldots \cap A_{m-2}) \cdot \\
& \quad \Pr(A_1 \cap A_2 \cap A_3 \cap \ldots \cap A_{m-2}) \\
& \quad \vdots \\
&= \Pr(A_m|A_1 \cap A_2 \cap \ldots \cap A_{m-1}) \cdot \\
& \quad \Pr(A_{m-1}|A_1 \cap A_2 \cap \ldots \cap A_{m-2}) \cdot \\
& \quad \cdots \cdot \Pr(A_2|A_1) \cdot \Pr(A_1)
\end{aligned} \tag{8.6}$$

For conditional simulation, we are after the joint ccdf of all random variables within the random function given all the (n) data, that is,

$$\begin{aligned} \Pr(Z(x_{n+1}) \leq z, Z(x_{n+2}) \leq z, \ldots, Z(x_m) \leq z | Z(x_1) = \\ z_1, Z(x_2) = z_2, \cdots, Z(x_n) = z_n). \end{aligned} \tag{8.7}$$

To draw from this conditional joint distribution to create realizations, sequential algorithms do a step-by-step updating of the ccdfs. The first ccdf, $F(Z(x_{n+1}) \leq z)$, is obtained through kriging at a location, with the conditional mean defined by the kriged value and the conditional variance defined by the kriging variance. It is drawn from to obtain a simulated value for that location. This simulated value is added to the pool of nearby data to provide data to kriging system at the next location, $F(Z(x_{n+2}) \leq z)|Z(x_{n+1}) \leq z_{n+1})$. At each step, the simulated value is added to the pool of data and previously simulated values, and steps continue until all cells in the grid have a value. The locations in the grid are visited in a random order. Each new realization is obtained by going through the procedure beginning from a different point in the simulation grid and drawing from the ccdfs randomly. The number of simulated cells, s, is much larger than the number of data, n. Because very close simulated cells screen data farther away [202], the conditioning data and ccdf are taken from a neighborhood surrounding the node to be simulated (referred to as the search neighborhood). The use of a search neighborhood instead of all the data to define the conditional cdfs has two advantages. It prevents the need to solve very large $(n+s) \times (n+s)$ kriging systems and it reduces the impact that the lack of second-order stationarity will have on the results [106].

When the kriging algorithm used to build the ccdfs is simple kriging (SK), each ccdf is defined by the SK mean and variance. Ordinary kriging (OK) is an approximation to SK, and may be used in its place. The SK and OK approaches result in what has been called sequential Gaussian simulation, a parametric approach [203]. Data that don't obviously fit a multi-normal model are often transformed to normal scores before applying the simulation algorithm. Realizations are then transformed back to the data distribution, though this does not ensure that a multi-normal model will be successful.

Conditional cdfs may also be obtained through indicator kriging, IK (chapter 6). This is a direct modeling of the ccdf without resort to any kriging variance or Gaussian hypothesis. Since the conditional expectation of the indicator variable identifies the ccdf value at threshold z_k, $E\{I(x; z_k)|(n)\} = \Pr(Z(x) \leq z_k|(n))$, where $I(x; z_k)$ is an indicator random variable at location x, an estimate of that indicator can be used to build a model for the z ccdf. Continuous ccdfs are effectively discretized by dividing the values of the primary variable into classes and transforming each class (k) to an indicator variable. Using the IK approach has been called sequential indicator simulation and can provide non-Gaussian ccdfs [217].

An algorithm closely related to sequential simulation is called probability field simulation. It is conceptually simpler and less computationally taxing than sequential methods. It represents a shortcut to these more rigorously defined methods; work remains to be done on the relationship of probability field simulation to the random function model, $Z(x)$.

Probability field simulation, first conceived by Srivastava [345] and Froideveaux [145], separates the steps of creating the local ccdfs and drawing from them. The ccdfs at every grid cell are defined in the first step, defined using any algorithm, be it simple kriging, ordinary kriging, indicator kriging etc. The second step is the selection of a value from every local ccdf to generate a realization. A "probability field" is a set (image) of values uniformly distributed between 0 and 1 used to draw from the field of ccdfs. A constant, "flat" probability field where all values are equal to 0.5 would result in the median value from each ccdf being selected at each location. The resulting realization would be smooth, like a kriged map, and not reflect the spatial pattern of the variable. To obtain realizations that represent the spatial pattern, probability fields must have a spatial pattern of their own. This pattern is characterized by the variogram of the uniform scores of the data (which have values between 0 and 1). Any simulation algorithm that can generate unconditional simulations can be used to create probability fields that reproduce this variogram [106]. Any number of unconditional simulations can be done to generate the probability fields; each is used to draw from the z ccdfs and yield a realization.

Drawing from ccdfs within the sequential or probability field algorithms employs the so-called quantile algorithm [182]. The quantile algorithm exploits the theorem that if u has a uniform distribution between 0 and 1, and $F(w)$ is a permissible

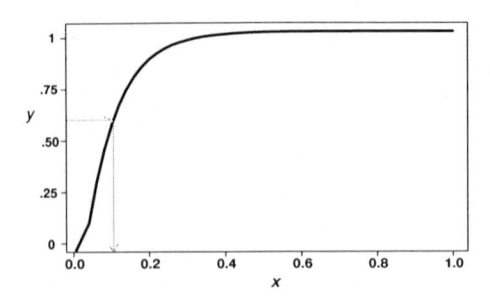

Figure 8.3: Illustration of how the quantile algorithm uses values of y, which are uniformly distributed, to draw values of x, which are distributed according to the cdf plotted here.

cdf, strictly increasing between 0 and 1, then the random variable w defined by $w = F^{-1}(u)$ is a continuous type random variable with distribution function $F(w)$. Therefore, u can be used in the inverse of the local ccdf to draw from that ccdf (Fig. 8.3).

8.3.3 Optimization

Simulated annealing has also been used to generate conditional simulations [105]. There are no Bayesian underpinnings to these optimization or "stochastic relaxation" procedures. Instead, an iterative approach is used to modify an initial image to gradually converge to realizations reproducing any target statistics (such as the univariate statistics or variogram). An objective function, $O(i)$, is defined as a function of the difference between the state of the image at the current iteration (i) and the desired state of a realization. At $i = 1$, the image is completely random; i.e. all pixels in the grid to be simulated are values independently drawn from the target histogram. At each iteration following, the image is slightly altered, for example through swapping two or more pixels. Convergence is achieved when the objective function reaches zero, or in practice some very small value. So, the required elements of the procedure include an objective function, a means of altering the current image, fast updating of the objective function, a decision rule about whether to accept or reject the current realization and a convergence criterion.

The decision rule is expressed as a function of a "temperature" parameter. Temperature, $T(i)$, is used to change the probability that an unfavorable image, with $O(i-1) \leq O(i)$, at iteration i will be accepted. Accepting unfavorable iterations guards against suboptimal minima; the larger the temperature the greater the probability of accepting such an unfavorable iteration. An example decision rule is defined by the probability of accepting the image on the ith iteration:

$$\text{Pr}(\text{accept } i\text{th image}) = \begin{cases} 1 & \text{if } O(i) \leq O(i-1) \\ e^{[O(i-1)-O(i)]/T(i)} & \text{otherwise} \end{cases} \quad (8.8)$$

$T(i)$ is made to decrease as i increases to avoid undoing an image that is getting close to being acceptable. The "art" of simulated annealing is to make sure the annealing schedule, as influenced by the rate of decrease of function $T(i)$, results in an acceptably small objective function in the smallest amount of time.

In comparison to the algorithms mentioned above, conditional simulation using optimization methods has a larger flexibility to include multiple criteria as different components of the objective function. Goovaerts [158] reviews the use of the univariate cdf, two-point statistics (such as the variogram), multiple-point statistics, correlation with other variables and cross-variograms with other variables as components in the objective function as well as weighting schemes for reproduction of these components.

8.4 Adding an ancillary variable

To this point, we have discussed only algorithms for univariate conditional simulation. To add information from ancillary variables, in particular from dense grids of spectral data, realizations must somehow be further conditioned to take into account this ancillary information. If the ancillary data are informative, they should act to reduce the variance of ccdfs modeled at each grid cell.

With two of the simulation strategies, adding back variability and the Bayesian approach, multivariable estimation (cokriging) lends itself to conditioning the ccdfs. Cokriging is considered to be a viable approach when the ancillary data are much more plentiful than the primary data—the accuracy is generally not improved if ancillary data is only available at the locations of primary variable measurements [216]. This implies that Cokriging is ideally suited to remote sensing applications, since spectral data are always far more plentiful than primary variable measurements. An alternative to traditional Cokriging that may be suited to conditional simulation using a dense grid of ancillary data is collocated cokriging [11, 158, 106]. This is an approximation to Cokriging whereby, at each grid cell to be simulated, only the ancillary datum at that cell is used rather than all neighboring ancillary data (i.e., only "collocated" ancillary information).

So, for example, either sequential Gaussian simulation or probability field simulation could be implemented using ccdfs defined using Cokriging estimation variances. If Cokriging variances are used, simulations should show less variation from realization to realization than those using just the primary variable approach given the same data, since as Myers [277] showed the Cokriging variances are always less than kriging variances. Adding ancillary information through an optimization technique can be done through the objective function—using, for example, an aspatial criterion such as the correlation coefficient between collocated primary and ancillary variables or a spatial criterion such as the cross-variogram [158].

An important question that must be answered before deciding to add an ancillary variable to a geostatistical analysis is, "How closely do the image data and ground data have to be related for the image data to reduce uncertainty about the primary variable?" Unfortunately, there is no analytical method for answering this question. The few that have done experiments with synthetic data give some clues.

Dungan [115] found that the correlation coefficient r between primary and ancillary variables had to exceed 0.3 for the root-mean-square error of Cokriging to be less than that from ordinary kriging. Kupfersberger and Bloschl [238] tested data sets with correlations of 0.4, 0.6 and 0.8 and found ancillary data with r values as low as 0.6 were significantly informative in estimation via cokriging.

Returning to the example from Fig. 8.1, let us assume that ground measurements of the primary variable are represented by the pixel values from band 5 at 150 locations. The "true" map of the primary variable is therefore represented by the band 5 image (Fig. 8.1c). Further, let us assume that the ancillary data are represented by band 3 – they are exhaustive and exhibit a correlation of .6 with the primary variable as judged by the 150 collocated data. Variograms from the ground samples, image data and their cross-variograms (Fig. 8.4) are required for cokriging.

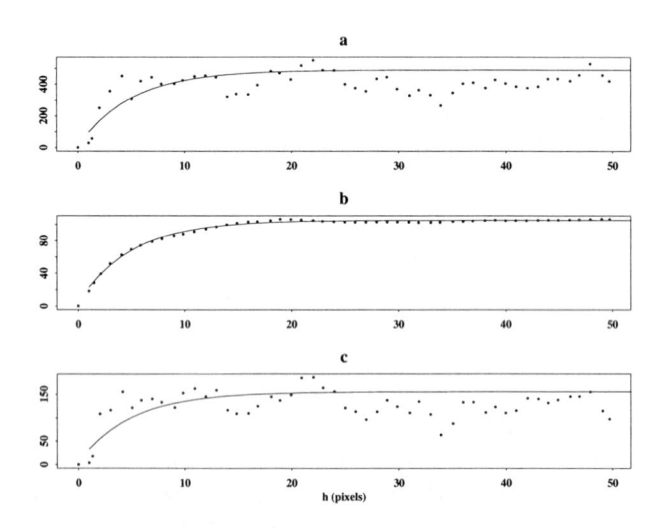

Figure 8.4: Variograms calculated (dots) and modeled (solid lines) from the example data. a) from 150 samples of the primary variable b) from the exhaustive band 3 image representing the ancillary variable and c) the cross-covariance between the primary and ancillary data at the 150 sample locations.

Given this model, and using a search neighborhood of 20 pixels, cokriging results in a map (Fig. 8.5a). Though the quality of the map estimated by cokriging appears far superior to that represented by kriging (Fig. 8.1b), it is still smoother than the "true" map (Fig. 8.1c). The cokriging variance map (Fig. 8.5b), as typical, reflects only the geometry of the sample locations and is unhelpful as a map of uncertainty.

The same primary and ancillary data are used in a sequential Gaussian simulation with the ccdfs defined by collocated cokriging [106]. A single realization from this simulation (Fig. 8.5c) shows the more realistic spatially variable pattern of the primary variable. The better reproduction of histogram and variogram by simulation than by cokriging is illustrated in Fig. 8.6. Exhaustive variograms from five

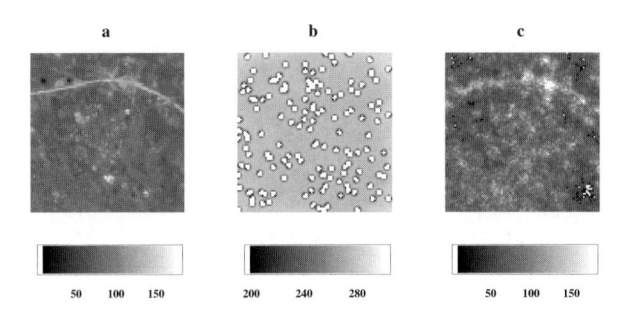

Figure 8.5: a) Map estimated using cokriging b) Cokriging variance map c) Single realization from sequential Gaussian simulation based on collocated cokriging.

realizations (Fig. 8.6b) are not identical to the model variogram used for cokriging (Fig. 8.4a), but are within an expected range of fluctuation. The images created by other simulation algorithms would not necessarily look like 5c, but should achieve similar reproductions of histogram and variogram.

8.5 Describing uncertainty

There is a variety of ways to use the multiple realizations resulting from conditional simulation. They may be summarized; subset using spatial or probability criteria; examined using spatial exploratory data analysis (EDA) techniques [171]; or each may be used as input to some other transform, such as a deterministic process model [126, 80]. The most obvious summary, the mean over all realizations, is called an E-type estimate, with "E" standing for expected value [221]. These estimates are similar to kriged maps, with the same smoothing effect. As such, the mean of the realizations is not itself a valid realization. Since one realization is no more likely to represent the actual spatial field of the primary variable than any other, the choice of a subset of realizations must be guided by other considerations. Map users may want to select realizations that represent widely different consequences in the analysis in order to explore the effect that extreme cases may have on contingent decisions.

Realization summaries that are more useful than the mean are higher order statistics at each grid cell, such as the variance of simulated values. Other measures of spread, such as the inter-quartile range or entropy measures have also been used [222]. Five realizations of the data used in the example show the variation among realizations (Fig. 8.7). The variance map calculated from 25 realizations (Fig. 8.7,

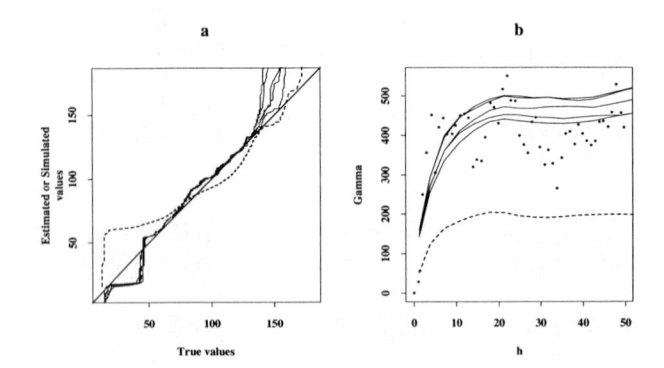

Figure 8.6: a) qq plot comparing distributions of estimated (cokriged) data (dotted line) and five simulated realizations (thin lines) versus the distribution of the true, primary variable. b) Variograms (thin lines) from five conditionally simulated realizations pass through the experimental variogram (dots) from the data. The variogram (dotted line) from the cokriged map does not.

lower right image) shows high values where there is large uncertainty and low values where there is less uncertainty.

Spatial exploratory data analyses are an increasingly important feature of geographical information systems [140]. An effective visualization of uncertainty is to create animations over all realizations [128]. Regions that change greatly during an animation are locations of high uncertainty – static regions have low uncertainty. Van der Wel et al. [375] suggest a similar interpretation for animations of probability values from classification of remotely sensed images. Srivastava [346] shows explicitly how probability field simulations can be ordered such that the realization-to-realization change in the animation is minimal. The eye can therefore follow patterns in the maps as they gradually appear, change shape and disappear. Image processing software that is designed for animation and multiple-image processing is well-suited to this kind of spatial EDA on realizations from simulation. One of the difficult practical issues of conditional simulation is to determine the number of realizations sufficient to characterize uncertainty. The number required should be reached when analyses based on the set of realizations no longer changes.

Using different algorithms will define different uncertainty "spaces." In other words, the set of realizations generated by one algorithm will not be identical to, nor necessarily result in the same summaries as, a set generated by another algorithm. In one of the few studies that employed more than one simulation algorithm, Goovaerts [160] found that for one specific data set, probability field simulation generated the largest uncertainty as measured by the variance of the results, followed by

Chapter 8: Conditional Simulation 151

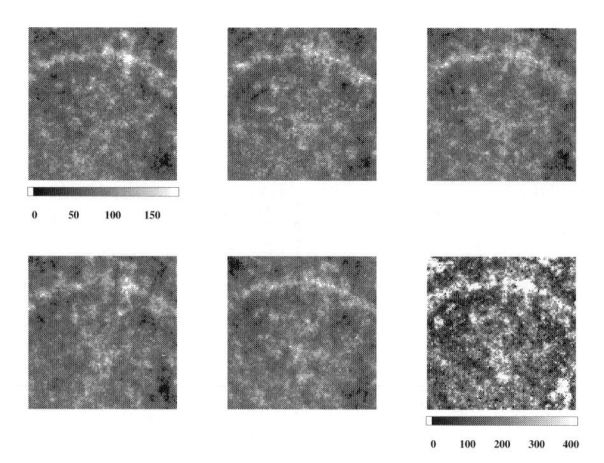

Figure 8.7: Five realizations from sequential Gaussian simulation based on collocated cokriging (top row and bottom left images). Variance map calculated from 25 realizations (lower right image).

indicator and Gaussian sequential simulation. The current state-of-the art is that algorithmically defined uncertainty must be relied upon.

8.6 The future of simulation in remote sensing

The application of conditional simulation methods to remotely sensing problems presents practical roadblocks and as well as significant potential. For example, the sizes of remotely sensed images are substantial, with most typically containing hundreds of thousands or millions of pixels. Computationally intensive geostatistical methods are challenging to implement on such large grids, since conditional simulation algorithms have been mainly developed on small grids (100-300 × 100-300). Of some compensating benefit, the relatively consistent spacing of image pixels allows one-time calculation of distances for variogram or spatial covariance calculations, reducing the computational burden. The exhaustive spatial sampling of a region makes variograms calculated from these data exact (chapter 5), reducing the uncertainty arising from sampling an ancillary variable.

With the existence of large grids from remote sensing comes an inevitable diminishment in the number of ground measurements relative to the number of image measurements. So, although several (e.g. [123]) have discussed the absolute number of primary variable measurements required for quantitative analysis, the density of samples is as important, at least for geostatistical methods. Maling [248] calls this the sampling fraction. Geostatistical methods rely on a sufficient number of nearby samples where each estimate will be made or simulated value will be drawn. Some sampling fractions in geostatistical studies are 0.01 [219], 0.075 [285] and 0.06 [202]. In remote sensing studies the fraction may be an order of magnitude or more smaller. A study of sampling density could provide information to create heuristics for the choice of models and methods.

The area on the ground represented by each resolution element is the support of image data. It is affected by the sensor and platform characteristics, point spread function and atmospheric conditions (chapter 7). It is a relatively consistent value across any given scene for many sensors, except for those with widely varying view angles (such as the NOAA Advanced Very High Resolution Radiometer) where resolution elements increase greatly in size outside the central swath. This consistency of support makes geostatistical models, which rely on fixed support, tenable. The flip side of this advantage is that for spaceborne sensors, supports are usually much larger than those of ground measurements. This support mismatch between image and ground measurements creates additional issues in estimation and simulation algorithms — the specific combination of supports for which they are developed must be explicitly accounted for. One of the longstanding grails in remote sensing is to apply methods developed with one sensor to data from a second sensor [204]. Change-of-support models are needed here, but they are complicated by the differences in radiometric characteristics among sensors. Conditional simulation has been used as a means of upscaling [314, 224] and may be more valid than estimation methods for this purpose.

This chapter has given but a broad overview of the purpose and methods of conditional simulation. It is currently an area of active research in geology, the field in which geostatistics originated, as well as in other environmental fields. As the uncertainties associated with mapping using remotely sensed data are increasingly appreciated, the advances made in mathematical geology for conditional simulation may begin to be exploited. Simulation models and methods are crucial to the "data expansion" role of statistics [219], in contrast to its usual data summary role.

8.7 Acknowledgements
The author thanks Ferenc Csillag, Lee Johnson, André Journel and Ramakrishna Nemani for reviewing this chapter.

Chapter 9

Supervised image classification

Ben Gorte

To monitor, analyze and interpret developments in our changing environment, spatial data are periodically collected and processed. Remote sensing is a valuable source for this purpose. Probabilistic methods can be used to extract thematic information from spectral data contained in multi-spectral satellite images of the earth surface. Such images consist of measurements of intensity of electro-magnetic radiation, usually sunlight, which is reflected by the earth surface. The area covered by a satellite image is subdivided into rows and columns, by which terrain elements are defined. The distance between adjacent rows (and columns) determines the resolution of the image. The measured reflection of the terrain elements is digitally stored in picture elements (pixels). When the pixels are displayed in rows and columns on a computer screen, the terrain becomes visible. In multi-spectral images each pixel consists of reflection measurements in several spectral bands, for example a number of visible colors and infrared.

To obtain information from multi-spectral imagery, the multi-dimensional, continuous reflection measurements have to be transformed into discrete objects, which are distinguished from each other by a discrete thematic classification. The relationship is not one-to-one. Within different objects of a single class, and even within a single object, different reflections may occur. Conversely, different thematic classes cannot always be distinguished in a satellite image because they show (almost) the same reflection. In such cases, deterministic methods are not sufficient. A probabilistic approach, however, is able to describe the spectral variations within classes and to minimize the risk of erroneous class assignments.

The remainder of this section is divided into three parts. First, standard image classification is introduced and put in a probabilistic framework. Next, refined probability estimates are presented. Finally, a quantification of the remaining classification uncertainty is described.

9.1 Classification

Classification chooses for each pixel a thematic class from a user-defined set. The choice is made on the basis of reflection measurements that are stored in the pixels of an image. The collection of measurements in one pixel is called measurement vector or feature vector. With M spectral bands the feature vector has M components and corresponds to a point in an M-dimensional feature space. The task of classification is to assign a class label to each feature vector, which means to subdivide the feature space into partitions that correspond to classes. This task can be achieved by *pattern recognition*.

9.1.1 Pattern recognition
In the description by Ripley [303] of a pattern recognition machine
> " ... we are given a set of N pre-determined classes, and assume (in theory) the existence of an oracle that would correctly label each example which might be presented to us. When we receive an example, some measurements are made, known as *features*, and these data are fed into the pattern recognition machine, known as the *classifier*. This is allowed to report
> 'this example is from class C_i'
> 'this example is from none of these classes' or
> 'this example is too hard for me'
> The second category are called *outliers* and the third *rejects* or *doubt* reports. Both can have great importance in applications."

Applied to satellite image classification, the presented examples are pixels in an image. The oracle in the above description is given by the terrain. It is assumed that each pixel can be classified by inspecting the corresponding terrain element.

Each pixel should be labeled as belonging to a class from a user-defined set of, for example, land-use of land-cover classes.

Images can be distorted in several ways and the results of classification can be disappointing. Reflections do not only depend on thematic class, but, for example, also on atmospheric conditions, soil moisture and illumination, where the latter is influenced by sun angle and terrain slope. Also noise, originating from the sensors and from analog-to-digital conversion, plays a role. In addition, some thematic classes consist of composites of land covers with different spectral properties. The class *residential area*, for example, contains houses (roofs), roads, private and public gardens etc., which cause different reflections. On the other hand it happens that terrain elements of different classes yield the same reflection measurements. Pixels with such reflection cannot be classified with complete certainty. These are the hard examples in Ripley's description, for which *doubt* reports might be generated.

This may also be the case in a mixed pixel, which contains objects of different classes, for example when objects are smaller or narrower than an element. This depends on image resolution. At any resolution, some terrain elements are intersected by a boundary between larger objects. Objects are usually not aligned with the boundaries of image pixels. Finally, some objects are not clearly delimited in the terrain. For example the boundary between forest and heath land, or between a built-up area and the surrounding agricultural land, cannot always be drawn clearly. There is a transition zone, which may be several terrain elements wide. The corresponding pixels are difficult to classify and may cause *doubt* reports. It may be argued, however, that in case of mixed pixels and fuzzy boundaries it is not the classifier which fails, but the oracle. In that view, a classifiers capability to give 'doubt' reports should be considered an achievement.

Ripley's *outliers* occur if the terrain contains objects with classes that do not belong to the user-defined set. Especially when classification has to determine the area covered by the various classes, it is necessary to take the so-called unknown class into account. For example, when crop areas are estimated by classification, agricultural survey data may be used to provide training samples for the various crops. Usually, such survey data do not contain samples of the other classes that may be present in the area such as villages, water bodies, forested areas, roads and farmhouses. Together these constitute the *unknown* class, which should be excluded from the crop area estimates.

9.1.2 Statistical pattern recognition

Often, statistics is used to tackle classification problems. It cannot prevent classification errors, but it can minimize their probability. Moreover, the probability for error can be established, even for each pixel separately. Bayesian classification calculates in each pixel p (with feature vector $\mathbf{x}_p$) the *a posteriori* probability $P(C_i|\mathbf{x}_p)$ for each class C_i, based on estimates of the probability density $P(\mathbf{x}_p|C_i)$ and the *a prior* probability $P(C_i)$, and then chooses the class with the largest probability. The calculation applies Bayes' formula

$$P(C_i|\mathbf{x}_p) = \frac{P(\mathbf{x}_p|C_i)\, P(C_i)}{P(\mathbf{x}_p)}$$

in which $P(\mathbf{x}_p)$ can be replaced by a normalization factor

$$P(\mathbf{x}_p) = \sum_{j=1}^{N} P(\mathbf{x}_p|C_j)\, P(C_j)$$

if no *unknown* class is present in the image.

When estimating probability densities, it is often assumed that reflections are drawn from a multivariate normal (Gaussian) distribution. The probability density that feature vector $\mathbf{x}_p$ occurs in class C_i is given by

$$P(\mathbf{x}_p|C_i) = (2\pi)^{-M/2} |V_i|^{-\frac{1}{2}} e^{-\frac{1}{2}(\mathbf{y}^T V_i^{-1} \mathbf{y})} \qquad (9.1)$$

with:
M	:	the number of spectral bands
V_i	:	the $M \times M$ variance-covariance matrix of class C_i
$|V_i|$	:	the determinant of V_i
V_i^{-1}	:	the inverse of V_i
$\mathbf{y}$	:	$\mathbf{y} = \mathbf{x}_p - \mathbf{m}_i$ ($\mathbf{m}_i$ is the class mean vector)
$\mathbf{y}^T$	:	the transpose of $\mathbf{y}$ (a row vector).

The distribution parameters are calculated from training data, samples of pixels of each class, indicated by the user in the image under consideration. For each class C_i, the training samples $T_i = \{<\mathbf{x}, C_i>\} \in T$ are analyzed. The feature vectors $\mathbf{x}$ in T_i are put into a sequence. The average of the vectors in this sequence yields the mean vector $\mathbf{m}_i$. Looking at the j-th component of each feature vector, the variance v_{jj}^i can be obtained. With $j \in [1..M]$, this gives the diagonal elements of V_i. Similarly, between the j-th and k-th components of all feature vectors in the training samples of C_i, the covariance $v_{jk}^i, (j, k \in [1..M], j \neq k)$ can be determined, to yield the off-diagonal elements of V_i.

Sometimes, the user can specify a *priori* probabilities as the expected share of each class in the total area. With correct a *priori* probability estimates, in addition to accurate probability densities, the highest overall classification accuracy can be expected. In certain cases it is better to assume equal a *priori* probabilities for all classes — although this is not beneficial for overall classification accuracy, the results are less biased. Consider the task to find roads in a forested area, where the proportion of forest pixels (say, 99 %) is used as prior probability for the class *forest*, leaving 1 % for *road*. In the result, *forest* may be assigned to all image pixels, giving the user a useless map with an overall accuracy of 99 %. If priors are not considered, the user can expect some forest to become classified as road, but at least most of the road will be found, too, The overall accuracy is expected to be lower, for example, 97 %. The user has to decide which of the classified *road* pixels are really road, but this is probably easy when looking at the spatial arrangement of these pixels.

Minimum Mahalanobis distance

For an efficient implementation, it is allowed to omit the constant factor $(2\pi)^{-M/2}$ from (9.1) and to take -2 times the logarithm of the remaining part. This yields for each class a value

$$D_i(\mathbf{x}_p) = \ln|V_i| + \mathbf{y}^T V_i^{-1} \mathbf{y}.$$

For each class, $\ln|V_i|$ and V_i^{-1} are calculated at the beginning of the classification, and no time-consuming exponentiations are necessary at each pixel. The decision function selects the class with the minimum D_i value.

Class prior probabilities $P(C_i)$ can be included by minimizing over

$$D_i'(\mathbf{x}_p) = D_i(\mathbf{x}_p) - 2\ln(P(C_i))$$

The decisions are the same as from maximum likelihood. Actual a *posteriori* probability values can be obtained, applying

$$P(C_i|\mathbf{x}_p) \sim e^{-\frac{1}{2}D_i'(\mathbf{x}_p)}, \tag{9.2}$$

followed by normalization.

A further reduction omits $\ln|V_i|$ and minimizes over the squared *Mahalanobis distances* M_i:

$$M_i^2(\mathbf{x}_p) = \mathbf{y}^T V_i^{-1} \mathbf{y}$$

between a feature vector $\mathbf{x}_p$ and a class mean vector $\mathbf{m}_i$.

Sometimes, V_i is replaced by V, the matrix of covariances between the M features over the entire image. Then, the differences between within-class variabilities are neglected, but feature space anisotropy is still considered.

Minimum Distance

If also V_i is omitted from the calculation (set to the $M \times M$ unity matrix), the decision is based on the minimum squared Euclidean distance E_i:

$$E_i^2(\mathbf{x}_p) = \mathbf{y}^\mathrm{T}\mathbf{y}.$$

Compared to the Euclidean distance $E_i(\mathbf{x}_p)$, the Mahalanobis distance $M_i(\mathbf{x}_p)$ between a feature vector $\mathbf{x}_p$ and a class mean vector $\mathbf{m}_i$ is weighed by the inverse of the variance-covariance matrix V_i. Therefore, distances become smaller as variances and covariances increase, and wide-spread classes seem 'nearer' than compact ones.

Similar to (9.2), probabilities could be re-constructed, now under the assumption that the variances are equal in all bands for all classes, while covariances are 0.

A final simplification might be obtained by using city-block distances

$$B_i(\mathbf{x}_p) = \sum_{j=1}^{M} |x_{p,j} - m_{i,j}|,$$

where $x_{p,j}$ and $m_{i,j}$ are the components of $\mathbf{x}_p$ and $\mathbf{m}_i$, respectively.

9.2 Probability estimation refinements

Statistical classification methods do not always attempt to estimate various probabilities as accurately as possible. Perhaps it is assumed that the largest a *posteriori* probability occurs at the 'right' class also when estimated rather roughly. Even the most advanced of the methods described above suffers from two drawbacks.

- The probabilistics in Bayes' formula pertain to the entire image, while a decision is taken for each individual pixel. Having only one set of class a *priori* probabilities, the class assigned to a pixel depends on the situation in the entire image and is, therefore, influenced by areas that are far away from the pixel under consideration and have no relation with it.
- The assumption that reflections can be modeled as Normal distributions may be unrealistic for certain land-use classes, such as *built-up* and *agricultural* areas, which may consist of several land covers with different spectral signatures in different (unknown) proportions. Also signatures of land-cover classes are influenced by soil type, soil moisture, sun incidence angle (on slopes) etc. and may be inadequately modeled by Gaussian densities.

9.2.1 Local probabilities

Estimates of the various probabilities can be refined by making them local, *i.e.* pertaining to parts of the area, instead of to the entire area ([353], [269]).

If the user is able to subdivide the image into regions, such that different class mixing proportions occur in different regions, and if these mixing proportions are known, then in each region the expected overall accuracy is higher with a region-specific set of priors (according to the mixing proportions) than with the 'global' set. Therefore, taking the results of all regions together, the overall accuracy in the entire image also increases.

The necessary subdivision of the image can be made with additional (map) data, which may be stored, for example, in a geographic information system. The basic idea is that when deciding upon an class in a particular pixel, statistical data related to, for example, the soil type in that element are more relevant than statistics for the entire area. Until now, the objection against this approach was that such detailed statistics are generally not available or not reliable. This information can be obtained from the distribution of reflections.

9.2.2 Local *a priori* probability estimation

In absence of detailed statistics concerning the relation between classes and ancillary map data, an iterative procedure may be used to estimate local prior probabilities $P(C_i)$ on the basis of the image data. This procedure builds on the assumption that the *a posteriori* probability for a class in a pixel is equal to the *expected contribution* of that pixel to the total area (measured in pixels) of that class. Then, area A_i for class C_i in a segment s equals the sum of the posterior probabilities $P(C_i|\mathbf{x}_p)$ over the pixels p in segments s (which contains A pixels):

$$A_i = \sum_{p=1}^{A} P(C_i|\mathbf{x}_p),$$

where, according to Bayes formula

$$P(C_i|\mathbf{x}_p) = \frac{P(\mathbf{x}_p|C_i) \; P(C_i)}{P(\mathbf{x}_p)}.$$

Note that $A = \sum_{i=1}^{N} A_i$.

The probability densities $P(\mathbf{x}_p|C_i)$ are estimated locally, as described below. The procedure is started with an initial set of equal prior probabilities $P(C_i) = 1/N, i \in [1..N]$. By normalizing A_i a new set is obtained

$$P(C_i) = A_i/A,$$

which is used in the next iteration.

It can be shown that the prior probabilities converge and that the result is correct if the probability density estimates are accurate. A prerequisite is, therefore, that a collection of representative training samples is available.

9.2.3 Local probability density estimation

If *a priori* probabilities in Bayesian probabilistics pertain to a sub-area, the probability densities must concern the same sub-area. When using densities concerning the entire image, one must assume that these are indeed independent of the position in the image, for example that reflections of a certain crop do not depend on soil type. As subareas get smaller, which is beneficial for the refinement of estimates, the assumption becomes more doubtful. Therefore, the next section shows how local probability densities are estimated on the basis of a single collection of training data. The latter is crucial when there are many subareas — separate collections for each sub-area cannot be obtained then.

In general, image classification would benefit from a universal model, describing class distributions that are valid for any image of a given satellite/sensor system. Such a model might be the result of collecting the training samples of many classifications However, such a model would have to take spectral variability across images into account, caused, for example, by different atmospheric conditions and sun angles, and by seasonal influences (soil moisture, crop growing stage etc.)

Therefore, in practice one uses a probability density model with class probability densities for one (entire) image. Consequently, supervised classification involves a training stage, in which the class probability densities for a particular image are established. Still, the spectral variability within each class in the image is influenced by, for example, differences in soil type, soil moisture and illumination (relief).

Taking this approach one step further, the method in this paper estimates local class densities on the basis of a subset of the training samples: only those that matter for a single region. Note, that this is also done in stratification. The difference is that we do not require that this (subset of the) training set is taken *inside* the region. Instead, we select the subset from the entire training set on the basis of the feature space. Only those training samples that are spectrally similar to the feature vectors in a region s are used to estimate class densities in s. For class C_i a subset T_i^s is selected from T_i, with feature vectors that are 'near' to any of the feature vectors in s. If the spectral variation of C_i in s is the same as in the entire image (Strahler's requirement), all T_i samples are used and $T_i^s = T_i$. In that case, the local density is equal to the global one. But if C_i is more homogeneous in s that in the entire image, T_i^s will be a proper subset of T_i, and the T_i samples outside T_i^s will not influence the density estimate.

As a simple example, consider a 'natural color' image, taken with an RGB-camera. The terrain contains the class *grass*, which is predominantly green, but sometimes yellow. Therefore, the global probability that grass is yellow is relatively small. If we look at a yellow region in the image, however, the probability for green grass is much smaller than the probability for yellow grass. To estimate these probabilities, we should only yellow training samples.

9.2.4 Non-parametric estimation

Local class density estimation can be elegantly combined with non-parametric methods such as k-Nearest Neighbor (k-NN). Generally, when classifying a pixel with feature vector $\mathbf{x}$, k-NN selects those k training samples that are nearest to $\mathbf{x}$ in the feature space. Next, those k samples are subdivided according to their classes, which yields a sequence $k_i(\mathbf{x})$, $i \in [1..N]$, with $\sum k_i(\mathbf{x}) = k$. Straightforward k-NN classification then assigns the class with maximum $k_i(\mathbf{x})$ to the pixel. A refinement of this algorithm allows for density estimation, by compensating $k_i(\mathbf{x})$ with the total number A_i of class C_i samples (the number of elements in T_i): $P(\mathbf{x}|C_i)$ is proportional to $k_i(\mathbf{x})/A_i$ [113]. Intuitively, this makes sense. If $k_i(\mathbf{x})$ is large, then the density of C_i at feature vector $\mathbf{x}$ is obviously high. But if we train the classifier with twice as many C_i samples, also $k_i(\mathbf{x})$ is twice as large (on average). To compensate for differences in training set sizes for the various classes, we have to divide by A_i (Figure 9.1).

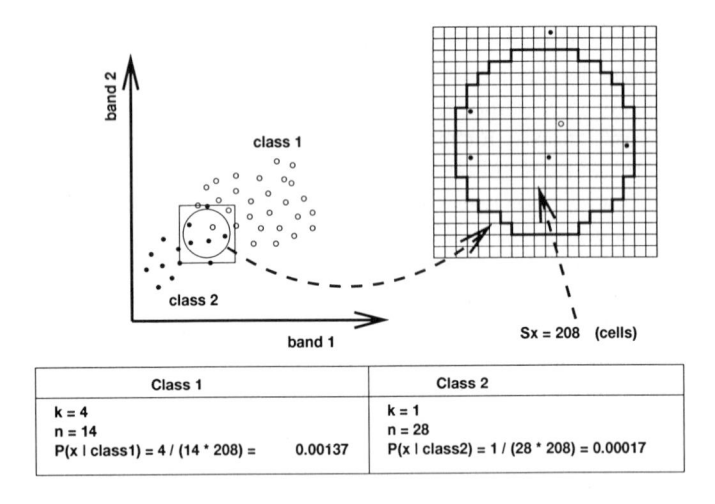

Figure 9.1: k-Nearest Neighbor estimation of probability densities

Note that straightforward k-NN classifiers, where such a compensation is not applied, are biased towards classes with many samples, which is similar to assigning large prior probabilities to such classes in a maximum likelihood classification. As a consequence, the effect of prior probabilities can be achieved in k-NN classification by *proportional sampling*. In that case, normalized k_i values serve as estimates for *a posteriori* probabilities:

$$P(C_i|\mathbf{x}) = \frac{k_i(\mathbf{x})}{\sum_{j=1}^{N} k_j(\mathbf{x})} = \frac{k_i(\mathbf{x})}{k}$$

If proportional sampling is not possible, compensation with N_i can be applied to get class densities. After that, Bayes formula may be applied to obtain *a posteriori* probabilities

$$P(C_i|\mathbf{x}) = \frac{\frac{k_i(\mathbf{x})}{A_i}P(C_i)}{P(\mathbf{x})},$$

where a normalization factor is substituted for $P(\mathbf{x})$.

When classifying a region s in the image, the k-NN algorithm can keep track of the training samples that get involved in the classification of the pixels in s. The

C_i samples that are used are the ones that belong to the 'nearest neighbors' of a pixel in s. Those samples become elements of T_i^s. After all pixels in s are examined, the number of elements in T_i^s are counted, giving A_i^s. For each pixel (with feature vector $\mathbf{x}$) in s, the density $P(\mathbf{x}|C_i)$ is estimated as

$$P(\mathbf{x}|C_i) \sim \frac{k_i(\mathbf{x})}{A_i^s}.$$

A user who incorporates additional data in this way can expect an increased classification accuracy. The largest a *posteriori* probability occurs with a different class than formerly, and this yields a better choice on average. It can be concluded that local estimates of probability densities and a *priori* probabilities lead to a *posteriori* probabilities that are more relevant.

Many kinds of additional data are suitable, such as soil maps, geological maps, elevation data and 'old' land use maps. The only requirement is that they subdivide the terrain in parts in which different mixing proportions of classes can be expected. These proportions do not have to be known beforehand, but are generated instead.

An example of classification improvement with local probabilities concerns the *Thematic Mapper* image of Twente, subdivided according to the (four digit) post code areas and using six classes: *city*, *residential* (suburbs and villages), *industrial*, *agriculture* (including grasslands), *natural vegetation* (forest and heather) and *water* [162].

Class-selection was according to land use. For example, there is only one class *agriculture*, without differentiation according to crops - also in practical cases such differentiation is often difficult to obtain, and perhaps not even required by the application. The class is spectrally heterogeneous. On the other hand, the classes *city*, *residential* and *industrial* are not only heterogeneous, but also have a large spectral overlap, whereas to distinguish them may be a user requirement.

The main purpose was to test the iterative classification method in a controlled experimental setup.

Table 9.1 compares different classifiers. The figures in the table are based on a single evaluation set. The k-NN classifications were made with $k = 11$. Local prior probabilities (per postal district) were estimated iteratively.

Table 9.1 allows for the following observations:
- The Gaussian method works better than the minimum Euclidean distance method (row 1 vs. 2). Due to differences in class variability, covariance matrices must be used and there are enough training samples to estimate them reliably.
- Straightforward k-NN classification (row 3) performs remarkably well. Proportional sampling was applied, such that a maximum a *posteriori* probability result is obtained.
- After compensating for differences in class training set sizes, k-NN gives maximum probability density classification (row 4) and can be compared with the Gaussian method using (row 2).
- Local probability estimation improves classification significantly.

Table 9.1: Comparison of classifiers for Twente data set

	probability density	prob. dens model	prior prob.	average accuracy	average reliability	overall accuracy
1	min. distance	global	-	70.99	47.79	65.93
2	gaussian	global	-	74.59	56.55	77.84
3	k-NN	global	inherent	75.31	69.85	86.23
4	k-NN	global	-	76.45	59.95	79.44
5	k-NN	local	-	74.68	58.83	78.46
6	k-NN	local	local	81.80	73.24	88.87

9.3 Classification uncertainty

When the maximum a *posteriori* probability of pixels after classification is less than 1, the correctness of the assigned label is uncertain. To evaluate classifications and decisions made on the basis of the results, a certainty measure is needed in addition to the classification itself. The per-pixel a *posteriori* probability vector contains all the information, but for interpretation, a single scalar number is preferable. Several scalar certainty measures can be conceived, such as the maximum probability value, which (by definition) indicates the probability that the classifier took the correct decision.

Entropy

To capture the entire probability vector, instead of only its maximum, *weighted uncertainty* measures can be used, such as the well-known *entropy* measure, originating from information theory ([324], [237]). The measure pertains to a statistical variable and to the uncertainties in its possible values, expressing the distribution and the extent of these uncertainties in a single number [157]. In the entropy measure, the uncertainty in a single value of a statistical variable is defined as the *information content* of a piece of information that would reveal this value with perfect accuracy. This quantity is weighted by the probability that the value occurs and summed over all values, which gives

$$e(\mathbf{x}_p) = \sum_{i=1}^{N} -P(C_i|\mathbf{x}_p) \log_2(P(C_i|\mathbf{x}_p)) \tag{9.3}$$

The classification example deals with six classes. If it is supposed for a moment that there are eight classes, then in case of complete certainty concerning class membership, three bits are needed to encode the information in each pixel: a class number between 0 (binary 000, corresponding to class 1) and 7 (binary 111 for class 8). With eight classes, the entropy measure (eq. 9.3 with $N=7$) yields a number between 0 and 3, which specifies how much of these three bits of information is still missing after classification has been completed. With only six classes, entropy is a number between 0 and $6 \times (-\frac{1}{6} \log_2 \frac{1}{6}) = 2.585$. (Figure 9.2). All a *posteriori* probabilities being equal to $\frac{1}{6}$ means that nothing is known about class membership, and the entropy value equals 2.585. If, on the other hand, one of the probabilities equals 1 (and the others 0), class membership is completely determined, which is reflected in entropy value 0. Between these two extremes is the situation with two probabilities equal to 0.5, the remaining four being 0. Then the entropy measure yields the value 1: one additional bit of information would be needed to change the complete ambiguity between two classes into a definite choice. Similarly, four times 0.25 and two times 0 gives entropy 2, the number of bits needed to choose one out of four.

Note that entropy expresses uncertainty **according to** the vector of a posteriori probabilities. It does not involve uncertainty **concerning** these probabilities: they are assumed to be correct. However, to estimate them correctly is exactly the problem in classification. As a consequence, the entropy measure cannot be used to compare classifiers that estimate probabilities in different ways.

Figure 9.2: Entropy measure for classification uncertainty

Chapter 10

Unsupervised class detection by adaptive sampling and density estimation[1]

Cees H.M. van Kemenade, Han La Poutré and Robert J. Mokken

Classification in remote sensing involves the mapping of the pixels of an image to a (relatively small) set of classes, such that pixels in the same class are having properties in common.

Until recently, most satellite imagery was at a relative low-resolution, where the width of a single pixel was between 10 m and 1 km. Imagery produced with new satellites have a higher spatial resolution. Such high-resolution images offer new opportunities for applications. In theory, a better classification is possible, as high-resolution images offer more information. In practice, many new problems appear. Due to the high spatial resolution, objects that are too small to locate in low-resolution images are visible in high-resolution images. As a result, the number of different classes that can be detected increases, and the discrimination between classes becomes more difficult. For low-resolution imagery one often assumes that a pixel value consists of the spectral vector of the underlying ground cover class plus a Gaussian distributed noise component. It is questionable whether this model is still applicable for high-resolution imagery. Therefore, in this chapter we use non-parametric models. These models are more flexible with respect to the shapes of the distributions modeled.

Apart from the spectral information, there is also spatial information available in the image. The human interpreter uses spatial information contrary to purely spectral classification methods, as these methods disregard the similarities between neighboring pixels during classification. Incorporation of the properties of the neighboring pixels means that we try to exploit the spatial structure present in the image.

10.1 Unsupervised non-parametric classification

We have developed an unsupervised non-parametric classification method. This classification method uses a clustering method to find the structure in the data.

---

1. Research supported under US Govt. European Research Office contract # N68171-95-C-9124.

A good introduction to clustering is given in [228] and in [303]. For a supervised classification method, a human interpreter has to pre-classify part of the image by marking a set of regions and classifying each marked region. Next, the supervised method can use these classified regions to obtain models that are used to classify the other parts of the image. During unsupervised classification we do not use a set of pre-classified training-samples. The method classifies the image based on the image itself, although some general assumptions about images can be used.

A clustering method can either be parametric or non-parametric. Parametric models usually assume a spherical or Gaussian model for the shape of a class. Let us assume for the moment that the noise contribution is mainly determined by small size objects with a spectral vector that differs significantly from the spectral vector of its surroundings. During classification, we have to detect the class of the most important ground cover for each pixel. For low-resolution imagery, a single pixel can contain many different small objects, and their combined surface usually covers a relatively small part of the pixel. If the sub-pixel objects belong to different classes, then a further cancellation of their spectral contributions is likely. The combined noise contribution can be modeled by a Gaussian distributed noise component. For high-resolution imagery, a pixel will only contain a few of these disturbing objects. It is then questionable whether the Gaussian noise model is applicable, and even if it is applicable, the variance is likely to be much larger as the disturbing objects are likely to cover a significant part of the pixel. A further complication is that for high-resolution imagery the number of detectable classes is larger, and thus one has to discriminate between more classes.

A parametric method will fit the free parameters to actual data. If the model matches, then these parametric methods can provide fast and efficient ways of obtaining the actual clustering of the data. If the model does not match, then the quality of the clustering can decrease severely. A method for this type of problem is the ISODATA clustering procedure [167, 168]. This method minimizes the sum of the squared distances between the points and the corresponding cluster center. This corresponds to the assumption that the clusters are approximately spherical.

More flexible modeling of data is possible when applying non-parametric clustering methods. We develop a non-parametric clustering method that is based on density-estimation [322, 327]. Non-parametric methods are more general in the sense that less prior information with respect to the actual shape of clusters is included. The resulting problem is more complex since the search for a good clustering of the data now involves both model selection and model fitting. Unfortunately, these two modeling phases cannot be performed separately. Model selection has to be performed before model fitting can take place. But model fitting is required to assess the quality of the selected model. In fact, model selection and model fitting often are so tightly integrated that one often cannot differentiate between these two any more.

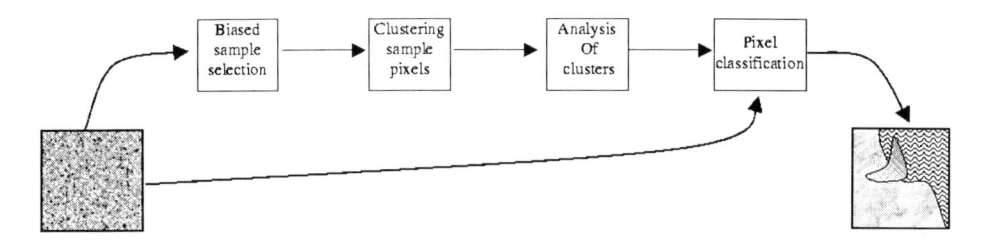

Figure 10.1: Schematic representation of the classification method

10.2 Outline of the method

We use a classification method based on density-estimation. The output of the method is a non-parametric clustering of the set of input points. A parametric clustering assumes a pre-specified model of the clusters, where an instantiation of this model can be described by a (small) number of parameters. Fig. 10.1 gives a schematic representation of the unsupervised classification method introduced in this chapter. On the left-hand side we see the unclassified image, on the right we have a classified image. The four steps of the method are:

Sample selection: A biased training-sample is selected from the image. The bias is used to reduce the amount of noise and the fraction of mixed pixels in the sample. During this selection, spatial information is used. The training sample consist of the spectral vectors of the selected pixels.

Clustering: An unsupervised non-parametric clustering method is used to find clusters. This clustering method is based on density estimation in the sample-space.

Analysis: The clusters are analyzed. For each cluster, the principal component is determined, and the distribution of the pixels when mapped at this principal component is investigated.

Pixel classification: Using cluster information and the original image, a classified image is generated.

10.3 Adaptive sampling of pixels

We develop an adaptive, biased sampling method, where the bias is directed towards the selection of pixels containing primarily one ground cover, and with little noise. We use it in our final algorithm inside a stratified random sampling framework (see chapter 13). For large images, we have to select for learning a limited sample of pixels from an image. We would like this learning-sample to be representative for the complete image. Therefore we select a stratified sample by means of the following procedure. To get a sample of size N, one partitions the complete image into a set of N rectangular regions of approximately equal size. Next a pixel is selected at random from each of these square regions. Application of such a procedure guarantees that the points in the learning-sample are distributed uniformly over the complete image. Especially if N is small, such a procedure is important.

The human eye is highly sensitive to spatial structure in an image. We almost never observe individual pixels. Instead we observe regions of pixels that are relatively similar. Within these regions the pixels have approximately the same spectral vectors, or the region shows a texture that is quite apparent. We call this property the local spatial homogeneity of images. Next, we show how to use the local spatial homogeneity to reduce noise in the sample.

If a region contains a single ground-cover, it is possible to get a sample with pixels that contain little noise by means of biased sampling. Here, the probability that a pixel is selected is determined by a similarity measure. We introduce a similarity measure that computes the similarity $sim(p)$ between a pixel p and the pixels in its spatial neighborhood $N(p)$. We show that the set of pixels with $sim(p) \leq \delta$, i.e. the pixels that are relatively similar to their neighbors, comprise a biased sample of pixels containing relatively little noise. Let the function $d(\vec{s}_1, \vec{s}_2)$ denote the spectral distance between two pixels. A similarity measure can be obtained by calculating $d(\cdot, \cdot)$ for all neighbors. Furthermore, the rank with respect to their distances is given by $r(i)$. The similarity measure we use is:

$$sim(p) = \sum_{i \in N(p)} W_{r(i)} d(\vec{s}_p, \vec{s}_i)$$

where p is the current pixel, $N(p)$ is its spatial neighborhood, and $\vec{W}$ is a weight vector such that $|\vec{W}| = 1$. The best choice of $\vec{W}$ depends on the structures one is interested in. If thin structures should be detectable, such as rivers or roads that have a width comparable to the width of a pixel, then W_1 and W_2 should be the only non-zero values. If one is not interested in such thin structures, then it is better to have $W_k > 0$ for higher values of k, as this puts a stronger emphasis on the similarity of a pixel with respect to its neighbors, and therefore a stronger emphasis on the homogeneity of the pixel. In the theoretical analysis that follows, we assume that we have a set of t pixels that are drawn according to the same distribution. From this set of pixels, the biased sampling method selects those pixels that contain relatively little noise, without computing the center of the distribution. This analysis is mainly a proof of principle for the biased sampling. For such a nice uni-modal distribution, it is also possible to estimate the location of the center by statistical methods, and select pixels that are close to this center. But such a simple approach fails in case of a multi-modal distribution. If all peaks of the multi-modal have approximately the same density, then it would be possible to select pixels in regions of high density, to get a set of relatively noise-less pixels. When densities of peaks differ strongly, a peak can only be recognized by the fact that its local spectral density is higher than the spectral density in a spectral neighborhood.

10.3.1 Theoretical analysis of a one-class problem

A theoretical analysis is possible for the case that $W_k = 1$, for a certain k. This corresponds to the case where $sim(p)$ equals the distance to the kth nearest neighbor. Let the spectral vector of a pixel be composed of the spectral vector of the primary ground cover within the area of this pixel and let it have Gaussian distributed noise

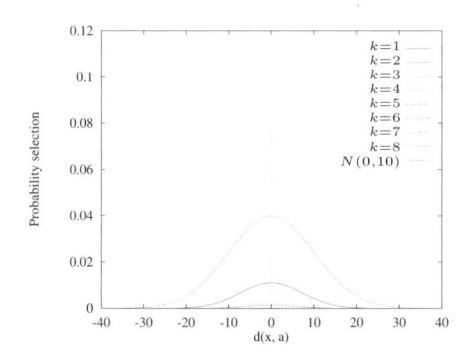

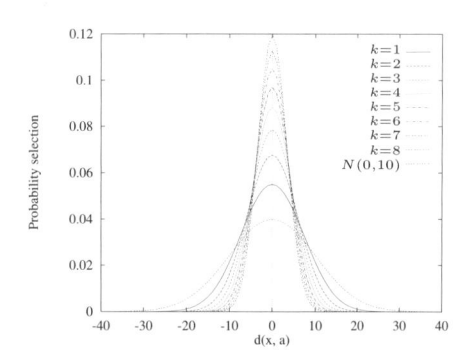

Figure 10.2: Density functions of samples obtained by using a series of values for k before (left) and after (right) normalization

with variance σ^2. Given a spatial region containing a set of pixels with the same primary ground cover, we can observe the distribution of these pixels in the spectral space. Let the distribution of pixels in the spectral space be given by the density function $f(\vec{s})$. We select pixels p such that $sim(p) \leq \delta$. Now the distribution $g(\vec{s})$ of the obtained sample is given by

$$g(\vec{s}) = f(\vec{s}) \sum_{j=k}^{t} B(t, j, p(\vec{s}, \delta))$$

where: $p(\vec{s}, \delta) = \int_{Sphere(\vec{s},\delta)} f(\vec{s}) d\vec{s}.$

Here $B(n, k, p)$ is the binomial distribution, with $0 \leq k \leq n$, t is the number of pixels in neighborhood $N(p)$, and $Sphere(\vec{s}, \delta)$ denotes a d-dimensional ball of radius δ around the point $\vec{s}$. Therefore after the sampling step we get the original density multiplied by a factor that depends on the averaged local density around spectral point $\vec{s}$. The value of k can vary between $k = 1$ and $k = t$. The case $k = 1$ corresponds to using the distance to the nearest neighbor, while $k = t$ corresponds to using the distance with respect to the most dissimilar neighbor. Large values of k correspond to a stricter criterion for homogeneity and therefore, for fixed δ, result in the selection of less pixels.

As an example, we computed the curves $g(\vec{s})$ for one-dimensional data and a Gaussian noise model with $\mu = 0$, $\sigma = 10$. The neighborhood $N(p)$ consist of the 8 adjacent pixels. Fig. 10.2 shows the unnormalized and normalized plots of the formula in one dimension for varying values of k and $\delta = 0.5$. Along the horizontal axis the distance between the actual value and the noise-less value $d(x, a)$ is shown. This value can be interpreted as the amount of noise in a pixel. The vertical axis corresponds to the probability that a corresponding pixel is selected. As k increases the probability of selection decreases rapidly, so the size of the sample gets smaller. When normalizing the results, it can be observed that the distributions tend to

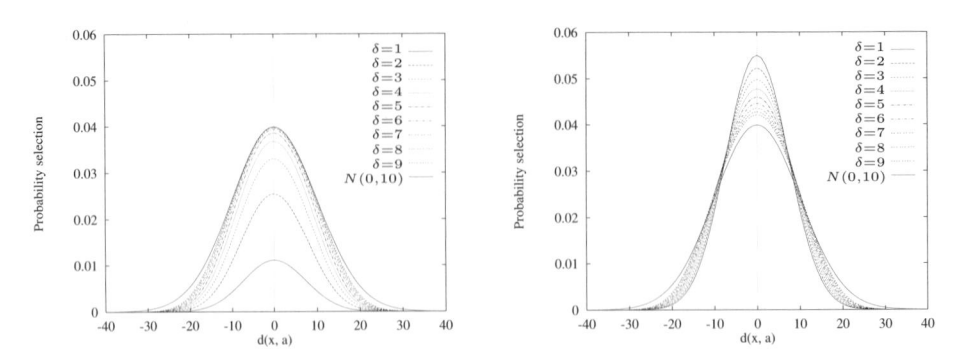

170 Cees H.M. van Kemenade, Han La Poutré and Robert J. Mokken

Figure 10.3: Density functions of samples obtained by using a series of values for δ before (left) and after (right) normalization

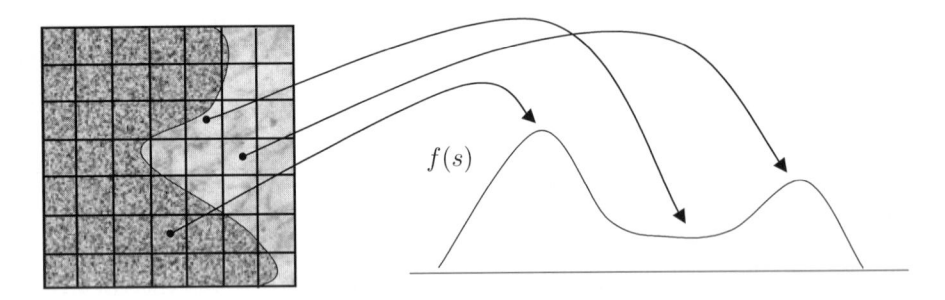

Figure 10.4: A square region containing two types of ground cover and the corresponding density function obtained when mapping all pixels to the spectral space.

be more strongly peaked as k increases. Therefore the selection is biased towards relatively noiseless pixels. The unsupervised classification process benefits from such a biased sampling step, as discussed in section 10.3.2.

Fig. 10.3 shows the unnormalized and normalized plots for different values of δ and $k = 1$. The threshold δ can vary between 0 and ∞. By increasing δ the selection criterion is weakened and more pixels are included. In the limit of $\delta \to \infty$ all pixels are selected, and the distribution converges to the original distribution of the pixels.

10.3.2 Discussion of a multi-class problem

In this section we show intuitively that the sampling procedure also works for multiple classes, and hence for a multi-modal density function of the pixels. Fig. 10.4 shows a region containing two classes, i.e. two types of ground cover. It also shows the density function obtained when mapping all pixels from the square patch to the spectral space. The arrows show the locations of three different pixels in the density-plot (of the spectral space). The peak on the left-hand side of the graph corresponds to the main ground cover within the square patch. The peak on the right corresponds to the other ground cover. The ridge in between these two peaks corresponds to the mixed pixels, that are located on the boundary of the two sub-regions.

Chapter 10: Unsupervised class detection 171

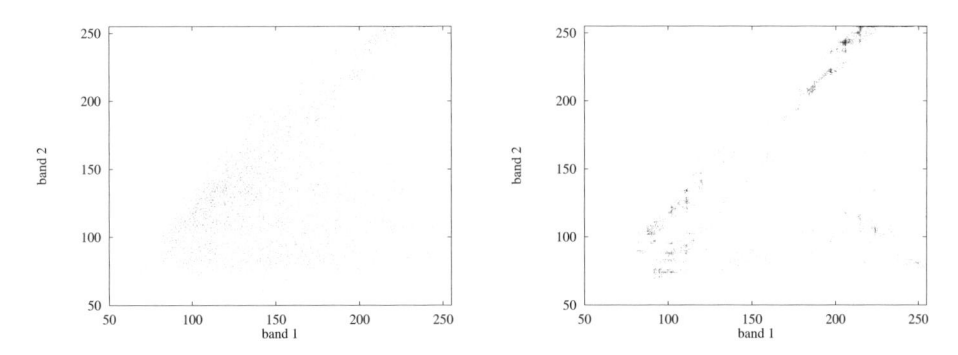

Figure 10.5: Results for a real image with stratified random sampling (left) and biased sampling with $k = 3$ (right)

When computing the similarity measure $sim(\cdot)$ for these three pixels, the two non-mixed pixels are likely to have a relatively low value of $sim(\cdot)$, as these pixels are surrounded by other pixels with the same primary ground cover. The mixed pixel probably has a high value of $sim(\cdot)$. Therefore the two pixels containing a single ground cover are more likely to be selected than the mixed pixel.

To summarize, the local spatial homogeneity is exploited by assuming that pixels in the spatial neighborhood $N(p)$ of point p likely belong to the same class as point p. Pixels with much noise and mixed pixels are both likely to have a large value of $sim(p)$. Therefore a biased sample selected according to the rule $sim(p) \leq \delta$ likely contains pixels with little noise and pixels containing only a single ground-cover.

Next, we apply the biased sampling method to a real image, and compare it with a stratified random sample. The image is a three-band 541 × 440 image, and the sample size is 4000. The stratified random sample is taken by dividing the image in a set of non-overlapping rectangular areas, and taking a random pixel from each area for the real images. The distribution in spectral space of such a random sample is compared to the distribution when using biased sampling. During the biased sampling a weight vector with $W_3 = 1$ is used. The value of δ was adapted such that exactly 4000 pixels were selected, so in fact we select the 4000 most homogeneous pixels. The result is shown in Fig. 10.5. The left graph corresponds to the stratified random sample (see chapter 13), and the right graph corresponds to the biased sample. The biased sample reveals more structure than the random sample. This example shows that biased sampling is able to highlight the cluster structure in the image-data.

10.4 Density estimation

As a basis for the clustering method we need a density estimation method [322, 327]. Assume that a set of points is drawn according to an unknown distribution. A density estimation method derives a density function on the basis of this data-set, that explains the distribution of the data-points. This density function is related to a distribution function, which should approximate the original, but unknown, distribution function that generated the data.

This method has to be able to handle multi-dimensional data. Furthermore, it is important that it operates fast, as the density estimation method is used as part of the clustering method. The most elementary method for density estimation is the histogram method. In a univariate case a range of data-values is split up into a number of disjoint bins, and the number of data-points in each of the bins is counted and used as an estimate of the local density. The width of the bins is called the window size (bin size). Histogram methods do not perform well in high-dimensional spaces, as the number of bins rises exponential with the dimension, so one either needs an excessively large number of data-points, or one should use a large window size, which results in a very coarse estimation of the density function.

Most modern density estimation methods are based on kernel estimators. The basic kernel estimator may be written compactly [322] as

$$\hat{f}(x) = \frac{1}{nh} \sum_{i=1}^{n} K\left(\frac{x-x_i}{h}\right) = \frac{1}{n} \sum_{i=1}^{n} K_h(x-x_i),$$

where $K_h(t) = K(t/h)/h$. In this formula x is the point where the density is estimated, n is the number of data points, h is the window size, x_i are the actual data points, and $K_h(t)$ is the kernel of the density estimation function. This kernel determines how points at distance t from the point x influence the density at point x.

We use a density estimation method with an adaptive window size h_x that is based on the distance to the kth nearest neighbor, the k-NN estimator [327]. The k-NN method and the kernel discrimination were first given by [130], according to [303]. This method is fast, it performs good in high dimensional spaces, and it fits well within the framework of the classification method described in section 10.7. Let the distance to the kth nearest neighbor be denoted by $d_k(x, \{x_i\})$, where $\{x_i\}$ denotes a list with all data-points. Now the local density at a query point x can be estimated by assuming a uniform density over a sphere centered at x with radius $d_k(x, \{x_i\})$, which leads to the formula

$$\hat{f}(x) = \frac{k}{d_k(x, \{x_i\})^d N V_d(1)},$$

where N is the sample size and $V_d(1)$ is the volume of a d-dimensional sphere with radius one. This formula fits within the kernel-based framework, mentioned earlier in this section, with an adaptive window width. The window width depends upon the local density. If the local density increases, then the window width decreases. Adaptive window widths become important as the dimension of the data increases. The k-NN estimator is seen to outperform the fixed kernel estimator when the dimension is larger than or equal to five [322].

The k-NN density estimator requires a method to find the kth nearest neighbor. To compute the kth nearest neighbor, we use an advanced data structure, viz., the

Chapter 10: Unsupervised class detection

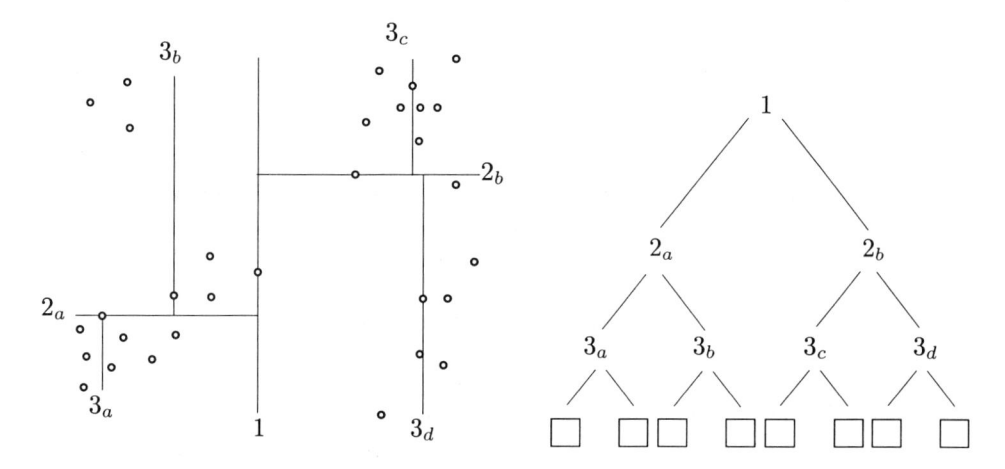

Figure 10.6: Example of a kd-tree on a two-dimensional dataset. On the left the data-points and the hyperplanes that subsequently split the dataset are shown, and on the right the corresponding kd-tree is shown.

kd-trees as introduced in [143], [36], and [144]. We use these trees because a brute-force computation of the nearest neighbors in a data set of size N will require $N-1$ distance computations, while by means of kd-trees the same computation will only take a number of steps proportional to $\log N$. The kd-tree is a multi-dimensional extension of the binary tree, where different levels of the tree uses a different dimension to obtain the discriminator. Fig. 10.6 shows an example of a kd-tree for a two-dimensional dataset. On the left we see the set of data-points, given by the circles and the partitioning hyperplanes, given as lines. Hyperplane 1 partitions the data-set in two sub-sets of equal size along the x-axis. These subsets are partitioned by the hyperplanes 2_a and 2_b along the y-axis. The third set of hyperplanes partitions the resulting 4 sub-sets along the x-axis again. On the right side of Fig. 10.6 we see the corresponding tree. In each node, the location of the hyperplane is stored. Given this kd-tree we can check in a number of steps proportional to $\log N$ whether a point is in the data-set, where N is the number of data-points. To perform this check one starts at the root of the tree, and compute on which side of hyperplane 1 the point is located. If the x-value of the point is smaller or equal to the value of hyper-plane 1, then the point is located in the left branch, otherwise it is located in the right branch. Assuming the point is in the right branch we go one level down in the tree and compare y-value of the point against hyperplane 2_b. As the depth of the tree is $O(\log N)$, a leaf is reached in a number of steps proportional to $\log N$. If the point is in the tree, then it is equal to the point located in this leaf.

Construction of the kd-tree takes $O(N \log N)$ time, where N is the size of the data set. A query for the kth nearest neighbor takes $O(\log N)$ time. For our application we used the optimized kd-tree [144]. In these trees the discriminating dimension is computed for each node separately, so different nodes at the same level can use a different discriminator, this contrary to the standard kd-tree. The optimized kd-tree performs better in practice, but the theoretical time-complexities for building the tree and searching an element are the same for both types of kd-trees.

10.5 Curse of dimensionality for density estimation

In this section we investigate the requirements for a cluster to be detectable by means of a k-NN density estimation method.

First, a one-class problem is studied. Let the pixels of the class be distributed according to the Gaussian distribution with standard deviation σ, denoted by $G(x, \mu, \sigma)$. This corresponds to a spherical symmetric distribution. Therefore the probability of finding a pixel with spectral vector $\vec{s}$ only depends on the distance $|\vec{s} - \vec{\mu}|$, where $\vec{\mu}$ is the noise-less spectral vector of the pure ground cover this class corresponds to. We compute the distance to the kth nearest neighbor in a d-dimensional space. To do so, we need the volume of a d-dimensional hyper-sphere of radius r, given by $V_d(r) = V_d(1) r^d$. Here $V_d(1)$ is the volume of the unit-sphere in d dimensions, which is given by the formula

$$V_d(1) = \frac{\pi^{\frac{d}{2}}}{\Gamma[\frac{d}{2} + 1]}$$

where $\Gamma[n]$ is the gamma function:

$$\Gamma[n] = \int_0^\infty x^{n-1} e^{-x} dx.$$

Because rescaling is possible, one can assume that $\sigma = 1$, and $\vec{\mu} = \vec{0}$, without loss of generality. So, the plots correspond to the scaled distance, where distances are scaled with respect to the actual σ. The fraction of pixels that fall within a sphere of radius r around the pure spectral vector $\vec{\mu}$ is given by

$$c_d(r) = \frac{\int_0^r G(x, 0, 1) \frac{\partial V_d(x)}{\partial x} dx}{\int_0^\infty G(x, 0, 1) \frac{\partial V_d(x)}{\partial x} dx}$$

The left graph of Fig. 10.7 shows the curves $c_d(r)$ for dimensions ranging from one to seven. This graph shows the curse of dimensionality. In a one-dimensional space, 50% of the points are located within a sphere with radius 0.68. In a seven dimensional space only 0.01% of the points is located within such a sphere, and a sphere of a radius of 2.6 is required to cover more than 50% of the points. Therefore, if the dimension increases, then the densities get lower and a larger fraction of the points is located in the tail of the distribution. We can also compute the average uniform density over sphere$(\vec{\mu}, r)$ by $\hat{f}_d(r) = \frac{c_d(r)}{V_d(r)}$. The right graph of Fig. 10.7 shows this density as a function of the radius of this sphere. These densities have been normalized, such that the integral over the densities is one. So, if a class covers only a fraction α of the image, then these normalized densities should be multiplied by α, to get the actual density.

Chapter 10: Unsupervised class detection 175

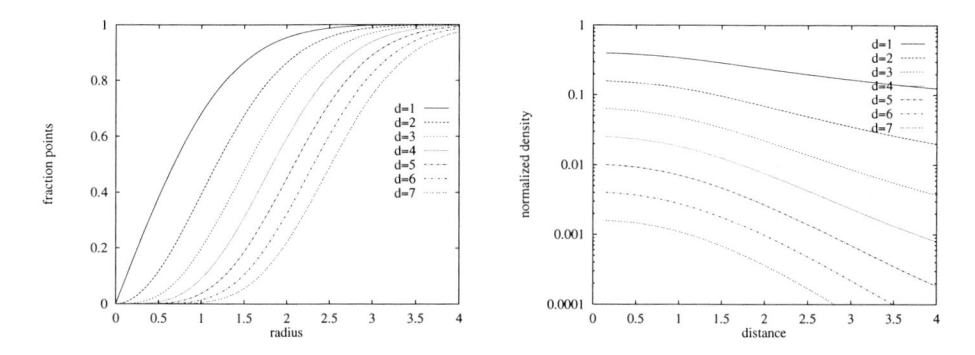

Figure 10.7: Fraction of points in sphere($\vec{\mu}, r$) (left) and average uniform density over this sphere (right) for different numbers of dimensions d.

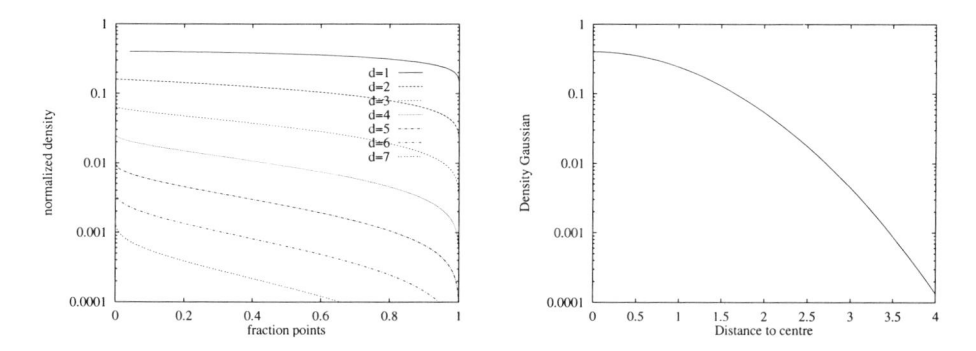

Figure 10.8: Density as a function of the fraction of points in the k-NN sphere for different numbers of dimensions d (left), and the function density as function of the distance $G(r, 0, 1)$ (right).

The k-NN density estimator uses the uniform density over a k-NN sphere as a density estimate. Let a class contains n sample points, and k be given. Now the k-NN sphere around the centre of the class contains a fraction k/n of the sample points that belong to this class. Using the above results, the estimated density over this k-NN sphere is computed. The results are shown in the left graph of Fig. 10.8. Note that the y-axis is logarithmic in both graphs. The density decreases rapidly as the dimension increases. Therefore, if the dimension increases, it becomes more difficult to detect a class. Furthermore, the curves show a downward slope. This effect gets stronger as the dimension increases. Classes with relatively few sample-points require a large k-NN sphere. Therefore such classes are more difficult to detect in high-dimensional spaces. Take for example a sample covering two classes, containing $10k$ points that belong to the first class and $5k$ points that belong to the second class. Thus the fraction of samples in the k-NN sphere is respectively 0.1 and 0.2. Now, the left graph in Fig. 10.8 is used to find the normalized densities of the peaks of the clusters that belong to these two classes. For a one-dimensional data-set the normalized peak-density 0.398 for the first class and 0.395 for the second class. Therefore, the non-normalized peak-densities of the corresponding clusters differ approximately by a factor two. For the seven-dimensional space the normalized peak-densities are 0.00056 and 0.00039. These densities differ more, and therefore the class with fewer sample points is more difficult to detect, and much more likely to disappear in the background noise.

Next, consider the separability of two classes. The right graph of Fig. 10.8 shows the normalized density as a function of the distance to the center of a cluster, denoted by $G(r, 0, 1)$. For two clusters, with the first containing n sample-points, and the second containing αn sample-points ($0 < \alpha \ll 1$) the left graph of Fig. 10.8 can be used to approximate the normalized density at the center of the second cluster. Multiplying the obtained value by α gives the relative density with respect to the first cluster. Now given this density, the right graph of Fig. 10.8 is used to determine the distance with respect to the center of the first cluster, where the tail-density of the first cluster corresponds to the peak-density of the second cluster. As a concrete example, imagine a 7-dimensional space where the first cluster contains $10k$ points, and the second cluster contains $5k$ points, hence $\alpha = 0.5$. With the left graph we find that the normalized density at the center of the second cluster is approximately $4 \cdot 10^{-4}$ and that of the first cluster $2 \cdot 10^{-4}$. With the right graph we see that the density of the first cluster is $2 \cdot 10^{-4}$ at a scaled distance of 3.9 from the center of this cluster. The same exercise for the 1-dimensional case shows that the comparable density is then attained at a scaled distance of approximately 1.2. Hence, if the dimension of the space increases, then a larger distance between the classes is required to separate them.

Density estimation becomes more difficult as the number of dimensions increases. Given a set of points drawn according to a Gaussian distribution, an increasing fraction of the points is located in the tails of the distribution as the dimension of the space increases. As a result densities decrease rapidly. When using a k-NN density estimator, a uniform distribution over a k-NN sphere is assumed. As a result the peak-density of small clusters will be underestimated, because the k points in the k-NN sphere form a significant fraction of all points in case of a small cluster. The underestimation of peak-density for small classes gets more severe as the dimension increases. As a result, small clusters are more likely to disappear in the tail of one of the larger clusters.

10.6 Density based sampling for Remote Sensing
10.6.1 Density estimation in Remote Sensing
By means of a simple fictitious example, we show that the number of pixels assigned to different types of ground cover, are likely to differ strongly. Consider an image consisting of 1000 × 1000 pixels, mainly covered with green grass. The image also contains a grey road with a width of 3 pixels and 10 red houses, each having dimension 10 × 10 pixels. Hence, the road covers 0.3%, the houses cover 0.1%, and the grass covers the remaining 99.6% of the image. In many applications of remote

sensing, it is important to detect such structures that cover only a very small part of the complete image.

With this image, the human eye can easily discover the structure. In a split second, we observe groups of pixels that are close together and have roughly the same color, even in the presence of noise. Therefore, the use of the local spatial structure in the image seems to be the key to the human visual recognition. If we map pixels to the spectral space, then the spatial information gets lost. In section 10.5, it was noted that detection of small clusters gets increasingly more difficult as the dimension of the space increases. Our method does a spectral clustering on a sample. During the sampling step, spatial information is available. Next, we show a sampling method that uses this spatial information to get a sample containing relatively many points that belong to the clusters that cover only a small part of the image.

10.6.2 Using local density estimates during sampling

We introduce a sampling method that incorporates spatial information. To do so, the method compares local and global density estimates. Pixels are selected on the basis of the ratio between the local and the global density estimates. The goal is to obtain a sample that contains a more even distribution of points over all classes. This means that pixels belonging to the classes that cover a small part of the image, should receive a relatively high probability of being selected.

To obtain the local density estimates, we use small patches from the complete image. Within such a patch the diversity is much smaller than in the complete image. This is a result of the limited number of pixels in the patch, but the effect is magnified by the local spatial homogeneity. As a result, it is relatively easy to discover the different classes in such a small patch. Furthermore, the ground-covers present in such a small patch, tend to cover a relatively large part of the patch. For a patch containing part of a house in the example image, the proportion filled by the house is likely to be much larger than the global proportion of image, which is 0.1%. To get the global density estimates, the method first draws a random sample S_g from the image. This sample will contain a high density in regions corresponding to spectral features of the most important ground covers in the image.

We select the learning sample by the following procedure. To select a single point we extract a $l \times l$ patch of pixels from the image. From this patch, a pixel is selected at random. Given the spectral vector of this pixel, the nearest peak in the density landscape has to be located. We have to find this peak in order to get reliable comparisons of the local and global density estimates. This peak likely corresponds to relatively noise-less pixels containing only a single ground cover. Therefore, locating this peak results in a reduction of noise. Furthermore, we need to locate this peak in order to get reliable estimates of the local and global density. The location of this peak is determined by the following procedure. The k_l-neighborhood of the point in the spectral space is taken, and the median point of the k_l pixels in this neighborhood is computed. A median point is determined by computing the median value for each dimension, so this median point does not have to correspond with a real point in the data-set. Next, the k_l-neighborhood of

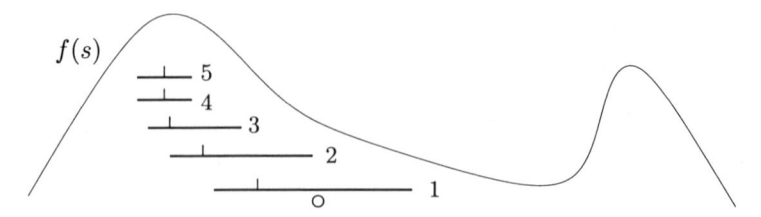

Figure 10.9: Search for high density region by iterated taking of k-neighborhoods around median points.

this median point is computed, resulting in a new median point. This process is continued until a k_l-neighborhood is obtained, containing only points that where visited already. The median point now is likely to approximate a local maximum in the density landscape. Fig. 10.9 gives a graphical example of this procedure. The curve represents the density in spectral space. The circle denotes the location of the random starting point in the spectral space, and the numbered lines below the figure correspond to the subsequent k_l-neighborhoods that are computed. The vertical marker on each line denotes the location of the median point of the sample. These median points are likely to be located on the side of the neighborhood that corresponds to the highest density. The fifth neighborhood does not contain any new points, and thus the iterated search is terminated. Now given this median point, we compute the the ratio between the local density, and the global density around this median point, both in spectral space. The local density is computed by means of the k_l^{th} nearest neighbor of the median point over all pixels in the patch, the global density is computed by means of the k_g^{th} nearest neighbor in sample S_g.

For the $l \times l$ patch, extraction of a random pixel, computation of the nearest peak, and estimation of the ratio between local and global density is repeated β times. The median point with the largest ratio is selected, and added to the learning-sample. The selected point typically has a ratio in the range 10^2 to 10^4. The whole procedure is embedded within a stratification framework. We stratify the image by covering the image with N non-overlapping rectangular regions, where N is the size of the learning sample. Each rectangular region provides one $l \times l$ patch, and therefore provides a single sample point.

A pixel belonging to a ground cover that covers a small part of the image has a small global density. Therefore, the ratio between the local and the global density is large. For a ground cover that covers large part of the image the global density is high, and therefore the ratio gets smaller. Noise reduction is obtained by searching for the peaks in the density landscape. It is possible that we select a noisy point that is located within a transitively closed neighborhood. The median point now is close to this point. Probably the local density is roughly equal to its global density and therefore the ratio is relatively small. As the typical ratios are larger than 10^2, such a noisy pixel is still unlikely to have the largest ratio amongst the β pixels that are selected from the patch.

Chapter 10: Unsupervised class detection   179

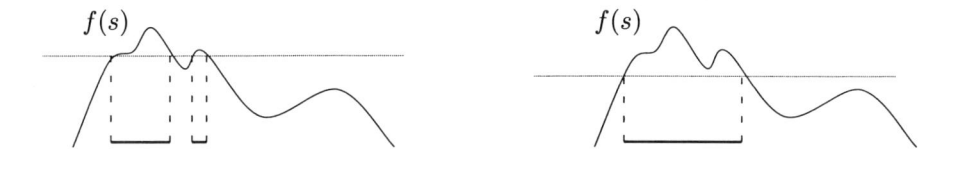

Figure 10.10: Density based hierarchical clustering

10.7 Hierarchical clustering

In this section we give a detailed description of the hierarchical clustering method we developed. This method takes a set of points as an input and produces a set of clusters. Simultaneously, it computes a measure for the separability of all the clusters. The full pseudo-code of the method is given in a technical report [230].

10.7.1 Water-level model

The hierarchical clustering method can be thought of as a method that counts the number of islands when the water-level of the lake is decreasing. If the water-level drops, then a new peak that surfaces increases the number of islands by one. If the region between two islands surfaces, then two islands are merged and the number of islands decreases by one. A graphical representation is given in Fig. 10.10. The left-hand side of this figure shows a density curve over a one-dimensional space. This density curve has three local maxima. The left peak corresponds to the highest density. The horizontal line in Fig. 10.10 represents the decreasing threshold, used by the method. On the left of the figure the threshold is still high. Each connected region above the threshold results in a separate cluster. The two clusters in the left-hand figure are denoted by the solid line-segments below the graph. On the right-hand side of Fig. 10.10, the threshold is lowered. The density in the region in between the two clusters is above the threshold. As a result the two clusters have been merged into a single cluster. If we lower the threshold even further, then the points corresponding to the third peak will be detected. In practice we do not know the density function of a data-set. We can approximate it by means of density estimation methods.

10.7.2 Algorithm

The hierarchical clustering algorithm keeps track of three lists. The first list contains all data-points, the second list contains all clusters, and the third list contains delayed cluster merge operations. Initially, the last two lists are empty. For each of the points the local density is estimated, and the list of points is ordered on decreasing density. Hence, a position closer to the start of this list corresponds to a higher density estimate. If the data are distributed according to the density given in Fig. 10.10, then the sample-points close to center of the left-hand cluster will be at the start of this list. If we process a sample-point s, then we set the current

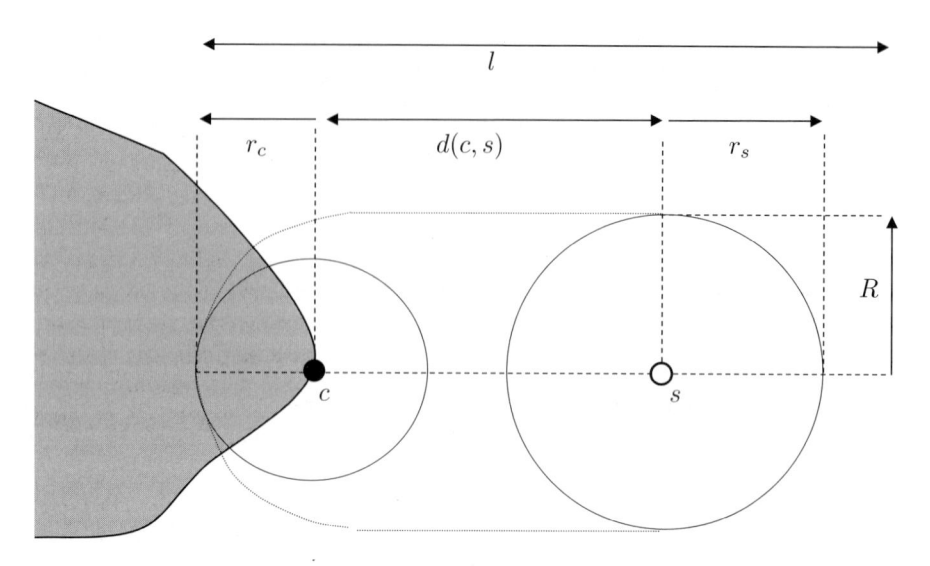

Figure 10.11: Estimating the combined density around two points by the construction of their cylindrical envelope.

threshold equal to its local density estimate. Before actually processing the point, we first process all delayed merge operations that should have been performed at this threshold. Once an operation is performed, it is removed from the merge operation list. Next, we generate a new cluster containing sample-point s only, and we add this new cluster to the list of clusters. Now, we have to determine when this new cluster can be merged with any of the other clusters. The computation for a single cluster C is visualized in Fig. 10.11. The new cluster consists of a single point denoted by the small circle. The cluster C is denoted by the grey region. The density at which the cluster containing s is merged with cluster C is computed as follows.

1. The nearest neighbor c of s in C is located.
2. The joint volume of the k-NN neighborhoods of point s and c is computed by taking a cylindrical envelope around both points with a shape resembling that of a pill.
3. Given the volume of this cylindrical envelope and given that it contains at least $2k$ points, approximate the joint density of point c and s.

The left-hand side of figure 10.11 corresponds to the situation where s is outside C. The right-hand side corresponds to the situation where the joint density of s and c is above the threshold, and therefore the cluster containing s is merged with cluster C. Using this approach, the separability of two clusters is determined by the density of the densest connection between these two clusters. The algorithm is sensitive to the minimal allowed density during the clustering process. If this density is too low, all clusters are likely to be merged into a single cluster, if it is too high, then low density clusters, corresponding to classes of ground cover that are relatively rare in the image, will not be detected. If the dimension of the

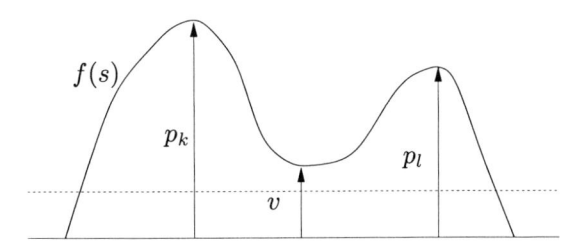

Figure 10.12: Separability of clusters

spectral space increases, then its sensitivity with respect to the minimal density will increase. Therefore, we introduce a second measure for the separability of two clusters, and use this measure to put an additional restriction on the merge of two clusters. The new measure is a separation measure, given by $\frac{v}{\min\{p_k, p_l\}}$, where p_i is the maximum of the density in cluster i, and v is the density in between the two clusters. This measure determines the density ratio between the peak density of the low-density cluster, and the density in between the clusters. An example is given in Fig. 10.12. This figure shows a density function over a one dimensional search space. The solid horizontal line denotes density 0. The dashed line denotes the current *densityThreshold* (which decreases during the run of the algorithm). The value of the separation measure is between zero and one. A value close to one means that the low-density cluster has a density that is close to the density of the region in between the clusters. Because estimated densities are used, it seems reasonable to merge the two clusters in this case. If the value is close to zero, then the low-density peak has a significant higher density than the region in between the two clusters, and it is likely that the low-density cluster corresponds to a separate class. Thus, no merge is performed. In the example in Fig. 10.12 a point in between the two peaks is considered because it is above the *densityThreshold*. The two clusters are now merged if $\frac{v}{p_l} > separation$.

10.8 Remote sensing application

The hierarchical clustering method produces a set of classes for the sample. Next, a pixel-classification method is needed to assign all pixels of the image to a class. The clustering method can produce classes that correspond to non-convex regions in the spectral space. Therefore, the pixel-classification method should be non-parametric. We use a nearest-neighbor classifier. A pixel is classified by finding the (spectral) nearest pixel in the learning-sample, and assigning the class of this pixel. We use a kd-tree to find the nearest neighbor of a pixel in a number of steps proportional to $\log N$, where N is the size of the learning sample. As this operation has to be repeated for each pixel, this pixel classification method turns out to be slow on large data-sets. As an alternative we implemented a method that does a principal component analysis, by means of singular value decomposition [213], on each of the clusters. Next each cluster is reduced to a line-segment. The direction of the line-segment is determined by the direction of its primary principal component of the cluster, the center of the line-segment is determined by the center of the cluster, and the length of the line-segment is set equal to two times the standard-deviation along the primary principal component [213]. Now a pixel is classified by mapping it on the line-segment of each of the clusters and selecting the class that is linked to the nearest line-segment. The position of the mapping of the point on the line-segment, is used as luminance-value in the output of the image. The time needed for the classification of a pixel now is proportional to the number of clusters.

We have applied the methods presented in this chapter to remote sensing data. We have tested the method on a high-resolution three-band aerial photograph of 500×538 pixels and on 7-band Landsat scene with 960×1130 pixels. For the Landsat images the sixth band, which corresponds to thermal emission, was removed from the data-set, following [302]. For both images we used a sample of 4000 points for learning. The tests were performed on a SUN workstation running at 180 MHz. The sampling method used during this experiment involves the computation of local and global density estimates. For the three-band image the generation of the biased sample took approximately 33 seconds, and a classification containing 13 clusters was obtained in approximately 16 seconds. For the six-band image the generation of the biased sample took approximately 95 seconds, and a classification containing 26 clusters was obtained in approximately 59 seconds. When using a sample containing 11,000 points for the 7-band image the sampling step takes 460 seconds, and the clustering step takes 421 seconds.

For the 7-band Landsat scene we also had a map, showing the results of a supervised classification of the land usage of part of this region. The resolution of the map and the Landsat scene were different, and geometric corrections were applied to the map. Therefore, we can only give a qualitative comparison between the map and the classification obtained by our tool. The types of ground-usage shown in the map are agriculture, industry, city, residential, water, and natural vegetation. When comparing the map to our results we observe that our method finds more classes. For example, we observe many different classes in the agriculture region. It is interesting to see that most of the regions found by our method are rectangular regions, that are aligned with the nearby regions. The shape, orientation, and size of these regions corresponds to the typical plots of land in agriculture regions. It seems that our method is able to discriminate between the different types of agricultural use of the land in this region. Also the water regions, that cover only approximately 0.6% of the total surface, come out clearly when using our method. We also find two classifications for the urban regions, the first class is mainly located near the center of urban regions, while the second class is located more towards the boundaries of the urban areas. This may correspond to the discrimination between city and residential area in the map. The boundaries between city and residential are different in our case, though we can easily imagine that these boundaries are not very well defined, and we see it as a promising result that our method already detects that you have different types of urban area's. There is only little industry in this region, and it seems like the industrial regions are classified as residential area's in our method. The area's with natural vegetation are split over two classes.

When doing the same analysis for a lower value of the separability parameter we get a more coarse grain classification. The agriculture regions are less diverse, and the city and residential area are merged in a single class. Natural vegetation is covered by a single class too.

10.9 Conclusions

We developed an adaptive, biased sampling method and a hierarchical clustering method. The sampling method exploits spatial information in order to select those pixels that correspond to a single ground cover, and contain relatively little noise. The sampling method has been tested by means of a theoretical model and on real data. In both cases, we observe that the clusters in the data-set are more clearly present when using a biased sample instead of a random sample.

The hierarchical clustering method is a fast, unsupervised clustering method that takes a set of points as an input and produces a set of non-parametric classes describing the input-data. The method is purely data-driven, and therefore the number of clusters obtained is dependent upon this data-set. In fact, the algorithm produces a whole range of clusterings simultaneously, and afterwards a number of clusterings can be extracted almost instantly. Apart from the sample-sizes and neighborhood sizes, the method uses a separability parameter. This parameter determines under what conditions two clusters can be merged into a single cluster, and therefore affects the final number of clusters. This parameter has an intuitive basis in terms of the ratio of the peak-densities of clusters and the density of the ridge connecting the clusters.

Anticipated further work is the development of non-parametric models out of the learning-data by means of radial basis neural networks, the use of evolutionary computation methods to search for models that allow a demixing of clusters consisting of multiple classes, and the usage of a Bayesian approach to exploit the spatial structure during the pixel classification. Spatial structure is exploited by computing prior probabilities over a spatial neighborhood, and use these to compute posterior pixel classification probabilities.

Chapter 11

Image classification through spectral unmixing

Freek van der Meer

The most widely used method for extracting surface information from remotely sensed images is image classification. With this technique, each pixel is assigned to one out of several known categories or classes through a separation approach. Thus an image is decomposed into an image containing only thematic information of the classes previously selected as the expected image elements. Two approaches are adopted which are described in the two previous chapters in this book: supervised classification (see Chapter 9) and unsupervised classification (see Chapter 10). In supervised classification, a training sample set of pixels is defined by the user to train the classifier. The spectral characteristics of each training set are defined through a statistical or probabilistic process from feature spaces, pixels to be classified are statistically compared with the known classes and assigned to the class to which they mostly resemble. In this way thematic information is obtained disregarding the mostly compositional nature of surface materials. In unsupervised classification, clustering techniques are deployed to segregate an image into a set of spectrally separable classes of pixels. Reflected radiation from a pixel as observed in remote sensing imagery, however, has rarely interacted with a volume composed of a single homogenous material because natural surfaces and vegetation composed of a single uniform material do not exist. Most often the electromagnetic radiation observed as pixel reflectance values results from the spectral mixture of a number of ground spectral classes present at the surface sensed. Singer & McCord [329] showed that, if the mixing scale is macroscopic such that photons interact with one material rather than with several materials, mixing can be considered linear and the resulting pixel reflectance spectrum is a linear summation of the individual material reflectance functions multiplied by the surface fraction they constitute. Mixing can be considered a linear process if: (1) no interaction between materials occurs, each photon sees only one material, (2) the scale of mixing is very large as opposed to the size of the materials, and (3) multiple scattering does not occur. Various sources contribute to spectral mixing: (1) optical imaging systems integrate reflected light from each pixel, (2) all materials present in the field of view contribute to the mixed reflectance sensed at a pixel, and (3) variable illumination conditions due to topographic effects result in spectrally mixed signals.

Rather than aiming at representing the landscape in terms of a number of fixed classes, mixture modeling and spectral unmixing [4] acknowledge the compositional nature of natural surfaces. These techniques strive at finding the relative or absolute fractions (or abundance) of a number of spectrally pure components or end-member spectra (shortly denoted as end-members) that together contribute to the observed reflectance of the image. Therefore the outcome of such an analysis is a new set of images that for each selected end-member portrays the fraction of this class within the volume bound by the pixel. Mixture modeling is the forward process of deriving mixed signals from pure end-member spectra while spectral unmixing aims at doing the reverse, deriving the fractions of the pure end-members from the mixed pixel signal. Spectral unmixing is becoming more readily applied for various applications dealing with multispectral image data. The interested reader is referred to some recent publications [25, 32, 135, 283, 320, 326, 373]. A review of mixture modeling and spectral unmixing approaches can be found in [200].

In this chapter, mixture modeling and spectral unmixing are described and two alternative methods for finding image end-members. The technique is applied to the LANDSAT Thematic Mapper image of the Enschede area (chapter 1).

11.1 Mixture modeling

Mixture modeling aims at finding the mixed reflectance from a set of pure end-member spectra. This is based on work by Hapke [170] and Johnson et al. [212] on the analysis of radiative transfer in particulate media at different albedos (i.e. overall reflectivity) and reflectances by converting reflectance spectra to single-scattering albedos. In remote sensing, spectra are measured in bidirectional reflectance (see chapter 3). The following expression can be established relating bidirectional reflectance, $R(i,e)$, (i.e., the radiant power received per unit area per solid angle viewed from a direction $e = cos^{-1}\mu_0$ on a surface illuminated from a direction $i = cos^{-1}\mu_o$) to the mean single-scattering albedo, W, as [275]

$$R(i,e) = \frac{w/4(\mu + \mu_o)}{H(\mu)H(\mu_o)},$$

where $\mu_o = cos(i)$, and $\mu = sin(e)$ and $H(\mu)$ is a function describing multiple scattering between particles that can be approximated by

$$H(\mu) = \frac{1 + 2\mu}{1 + 2\mu(1-W)^{0.5}}.$$

The mean single-scattering albedo (see also Chapter 3) of a mixture is a linear combination of the single-scattering albedos of the end-member components weighted by their relative fraction as

$$W(\lambda) = \sum_{j=1}^{n} W_j(\lambda) f_j,$$

Chapter 11: Spectral unmixing

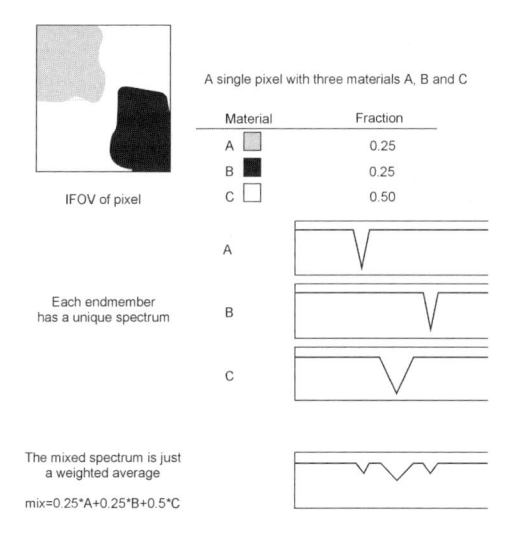

Figure 11.1: Diagram of linear spectral unmixing. A IFOV of a pixel is composed of three unique end-members with known spectra A, B, and C. The mixed spectrum to be found at the pixel is the linear weighted average of the three end-members. Spectral unmixing uses the three end-member spectra and the mixed pixel spectrum to derive the relative fractions of A, B and C in the volume bound by the pixel.

where λ is the spectral waveband, n is the number of end-members and f_j is the relative fraction of component j. Further, f_j is a function of the mass fraction M_j the density, ρ_j, and the diameter, d_j, of the particles of the end-member material, j, as

$$f_j = \frac{M_j/\rho_j d_j}{\sum_{j=1}^{n} M_j/\rho_j d_j}.$$

These equations allow therefore to model reflectance properties and mean single-scattering albedos of mixtures and form the basis of mixture modeling studies.

11.2 Spectral unmixing

Spectral unmixing is a deconvolution technique that aims at estimating the surface fractions of a number of spectral components or (end-members) together causing the observed mixed spectral signature of the pixel (Fig. 11.1). A linear combination of spectral end-members is chosen to decompose the mixed reflectance spectrum of each pixel, R_i, into fractions f_j of its end-members j in the i bands, Re_{ij}, by

$$R_i = \sum_{j=1}^{n} f_j \underset{ij}{Re} + \varepsilon_i \quad \text{and} \quad 0 \leq \sum_{j=1}^{n} f_j \leq 1,$$

where each of the n image end-members has an residual error ε_i which is approximated by the difference between the measured and modeled digital number (DN) in band i. A unique solution is found from this equation by minimizing the residual error, ε_i, in a least-squares solution. This residual error should in theory be equal to the instrument noise in case that only the selected end-members are present in a pixel. Residuals over all bands for each pixel in the image can be averaged to give a root-mean square (RMS) error, portrayed as an image, which is calculated from the difference of the modeled, R_{jk}, and measured, R'_{jk}, pixel spectrum as

$$\text{RMS} = \frac{1}{m} \sum_{k=1}^{m} \sqrt{\sum_{j=1}^{n} (R_{jk} - R'_{jk})^2 / n},$$

where n is the number of spectral bands and m the number of pixels within the image. The solution to unmixing is found through standard matrix inversion such as by Gaussian elimination. Boardman [44] suggested the use of singular value decomposition of the end-member matrix, having the advantage that singular values can be used to evaluate the orthogonality of the selected end-members. If all end-members are spectrally unique, all singular values are equal. For a degenerate set of end-members all but one singular value will be equal to zero.

A simple example to illustrate this is the following. Given that x represents a component (e.g., soil, water) and y is a pixel where the pixel is not pure and the various mixing components are correlated (as is inevitable in nature) then the mixing problem becomes

$$y = \sum \beta x + \varepsilon.$$

The solution to find the end-member abundances is

$$\widehat{\beta} = [X'X]^{-1} y.$$

When the mixture components, x, are correlated it means that the matrix $[X'X]$ is non-orthogonal. Therefore the inversion of $[X'X]$ will generate computational error (the error term, ε) which will increase as the condition number λ_1/λ_η of $[X'X]$ increases (with the λ's being the eigenvalues).

A solution to mixing as found by a matrix inversion is given by [323]. They define a vector of expected pixel signals $\mu_i = \{\mu_{i1}, \mu_{i2} \ldots ..\mu_{ic}\}^T$ for the c ground cover classes giving an expected mixed pixel signal under strictly linear conditions as

$$f_1 \mu_1 + f_2 \mu_2 \ldots f_c \mu_c = M \cdot f.$$

The columns of the matrix, M, are the vectors, μ_i, which are the end-member spectra. The observed signal of pure pixels exhibits statistical fluctuations due to sensor noise characterized by a noise variance-covariance matrix, N_i. Therefore pixels with the mixture will exhibit fluctuations around their mean value characterized by the noise covariance matrix, $N(f)$, given by

$$N(f) = f_1 N_1 + f_2 N_2 \ldots + f_c N_c.$$

Chapter 11: Spectral unmixing 189

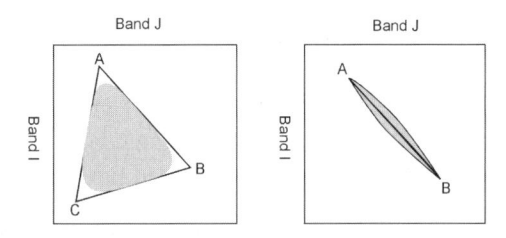

Figure 11.2: Selection of end-members using the extreme pixels bounding the cluster of pixels in a feature space in case of a two component mixture (top) and a three component mixture (bottom).

Generally, to find a solution to the mixing problem c has to be 1 less than the number of spectral bands in the data set (thus for Landsat TM $c = 5$ assuming that the thermal channel is excluded from the analysis). If $N(f)$ is independent of f, thus when the noise components are un-correlated, the linear model can be defined according to [323] as

$$x = Mf + e,$$

where e is the vector of errors satisfying

$$E(e) = 0 \text{ and } E(ee^T) = N.$$

This signifies that the expectation (i.e., variance) of the noise component is close to the sensor noise component.

The spectral unmixing model requires a number of spectrally pure end-members to be defined which cannot exceed the number of image bands minus 1 to allow a unique solution to be found that minimizes the noise or error of the model. Furthermore, constraining factors are that the fractions sum to unity and that the fractions for the individual end-members vary from 0 to 100%. When adopting these assumptions, constrained unmixing is applied in other cases we refer to unconstrained unmixing.

11.3 End-member selection

A set of end-members should allow to describe all spectral variability for all pixels, produce unique results, and be of significance to the underlying science objectives. Selection of end-members can be achieved in two ways:
1. from a spectral (field or laboratory) library, and
2. from the purest pixels in the image.

The second method has the advantage that selected end-members were collected under similar atmospheric conditions. End-members resulting through the first option are generally denoted as 'known', while those from the second option results are known as 'derived' end-members. Identification of the purest pixels in the scene is done through compression of the data using principal component analysis (PCA) following [339]. Spectrally pure pixels are found at the vertices of the polygon that bounds the data space of the principal components (Fig. 11.2). In case of LANDSAT TM data, over 90% of the spectral variability is mapped into PC1 and PC2, thus spectral end-members are usually selected from a scatterplot of PC1 and PC2. Selection of spectral end-members often is an iterative process where additional end-members are selected on the basis of clearly visible spatial patterns in the root mean squared error image until the RMS error obtained does not show any obvious systematic spatial patterns of error distribution. Boardman et al. [45] introduced an algorithm for finding the most spectrally pure pixels in an image based on the method of Smith et al. [339]: the pixel purity index (PPI). The PPI is computed by an iterative process of projecting n-dimensional scatterplots (where n is the number of bands) onto a random unit vector. The extreme pixels in each projection are recorded and the total number of times each pixel is marked as extreme is noted. A "Pixel Purity Image" is created in which the DN of each pixel corresponds to the number of times that pixel was recorded as extreme. The technique runs through a number of iterations (typically hundreds) that are defined by the user until the process converges to a maximum number of pure pixels. A threshold value in data units is used for extreme pixel selection. For example, a threshold of 2 will mark all pixels exceeding two digital numbers (DN), both high and low, as being extreme. This threshold selects the pixels on the ends of the projected vector. The threshold should be approximately 2-3 times the noise level in the data. For TM data, which typically has less than 1 DN noise, a threshold of 2 or 3 is considered suitable. Larger thresholds will cause PPI to find more extreme pixels, but they are less likely to be pure end-members.

11.4 Application to the Enschede image

The first step in the application of spectral unmixing to the LANDSAT TM image of Enschede was to calibrate the image data set to reflectance using the LANDSAT TM calibration described in chapter 3. Using the calibration coefficient found in tables 1-3 of chapter 3, the raw DN values are converted to radiance and subsequently to at-sensor reflectance (see Fig. 3.10) to retain spectroscopic image data that can be compared with other sources of spectral data such as field- or laboratory spectral libraries.

End-members to be used in the spectral unmixing process are selected using the PPI approach outlined in the previous paragraph, a threshold of 2 is used and the process is run through a series of 100 iteration. Note that after 6 iterations the process converges slowly to around 2000 pure pixels, but that after some 95 iterations the curve again shows a pronounced jump toward 3500 extreme pixels prior to stabilizing. A plot showing the number of iterations in the calculation of the PPI versus the number of extreme pixels is shown in Fig. 11.3. The PPI image for the Enschede LANDSAT TM data set is shown in Fig. 11.4. The final selection of end-members from the pixel purity image, which typically results in 3500 extreme pixels, is done by expert opinion. Alternatively, we could apply a statistical clustering technique to identify spatial clusters of spectrally extreme pixels. Here we

Chapter 11: Spectral unmixing

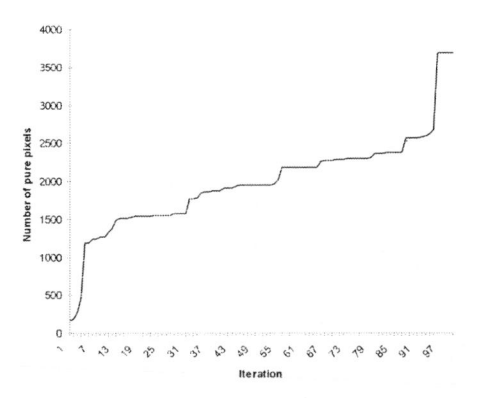

Figure 11.3: Number of extreme pixels found during the iterative process of calculating the pixel purity index. The number converges after a while to a maximum.

selected the maximum number of 5 end-members possible for the analysis: two vegetation spectra, two soil spectra and a water spectrum (Fig. 11.5). Unconstrained unmixing was used, the resulting fraction images for the five selected end-members and the calculated RMS image are shown in Fig. 11.6. Derived fractions (unconstrained unmixing) of the end-members for three selected pixels representing water, vegetation and urban area are given in Fig. 11.7. These not only demonstrate the derived output of the unmixing analysis at a pixel (or sub-pixel) level, but also show that unexplained spectral variability will contribute locally to negative fractions and pixels for which the sum of the fractions do not reach unity. Constraining the data through the restrictions opposed by the constrained unmixing approach will overcome this in a mathematical way, however often leading to undetectable errors in the resulting products. Most likely the negative fractions and non-unity of the resulting pixel response may partly be explained by the fact that the most spectrally pure end-members which were included in the analysis in reality are of a mixed nature. Given the large (approximately 30 m) spatial resolution of the Landsat TM sensor, it seems unlikely to find pixels that are truly composed of one single, homogenous and spectrally pure material. It is more likely that pixels are composed of mixtures of various materials together constituting the observed pixel reflectance. Furthermore, the selected end-members can hardly be regarded as statistically independent and in some cases (i.e., the two soil end-members and the two vegetation end-members) are linearly scaled versions of each other. In a complex system of linear equations such as presented by the spectral unmixing technique, such linear scaling will result in un-precise estimates of fractions and in the extreme case in singular matrices. Here we can use the singular values to evaluate the orthogonality of the selected end-members. In this example, indeed, the singular values show that the two soil and the two vegetation end-members cannot be sufficiently deconvoluted thus leading to errors in the estimates of their fractions.

Figure 11.4: Pixel purity index image for the Enschede scene showing (in white) the purest pixels in the data set.

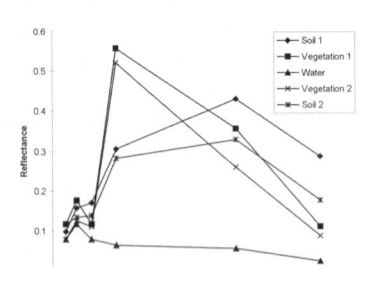

Figure 11.5: Reflectance spectra for selected end-members in the unmixing process derived from the PPI approach for the Enschede data set.

Chapter 11: Spectral unmixing 193

Figure 11.6: Fraction images for the five selected end-members and the RMS image.

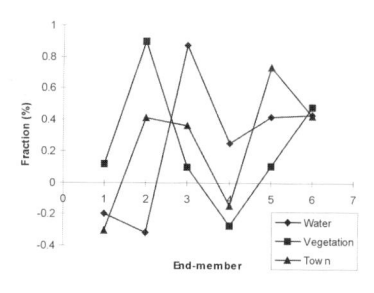

Figure 11.7: Derived fractions (unconstrained unmixing) of the end-members soil 1 (1), vegetation 1 (2), water (3), vegetation 2 (4) and soil 2 (5) for three selected pixels representing water, vegetation and urban area.

PART III

Chapter 12

Accuracy assessment of spatial information

Andrew K. Skidmore

The main product from a geographic information system (GIS), as well as from remotely sensed images, are maps (or a summary of information from maps). Possibly the most important question to be asked by a user is 'how accurate is the spatial information?' The aim of this chapter is to review the sources of error in spatial data, to discuss methods of assessing mapping accuracy, and to evaluate the accumulation of thematic map errors during GIS processing.

The chapter is divided into three sections. The first section describes the error inherent in raster and vector spatial data. Various methods of quantifying errors are then presented in the second section. The third section deals with attempts to calculate the accumulation of error in spatial data.

12.1 Error inherent in raster and vector spatial data

12.1.1 Raster Image errors

Raster images may be obtained from remote sensing sensors carried by aircraft or spacecraft platforms, or by converting an existing line map (vector data structure) to a raster data structure. A raster image is usually constructed from a regular grid of adjacent rectangular cells or pixels (i.e. a rectangular tessellation), though other tessellation's based on triangles and hexagons have been devised [243]. Two types of error are inherent in remotely sensed images, *viz.* geometric and radiometric or thematic errors. These error sources are addressed in detail in numerous monographs and papers including [74] and [302].

Geometric error in a remotely sensed image is caused by 1) movement in the remote sensing platform, 2) distortion due to earth curvature and terrain, 3) different centrifugal forces from earth affecting spacecraft movement, 4) earth rotational skew, and 5) distortions introduced by the remote sensing device itself including systematic distortions caused by sampling sequentially from each detector, and non-linear scanning [6]. Geometric error causes a point on the remotely sensed image to occur in the wrong position relative to other points in the image.

Correction of geometric errors in remotely sensed data is now a routine aspect of preprocessing remotely sensed data. The map or image is usually 'rubber sheet stretched' to an appropriate base map of known projection. Corrected images with geometric errors less than 0.5 pixel are obtainable and acceptable [137, 120, 333]. Base maps from which control point information is derived, however, may be of poor quality. For example, Bell [34] reported that maps used to geometrically correct images of the Great Barrier Reef contained errors of up to 1 km. The accurate selection of control points is crucial to obtaining acceptable results. Increasingly, global positioning satellite receivers are used to accurately locate field locations for the geometric correction of images [16, 27].

Locational accuracy is sometimes expressed as root-mean-square (rms) error, which is the standard error (of the difference between the transformed GCPs and the original GCPs) multiplied by the pixel size. The rms error is defined by the American Society for Photogrammetry and Remote Sensing (ASPRS, 1988) as

$$rms_x = \sqrt{\frac{D^2}{n}},$$

where

$$D^2 = d_1^2 + d_2^2 + ... + d_n^2 = \sum_{i=1}^{n} d_i^2,$$

where d_i = discrepancy in the x-coordinate direction (i.e. $x_{map} - x_{check}$), $i = 1, ..., n$.

Points within a rubber stretched image will no longer be on a regular grid as they have been warped to fit into the projection defined by the GCPs. To obtain a regular grid, an interpolation method is employed to nominate a value for a regular grid point, which falls between the points in the rubber stretched image. Lam [239] provides a review of other interpolation methods including splines, finite difference and kriging (see Chapter 6).

Radiometric errors in remotely sensed images occur as a result of differential scattering of electromagnetic radiation at varying wavelengths, sensors that have poorly calibrated multiple detectors within a band, sensor calibration error, signal-digitization error, and scene specific error such as off-nadir viewing, irradiance variation, and terrain topography [302] (see Chapter 3). Correction for band to band distortion is performed using image histograms (shifted to the origin to remove atmospheric scattering effects), while line striping effects are reduced by calibration of detectors, or by matching detector statistics during computer processing [356].

A final type of error in remotely sensed images may be caused by a time lag between ground checking (truthing) and image collection. In this case, a point on a map may be noted as incorrect during field checking, when in fact it was correct at the time that the map was made.

12.1.2 Error In Vector Spatial Data

In contrast to remotely sensed images, vector images have been traditionally recorded and stored as maps. Maps are subject to many errors [358] . Errors are introduced during data collection, as well as during the creation of the map. Thapa and Bossler [358] outline many of these errors, ranging from personal, instrumental and environmental errors during primary data acquisition activities in photogrammetry and surveying, through to errors in the map making process including plotting control, compilation error, the original line smoothing by draughtsmen which may not follow the true isolines on the ground [63], scaling problems, digitization errors [46], mislabeling, misclassification, lack of an independent survey for ground control points [46] etc. Other errors may be associated with the physical medium used to store the map (for example paper stretch and distortion).

Maps may be represented in computer GISs by a variation of the vector data structure [297], or converted to a raster data structure (see Chapter 4). Vector maps may be input to a computer by digitization. Digitization, however, introduces a number of errors. Varying line thickness on the original map requires automatically scanned vector lines to be thinned. During manual digitization the center of the map line must be followed if the map lines vary in thickness [298]. This requires very careful hand digitizing, or high accuracy automatic scanners. The number of vertices (points) used to approximate a curve is also critical [10]. Too few vertices will result in the line appearing stepped, while too many vertices create large data volumes. Thus, even with extreme care, error is introduced during digitization.

As for raster images, the main method of correcting geometric error in vector images is by using ground control points from a base map of better quality to transform the vector image to a known projection (see previous section for details).

12.2 Methods of quantifying errors
12.2.1 The error matrix

Quantifying error in maps is based on various statistics derived from the error matrix (also called a contingency table or confusion matrix) concept, first expounded for remotely sensed data in the 1970s (for example, [181]). The aim of the error matrix is to estimate the mapping accuracy (i.e. the number of correctly mapped points) within an image or map. An error matrix is constructed from points sampled from the map. The reference (also called ground truth or verification) data is normally represented along the columns of the error matrix, and is compared with the classified (or thematic) data represented along the rows of the error matrix. In most sampling schemes the reference data must be obtained by field inspection. The major diagonal of the matrix represents the agreement between the two data sets (Table 12.1); the overall classification accuracy is calculated by the ratio of the sum of correctly classified pixels in all classes to the sum of the total number of pixels tested.

To check every point, pixel or grid cell for correctness would be impossible except for the smallest map, so various sampling schemes have been proposed to select pixels to test. The design of the sampling strategy, the number of samples required, and the area of the samples has been widely debated.

Table 12.1: Typical error matrix, from [332]. Table Legend: I = Yertchuk, II = Gum/Stringybark, III = Silvertop Ash, IV = Blue-leaved Stringybark, V = Clear-cut/road, VI = Tea Tree, VII = Gum/Silvertop Ash, VIII = Black Oak, IX = unclassified

	Class	\multicolumn{9}{c	}{Reference or ground truth class}
		I	II	III	IV	V	VI	VII	VIII	IX	Total
	I	14	3	7	2		5		7		38
	II		14	3				3	4	1	25
	III		1	16	1		1	1	1		21
Thematic	IV	4		8	3		2	1	3		21
or	V	2				16				1	19
classified	VI			1		1					2
class	VII		1	3				3			7
	VIII								2		2
	IX										
Total		20	19	38	6	17	8	8	17	2	135

Overall classification accuracy: 50.4%

12.2.2 Sampling design

As with any sampling problem, one is trying to select the sampling design which gives the smallest variance and highest precision, for a given cost [69]. A number of alternative designs have been proposed for sampling the pixels to be used in constructing the error matrix (see Chapter 13). Berry and Baker [37] commended the use of a stratified systematic sample where each stratum has an unaligned systematic sample. Simple random sampling in land evaluation surveys emphasizes larger areas, and undersamples smaller areas [409]. Zonneveld [409] suggested a stratified random sample was preferable, and Van Genderen [377] agreed that a stratified random sample is 'the most appropriate method of sampling in resource studies using remotely sensed data'.

Rosenfield et al. [308] and Berry and Baker [37] suggested a stratified systematic unaligned sampling procedure (i.e. an area weighted procedure) as a first stage sample to assist in identifying categories occupying a small area, followed by further stratified random sampling for those classes with fewer than the desired minimum number of points. Todd et al. [363] argued that single stage cluster sampling is the cheapest sampling method, as multiple observations can be checked at each sample unit on the ground.

Congalton [78] simulated five sampling strategies using different number of samples over maps of forest, rangeland and grassland. The aim of the study was to ascertain the effect of different sampling schemes on estimating map accuracies using error matrices. He concluded that great care should be taken in using systematic sampling and stratified systematic unaligned sampling because these methods could lead to bias (i.e. overestimate population parameters). Congalton [78] also stated that cluster sampling may be used, provided a small number of pixels per cluster are selected (he suggests a maximum of 10 sample pixels per cluster). Stratified random sampling worked well and may be used where small but important areas need to be included in the sample. However, simple random sampling may be used in all situations.

12.2.3 Number of samples for accuracy statements

The number of samples may be related to two factors in map accuracy assessment: (1) the number of samples that must be taken in order to reject a map as being inaccurate; and (2) the number of samples required to determine the true accuracy within some error bounds for a map.

Van Genderen [377] pointed out that we wish to know, for a given number of sample pixels, the probability of accepting an incorrect map. In other words, when a high mapping accuracy is obtained with a small sample (e.g. of size 10), there is a chance that no pixels which are in error may be sampled, a so-called type II error α (or producer's risk) is committed. The corollary as stated by [150] is also important, that is the probability β of rejecting a correct map (i.e. committing a type I error, or a consumer's risk) must also be determined.

Van Genderen and Lock [376] and Van Genderen et al. [377] argued that only maps with a 95 per cent confidence intervals (i.e. $\beta = 0.05$) should be accepted and propose a sample size of 30. Ginevan [150] pointed out that [377] made no allowance for incorrectly rejecting an accurate map. The trade-off one makes using the approach of [150] is to take a larger sample, but in so doing reduce the chance of rejecting an acceptable map: that is Ginevan [150] proposed a more conservative approach. Hay [172] concluded the minimum sample size should be 50. His minimum sample size is larger than that of [376]: he stated that 'any sample size of less than 50 will be an unsatisfactory guide to error rates'.

The methods discussed above calculate sample size based on confidence limits or acceptance testing. Thomas and Allcock [359] described a method for calculating the confidence intervals about a mapping accuracy statement, for a sample of a size specified by the user. Their technique is based on using binomial distribution theory. An assumption is that the minimum number of samples should be greater than 50, with a sample in the hundreds being more acceptable. The lowest 99.9 per cent confidence level for a map's accuracy is calculated as follows:

$$99.9\%\text{CL} = (m - 3e_m) - 3(s + 3e_s) \qquad (12.1)$$

where N =number of samples taken, p = fraction of samples that have been correctly classified, $q = 1 - p$, $m = N \cdot p$, $s = \sqrt{N \cdot p \cdot q}$, $e_m = s/\sqrt{N}$, $e_s = s/\sqrt{2N}$.

To calculate the 99 and 95 per cent confidence levels, the value 3 in Equation 12.1 can be replaced with 2.33 and 1.65, respectively. Note that this technique can be used on a class by class basis, or for the whole image.

12.2.4 Sample unit area

The number of samples must be traded-off against the area covered by a sample unit, given a certain quantity of money to perform a sampling operation. 'Should many small-area samples or a few large-area samples be taken?' was a question posed by [86].

In the context of remote sensing images, [396] investigated inter-pixel variation in remotely sensed images. If the spatial resolution of a sample is considerably finer than the objects (or ground resolution elements) in the scene, then pixels will be highly correlated with their neighbors and inter-pixel variation will be low. However, the intra-pixel variance will be high, as multiple ground resolution elements occur within the sample (a situation Cochran [69] states will lead to a biased result). As resolution of ground objects becomes closer to pixel resolution, the likelihood of neighbors being similar decreases, and inter-pixel variation becomes high. At the same time, intra-pixel variation will remain relatively low because there is still less than one ground resolution element coinciding with the sample. When the ground resolution elements are much smaller than the samples, inter-pixel variance is low as many elements are averaged over one sample. Intra-pixel (or within sample) variation would increase, however, because many elements contribute to the variance of the pixel [104]. These relationships are generalized in Fig. 12.1.

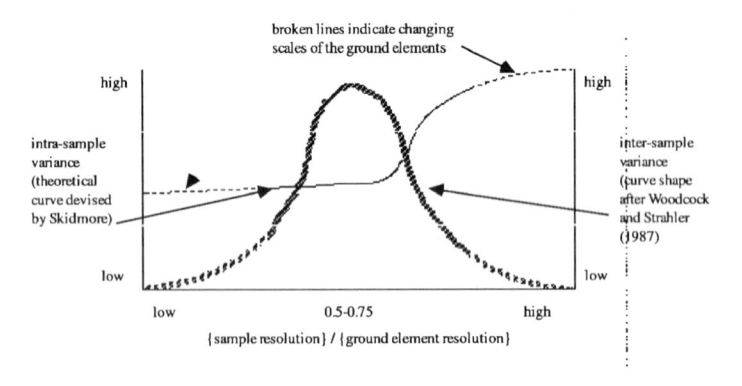

Figure 12.1: Change in intra-pixel and inter-pixel variance as the resolution of the ground elements decreases relative to the image pixel resolution.

Woodcock and Strahler [396] made two interesting observations applicable to the problem of determining sample size when they degraded image (or sample) resolution for various scenes, and noted changes in inter-pixel (or local) variance within a three-by-three moving window. Firstly, the peak in between-sample variation occurred when the sample resolution was 0.5 to 0.75 of the size of the ground resolution elements (note that there is no absolute scale on the x axis for Figure 1). A second observation by [396] is related to the first: the scale of the ground resolution elements is important in determining the sample variation.

There is other evidence to support these ideas. Generally, mapping of heterogeneous classes such as *forest* and *residential* is more accurate at 80 m resolution than at finer resolution such as 30 m [364]; however more homogeneous classes such as agriculture and rangelands are more accurately mapped at 30 m than 80 m [364]. Similarly, [7] reported lower mapping accuracy of forests at 30 m compared with 80

m.

Curran and Williamson [86] stated that it is preferable to have a few highly accurate ground measurements, rather than a greater number of less accurate measurements. In other words, measurement quality at a sample point is more important than the quantity of collected samples. This conclusion was made after they found that the error in measuring a variable or class on the ground was much higher than the error in the map or image samples.

In conclusion, the ideal solution is to minimize intra-pixel and inter-pixel variation (see Figure 1). The optimal sample area size depends on the cover type being mapped - in other words the size of the ground resolution elements as well as the level of heterogeneity within the cover class. As the ground resolution element becomes larger, so should the area (size) that the sample unit covers (that is, imagine sampling a tree compared to a grass).

12.2.5 Measures of mapping accuracy generated from an error matrix

As introduced above, a commonly cited measure of mapping accuracy is the *overall accuracy* which is the number of correctly classified pixels (i.e. the sum of the major diagonal cells in the error matrix) divided by the total number of pixels checked. A typical error for a field checked remotely sensed image is shown in Table 12.2, taken from a paper by Congalton and Mead [1] [76]. Anderson et al. [12] suggested the minimum level of interpretation accuracy in the identification of land-cover categories should be at least 85 per cent.

Table 12.2: Error matrix and associated accuracy statistics taken from Congalton and Mead (1983).

		\multicolumn{5}{c}{Ground truth class}	Commission	User
		A	B	C	D	total	Error	Accuracy
Image	A	35	14	11	1	61	43	57
or map	B	4	11	3	0	18	39	61
class	C	12	9	38	4	63	40	60
	D	2	5	12	2	21	90	10
total		53	39	64	7	163		
Omission error		34	72	41	71			
Producer accuracy		66	28	59	29			
total of diagonals				86				
total				163				
Map accuracy				52.8				

In Table 12.2, user (consumer) and producer accuracy is calculated for the error matrix; terms which are equivalent to type I and type II errors and to commission

---

1. Hudson and Ramm [184] show that the original paper by Congalton and Mead [76], which is based on work by Cohen [70], contained a numerical error as well as an error in the formulae for calculating the variance of the Kappa coefficient. Though the error introduced is small, the mistakes were adopted by a number of subsequent authors.

and omission error, respectively.

Overall classification accuracy is the ratio of the total number of correctly classified pixels to the total number of pixels in each class; this measure strongly overestimates the positional class accuracy [227] Kalensky and Scherk [227] went on to define the accuracy of each class or "class mapping accuracy" as:

$$M_i\% = \frac{N_i}{N_i + E_i},$$

where $M_i\%$ = mapping accuracy of class i, N_i = number of correctly classified pixels in class i and E_i = number of erroneous pixels in class i (i.e. sum of omissions and commissions).

Cohen [70] and Bishop et al. [43] defined an estimate of a measure of overall agreement between image data and the reference (ground truth) data called κ. It can be estimated by

$$\widehat{\kappa} = \frac{N \sum_{i=1}^{r} X_{ii} - \sum_{i=1}^{r} X_{i+}X_{+i}}{N^2 - \sum_{i=1}^{r} X_{i+}X_{+i}} = \frac{\theta_1 - \theta_2}{1 - \theta_2},$$

where

$$\theta_1 = \sum_{i=1}^{r} \frac{X_{ii}}{N} \quad \text{and} \quad \theta_2 = \sum_{i=1}^{r} \frac{X_{i+}X_{+i}}{N^2}.$$

Note that X_{i+} is the sum of the ith row, X_{+i} is the sum of the ith column, and X_{ii} is the count of observations at row i and column i. Note also that r is the number of rows and columns in the error matrix, while the total number of observations is N. Using this notation for the Table 12.2 example, the row total for class A is $X_{1+} = 61$, the column total for class C is $X_{+3} = 64$, and the number of counts at the class A row and the class A column is $X_{ii} = 35$.

The asymptotic variance of $\widehat{\kappa}$, $\widehat{\sigma}^2(\widehat{\kappa})$ equals:

$$\widehat{\sigma}^2(\widehat{\kappa}) = \frac{1}{N}\left(\frac{\theta_1(1-\theta_1)}{(1-\theta_2)^2} + \frac{2(1-\theta_1)(2\theta_1\theta_2 - \theta_3)}{(1-\theta_2)^3} + \frac{(1-\theta_1)^2(\theta_4 - 4\theta_2^2)}{(1-\theta_2)^4}\right),$$

where

$$\theta_3 = \sum_{i=1}^{r} \frac{X_{ii}(X_{i+} + X_{+i})}{N^2} \quad \text{and} \quad \theta_4 = \sum_{i,j=1}^{r} \frac{X_{ij}(X_{i+} + X_{+i})^2}{N^3}.$$

Note that X_{ij} is the value in the error matrix at the intersection of the ith row and the jth column. For example, in Table 12.2, X_{23} has a value of 3. Using the above error matrix (Table 12.2), the value of κ may be estimated using a simple method:

$$\widehat{\kappa}_S = \frac{Nd - q}{N^2 - q} \div 100.$$

The variable d is the sum of the diagonals, that is $d = \sum_{i=1}^{r} X_{ii} = 35 + 11 + 38 + 2 = 86$. The variable q is the number of cases in diagonal cells due to chance, that is $q = \sum_{i=1}^{r} X_{i+} X_{+i} = 3233 + 702 + 4032 + 147 = 8114$, calculated as:

$q(AA)$ = Total row A × Total column $A = 61 \times 53 = 3233$
$q(BB)$ = Total row B × Total column $B = 18 \times 39 = 702$
$q(CC)$ = Total row C × Total column $C = 63 \times 64 = 4032$
$q(DD)$ = Total row D × Total column $D = 21 \times 7 = 147$

Therefore, $q = q(AA) + q(BB) + q(CC) + q(DD) = 8114$.

Finally, the variable N is the total number of map cells or image pixels tested viz. $N = 163$. Thus

$$\hat{\kappa}_S = \frac{Nd - q}{N^2 - q} \div 100 = \frac{163 \times 86 - 8114}{163^2 - 8114} \div 100 = 0.3199.$$

Note that κ may also be estimated using the formulae given by [70] and [43] above. Then:

$$\hat{\kappa} = \frac{\theta_1 - \theta_2}{1 - \theta_2} = \frac{0.5276 - 0.3054}{1 - 0.3054} = 0.3199.$$

Values of κ ranges in value from 0 (no association, that is any agreement between the map and the ground truth equals chance agreement) through to 1 (full association, there is perfect agreement between the two images). Less than chance agreement leads to negative values of κ. The value of κ calculated above ($\hat{\kappa}_S = 0.32$) indicates that a little less than one third of the image and test grid cells are in agreement.

The estimated asymptotic variance of κ for the data in Table 12.2 equals 0.0027. The variance of κ is used in the next section to calculate whether there is a significant difference between maps.

12.2.6 Change detection - testing for a significant difference between maps

Testing for a statistically significant difference between maps is important when analyzing change. In addition, a user may wish to check differences in the type of classifier [307, 184, 332], type of imagery (i.e. spatial or spectral resolution), date of imagery [77], and film/filter combinations for photointerpretation [76]. These factors may all be important in determining the utility of one remotely sensed product over another.

Congalton et al. [77] and Skidmore [333] applied discrete multivariate analysis techniques developed by [70] and [43] to test whether two error matrices were significantly different. In these studies, the type of classifier was varied, while other factors such as date of image collection, training areas etc. were held constant.

To test for a statistically significant difference between two independent error matrices, [70] proposed using the κ-values (e.g. κ_1 and κ_2 representing images 1 and 2 respectively) and their associated variance by evaluating the normal curve deviate:

This test statistic may be applied to paired combinations of error matrices, to ascertain whether the error matrices are significantly different. Obviously only one image factor should be changed at any time, such as classifier type or date of image collection. A null hypothesis can be set up to test whether the κ values for the two images differ;

$$H_o : \kappa_1 = \kappa_2 \text{ versus } H_a : \kappa_1 \neq \kappa_2.$$

The null hypothesis is rejected using the normal curve deviate statistic (z) for $\alpha = 0.05$ if $z_t > 1.96$ (i.e. $z_{\alpha=0.05} = 1.96$). Note that any other rejection region can be used e.g. $\alpha = 0.01$ or $\alpha = 0.001$.

The data in Table 12.2 was compared with a second error matrix. Let $\widehat{\kappa}_1$ be the estimated kappa value for the map produced by interpreter 1 and $\widehat{\kappa}_2$ that by interpreter 2. With $\widehat{\kappa}_1 = 0.3199$, $\widehat{\kappa}_2 = 0.5004$, $\widehat{\sigma}(\widehat{\kappa}_1) = 0.002781$ and $\widehat{\sigma}(\widehat{\kappa}_2) = 0.001802$ we find that

$$Z = \frac{\widehat{\kappa}_1 - \widehat{\kappa}_2}{\widehat{\sigma}(\widehat{\kappa}_1) - \widehat{\sigma}(\widehat{\kappa}_2)} = 2.67.$$

This value exceeds $z_t = 1.96$ (at $\alpha = 0.05$). Therefore reject the null hypothesis and conclude that there is a significant difference between the 2 photo-interpreters.

Rosenfield and Fitzpatrick-Linz [309] discussed the Cohen κ-coefficient [70] as a relation to a family of coefficients which correct for chance agreement between two error matrices (or contingency tables). They commended the Cohen κ coefficient statistic because it considers within class correlation as well as overall image correlation: in other words all cells in the error matrix are considered.

A significant difference between maps may be also calculated using traditional statistics. Rosenfield and Melley [307] compared mapping accuracy obtained from Landsat MSS imagery and high altitude aerial photographs using one-way statistical tests. Test sites were compared on both imagery types as being correctly or incorrectly classified. These test sites were paired, so a research hypothesis could be stated that there was no difference in the number of correctly classified sites for the two images. This hypothesis was tested using the Student t-test as well as nonparametric tests (Sign test and Wilcoxon U-test). In addition, a two way analysis of variance was performed to test whether there was a significant difference between imagery as well as a significant difference between the mapping accuracies of the classes.

12.2.7 Methods of quantifying errors in a vector data layer

The methods discussed above for quantifying error in raster images are equally applicable to quantifying error in vector polygons. Instead of checking whether an image pixel or grid cell is correctly classified, a point within the polygon is verified against the ground truth information. A particular problem of vector images are ground truth samples which occur across boundary lines; in this case the class with the largest area within the sample area may be selected to represent the vector map image.

A method of assessing map accuracy based on line intersect sampling was described in [335]. Line intersect sampling is used to estimate the length of cover class boundaries on a map that coincide with the true boundaries of the cover classes on the ground. Maps may be either traditional cartographic maps, or maps generated from remotely sensed imagery. A ratio of coincident boundary to total boundary is proposed as a measure of map accuracy and this ratio is called the *boundary error*. Though this technique has been developed for vector maps, it is equally applicable to raster maps.

Skidmore and Turner [335] found that the true boundary lengths were not significantly different from the estimated boundary lengths sampled using line intersect sampling, with $\alpha = 0.05$. The estimated boundary accuracy (64%) was extremely close to the true boundary accuracy (65.1%), and there was no significant difference between the true and estimated boundary accuracy.

12.3 Error accumulation in GIS overlay

Maps of the same region may be digitized, geographically rectified, and stored as a series of layers in a GIS. GIS overlaying techniques (see Chapter 4) may produce many apparently interesting products, but if the input data are erroneous, then any derived map will also contain error.

GISs often contain nonspatial or text data which will contribute to error accumulation, such as a misspelled name (ie. Smith or Smyth) or incorrect label (ie. < 10 rather than < 100). Non-spatial data may also include knowledge or rules used by expert systems [333]. The inaccuracies of polygon boundaries and error in geometric rectification of data layers are also contributing factors to error accumulation when overlaying GIS data layers.

The method(s) by which errors accumulate during overlaying is important for modeling error in the final map products. The first necessity for modeling map error accumulation is to quantify the error in the individual layers being overlaid.

12.3.1 Reliability diagrams for each layer

Wright [401] suggested that reliability diagrams should accompany all maps. He emphasized that the sources used to generate different regions of the map have varying accuracy, and these sources should be clearly stated on the map. For example, one region may have been mapped using low altitude aerial photography and controlled ground survey, and would therefore be more accurate than another region mapped using high altitude photography and only reconnaissance survey. This theme was taken up in [63] and [247], who suggested such a reliability diagram showing map pedigree should be included as an additional layer accompanying each map layer in a GIS. In the 1980s, the term *data dictionary* became widespread as a method for describing and pointing to data in a database, and a number of standards were proposed for data dictionaries. A part of a data dictionary, or data base, could include information pertaining to the accuracy of a GIS data layer, including the source of the data, method of conversion from analogue to digital form, quietly of the original field work (ie. were boundaries field checked) etc. The data dictionary concept evolved into the currently favored term of *metadata*, that is information about the data is stored within the data. However, for the purposes of error accumulation modeling, reliability diagrams do not provide a quantitative statement about the accuracy (or error) of the map.

A second technique that can be used to estimate data reliability is Kriging interpolation [239] — see also Chapter 6. It assumes that the distance between samples is related to sample values: sample values are correlated and the dependence between sample values decreases with distance (see Chapter 5). Assuming that the samples are correct, the spatial variance from the semi-variogram is used to generate a map of *reliability*, given by the kriging variance. Caution must be exercised, however, in using the Kriging variance map as a reliability map that indicates error. The kriging variance represents uncertainty in interpolation and not an error in mapping. In addition, most successful Kriging applications have been at a local scale for soil and geological property mapping (ie. soil pits at about 100 m intervals) [337, 349].

Thirdly, reliability layers may also be generated using image processing and GIS classifiers. The supervised nonparametric classifier described in [336] classified remotely sensed and GIS digital data. The classifier outputs for all cells and the empirical probability of correct classification for each class according to the training area data, and thereby gives an indication of map accuracy. Similarly, the well known maximum likelihood classifier [302] also produces a posterior probability for each cell or pixel. This may also be viewed as the empirical probability of correct classification, as based on the training area data.

12.3.2 Modeling error accumulation

The few methods proposed for modeling error accumulation are limited in their application. Working with ideal data, these methods may allow some conclusions to be drawn about error accumulation during GIS overlay operations. However, the methods break down when used with map layers created under different conditions than assumed by the methods.

Using the statistics of [294] with empirical data, Burrough [60] concluded that with two layers of continuous data, the addition operation is relatively unimportant in terms of error accumulation. The amount of error accumulated by the division and multiplication operations is much larger. The largest error accumulation occurs during subtraction operations. Correlated variables may have higher error accumulation rates compared with non-correlated data, because erroneous regions will tend to coincide, and concentrate error rates there.

Newcomer and Sjazgin [280] used probability theory to model error accumulation. However, they assumed that the data layers are dependent, which is problematic if layers are related as no additional information is generated when the layers are combined.

Heuvelink and Burrough [179] modeled the accumulation of error in Boolean models, using surfaces interpolated by Kriging as the estimated error source. The technique is limited to a local scale and applications suited to Kriging interpolation.

Skidmore [334] developed a method for modeling the accumulation of error in GIS based on the theory of central concept [60, 315]. The distance of unknown points to well defined central points (such as a soil pit or a forest plot) are measured in environmental space. This distance is used as a measure of accuracy of the unknown points.

12.3.3 Bayesian overlaying

The use of Bayesian logic for GIS overlaying is explained in [333]. As with Boolean, arithmetic and composite overlaying, there is inherent error in the individual data layers when overlaying using Bayesian logic. In addition, Bayesian overlaying uses rules to link the evidence to the hypotheses: the rules have an associated uncertainty representing an additional source of error.

12.4 Conclusions

Maps and images are inherently erroneous; these errors have been well documented in the literature. In this chapter, methods are described to estimate the accuracy of a map, as well as attempts to measure error accumulation during map overlay. The main conclusions from this chapter are:

- Many sampling units increase the precision (variance) of the estimate of map accuracy.
- Random or stratified random sampling maximizes precision and accuracy (albeit at a higher cost compared with cluster sampling or systematic sampling).
- Cluster sampling offers reduced sampling costs, but is dependent upon low intra-cluster variance to be effective.
- Systematic sampling schemes may lead to bias in parameter estimation if periodic errors align with the sampling frame
- During sampling, field parameters should be measured accurately.
- Though modeling the accumulation of error during GIS overlay analysis is still in its infancy, some methods for measuring error accumulation during GIS analysis are presented.
- Any procedure to reduce mapping error in individual layers in a GIS will improve the mapping accuracy of an overlay generated from the GIS.
- Until better error modeling techniques are developed for GISs, accuracy statistics should be calculated for each layer in a GIS, as well as for each layer produced by a GIS.

Acknowledgments

Xuehua Liu assisted in preparing the correct version of the Kappa statistic formulae and example.

Chapter 13

Spatial sampling schemes for remote sensing

Jaap de Gruijter

The objective of this chapter is to communicate to researchers some basic knowledge of sampling for applications in remote sensing projects. As a consequence, all theory, formulae and terminology not essential for application in remote sensing or for basic understanding, is omitted. Reference is made to the sampling literature where possible. However, in one way this chapter is broader than usual texts on sampling, because full attention is paid to how sampling theory can be employed and embedded in real-life research projects. Thus it is discussed in detail how an efficient and effective sampling scheme can be designed in view of the aims and constraints of the project at large.

The chapter also discusses the differences and the choice between the design-based and the model-based approach to sampling, as much confusion around this issue still exists in the applied literature. More text is devoted to design-based strategies than to model-based strategies, not because they are more important but because there are more of them, and reference to existing literature is often problematic. In the general sampling literature design-based strategies are mostly presented in a non-spatial finite population framework, and 'translation' into the spatial context is in order.

13.1 Designing a sampling scheme
13.1.1 Towards better planning
Research projects which involve sampling usually include the following activities:
- Planning of the fieldwork: given the purpose of the project, the budget and possible logistic constraints, it is decided how many, where, how and when samples and/or field measurements are to be taken;
- Fieldwork: taking samples and/or field measurements;
- Laboratory work: sample preparation and analyses (optional);
- Data recording;
- Data processing;
- Reporting.

Roughly speaking, the above activities can be thought of as consecutive stages of a project, but obviously the activities can overlap in time as, for instance, data recording and fieldwork are often done simultaneously. Also, there can be switching back and forth between activities. For instance, if during data processing some deficiency is discovered, it may be needed to do additional fieldwork.

The main purpose of this section is to argue that, although the above sequence of activities seems logical, it is not a good sampling practice. This is because an essential element is missing at the beginning: the element of planning the whole chain of activities, including the statistical procedure of data processing. Careful planning of the entire project is a pre-requisite of good sampling practice and should precede any other activity. Researchers usually put enough effort and ingenuity in deciding how and where to take samples. That is not the problem. Very often, however, the ideas about how to analyze the data remain rather vague until the data are there and crisp decisions must be made about what to do with them. In that case, more likely than not, data analysis and data acquisition will not be properly tuned to each other. Due to this mismatch, the potential qualities that a data acquisition plan might have are not fully exploited, and sub-optimal results are obtained. One example is where a stratified random sample has been taken, but this sample is analyzed as if it were a simple random sample. Another example is where the data are to be analyzed by some form of kriging, but it appears that the variogram needed for this can not be reliably estimated from the data. Finally, a situation often encountered is where the conclusions to be drawn from the sample data can only be based on questionable assumptions because the sample was not properly randomized. These examples will become more clear in the next sections.

In conclusion, it is recommended that not only the fieldwork is planned but the entire project, with special attention for the tuning of data acquisition with data processing and vice versa. Proper planning of the entire project will always pay itself back by increased efficacy as well as efficiency. The plan itself is referred to as the 'sampling scheme'. Used in this broad sense it covers much more than just a layout of sample locations in the field. The sampling scheme captures all the decisions and information pertinent to data acquisition, data recording and data processing [109]:

a. Purpose of the sampling: target area, target variable(s), target parameter(s);
b. Constraints: financial, logistic, operational;
c. Method of taking samples: dimensions of sample elements and sampling devices;
d. Method(s) of determination: field measurements and/or laboratory analyses;
e. Sampling design: this specifies both the sample size and how the sample locations are to be selected;
f. The actually selected sample points;
g. Protocols on data recording and fieldwork;
h. Method(s) of statistical analysis;
i. Prediction of operational costs and accuracy of results.

Apart from tuning data acquisition to data processing and vice versa, there is a more general reason why the project should be planned as a whole rather than to optimize parts of it in isolation from each other: the consequences of a decision about a single issue, in terms of quality and costs, depend on the decisions taken on other issues. A simple example is where two methods of determination are available for the target variable: a cheap but inaccurate method and an expensive but accurate method. The choice between either has an effect on both the costs and the accuracy of the final result, and these effects depend on the sample size. Given a fixed budget, choosing the cheap method implies that a larger sample size can be afforded. Whether or not this leads to a better result depends on various factors. How to design a sampling scheme is dealt with in the next sections.

13.1.2 A guiding principle in designing sampling schemes

A safe way to a good sampling scheme is this principle: 'Start at the end, then reason backwards'. This means that one should first determine precisely what type of result is demanded. Only when the type of result is defined it becomes useful to search for a sampling scheme that leads to that result in an efficient way. The reason for this is that different types of results ask for different sampling schemes. Although this is an extremely important fact in sampling, it is not always clearly realized.

For instance, if the spatial mean of a region must be estimated, other, less expensive sampling schemes are needed than for local estimation at points, as for mapping. Another example is that data needs for generating hypotheses are totally different from those of testing hypotheses. The same is true for estimation of model parameters, for instance of variograms, compared with model validation.

Types of results can be divided in three broad groups. Firstly, the purpose of sampling may be estimation of the frequency distribution of a variable, or one or more parameters of that distribution. Examples are 'location' parameters such as the mean, quantiles (e.g. the median) and the mode, or 'dispersion' parameters such as the standard deviation, the range and tolerance intervals. These results are related to the area as a whole; they have no geographical coordinates. Secondly, the purpose may be some kind of description of the spatial distribution of the variable within the area. Examples are: prediction of values at points, estimation of means within parts of the area, or construction of contour maps. As opposed to the first group, these results contain geographical coordinates. Thirdly, there is a miscellaneous group of special purposes such as estimation of model parameters, model validation, generating hypotheses and multivariate statistics, including classification.

In principle, different types of results ask for different sampling schemes, because a given scheme may not yield the type of result that is required, or if it does, it may do so in an inefficient way. In conclusion, a good way of designing a sampling scheme is by reasoning backward through the following steps:
1. Decide precisely what type of result you want to end with. For instance, a map of a given variable, at a given scale and with a given accuracy. Or testing of a given hypothesis, at a given significance level and with a given power.
2. Determine what kind of data analysis leads to that result.

3. Determine what the data needs are for this analysis.
4. Search for a sampling scheme to get those data in the most efficient way.

To aid the search for a good sampling scheme some practical, scientific as well as statistical issues are discussed in the following sections.

13.1.3 Practical issues
Avoid undue complexity
Researchers often know much about the processes that have generated the spatial pattern of soil or vegetation properties in the study area. They may be tempted to express all this knowledge in detail in the form of a highly complex sampling design. Albeit understandable, this attitude entails two risks which are easily underestimated. Firstly, due to unforeseen operational difficulties during fieldwork, it may prove impossible to carry out the design in all its complexity. The fieldwork must then be adjourned until the design is re-adjusted. This may be time consuming and is likely to cause unwanted delay. Secondly, the complexities are introduced to increase the efficiency, but they may make the statistical analysis much more intricate and time consuming than expected. In conclusion, it is usually wise to avoid highly complex sampling designs, because the theoretical gain in efficiency compared with simpler solutions is easily overridden by practical difficulties.

Allow for unexpected delay in fieldwork
Even if the researcher is familiar with the circumstances in the terrain, there can be factors beyond his control that prevent the fieldwork to be completed within the available time. Clearly, unfinished fieldwork may seriously harm the statistical potential of the design. It is therefore prudent to allocate spare time in the scheme for contretemps, say 20 % of the total time for fieldwork, and to include a number of optional sample points to be visited as far as spare time allows.

Include a test phase if necessary
If there is significant uncertainty about the logistics of the fieldwork or the spatial variability, a preliminary test phase is always worth the extra effort. The information accruing from even a small sample collected prior to the main sample enables the latter to be optimized more precisely and reduces the risk that the project will not meet its goal at all. In the final statistical analysis the sample data from the test phase are combined with the main sample, so the additional effort is limited to travel time and statistical analysis.

Evaluate the scheme beforehand
It is good practice to quantitatively predict the cost of operation of the scheme and the accuracy of the result, prior to the fieldwork. Predicting cost and accuracy can be done in sophisticated ways, using mathematical models [109], or more globally, using experiences from similar projects, rules-of-thumb and approximations. A test phase will of course improve the prediction of cost and accuracy.

Explicit evaluation ex ante in terms of cost and accuracy is not only a final check of whether the scheme can be trusted to lead to the goal, it also enables comparison

Chapter 13: Spatial sampling schemes 215

with evaluation ex post, i.e. after the project is finished. If this reveals significant discrepancies, the causes should be analyzed. This may provide a ground for better planning of future projects.

13.1.4 Scientific issues
Protocol for fieldwork
Rules for fieldwork will usually concern the physical act of taking samples and/or measurements in the field, but they should also tell what to do if a sample point is inaccessible or if it falls outside the target area. An example of the latter in vegetation sampling is where, on inspection in the field, it turns out that at the given point there is no 'vegetation' according to a given definition.

A poor protocol may seriously affect the quality of the results. Obvious requirements for a protocol are that it is complete, unambiguous, practically feasible and scientifically sound. The scientific aspect plays a role, for instance, when a rule says that an inaccessible sampling point is to be shifted to a nearby location in a certain way. In principle this leads to over-representation of boundary zones and, depending on the kind of design and the statistical analysis, this may result in biased estimates.

Protocol for data recording
As for fieldwork, there should be sound rules for data recording. These rules should not only cover regular recording but also prescribe different codes for when a sampling point falls outside the target area, for when it is inaccessible, for when a value cannot be measured because it is too large or too small ('censoring' in the statistical sense), and for when the property cannot be measured for other reasons.

13.1.5 Statistical issues
Prior information on spatial variability
All prior information about the spatial variability in the area should be employed in the search for an efficient sampling design. Examples of prior information are satellite images, aerial photographs, thematic maps (e.g. vegetation and soil maps) and theory about the genesis of the spatial patterns. Images, photographs and maps may provide a useful stratification of the area. In that case the area is split into a number of relatively homogeneous sub-regions (called 'strata'), which are then sampled independently from each other (Section 13.3.3). Genetic theory may enable intelligent guesses about the spatial correlation. For instance, eolian deposition of parent material in the area may be known to have resulted in little short-range variation of texture. Then, if the target variable is closely related to texture, it will be important for efficiency to avoid sampling at points close to each other.

If prior information on the spatial variability is captured in the form of a variogram, this variogram can be used to predict the sampling variance for a given design [109] (see also Section 13.3.8 below). If in addition a model for the costs is available then it is possible to optimize the sampling design in a fully quantitative way [110].

Modes of sample point selection

Three possible modes of sample point selection can be distinguished: convenience sampling, purposive sampling and probability sampling. The concept of convenience sampling is self-explanatory. An obvious example is when sampling is limited to road sides or other easily accessible spots. The advantage of this mode is that it saves time and cost. The disadvantage is that the statistical properties are inferior compared to the other modes. For instance, estimates from a convenience sample have to considered as biased unless one is willing to accept specific assumptions about the sampling process and the spatial variation. These assumptions are often debatable, and this may or may not be acceptable, depending on the context of the project.

Purposive sampling tries to locate the sample points such that a given purpose is served best. A well known example is the 'free survey' method of mapping soil classes, whereby the surveyor locates the sample points where they are expected to be most informative with respect to soil class delineation. In this example the points are selected in a subjective manner, using experience, visible landscape features and pedogenetic hypotheses. However, purposive sampling may also proceed by formally optimizing an objective function related to the purpose. For instance, if the purpose is to map a spatial distribution by kriging and if geographical boundary effects are disregarded, then it can be shown that the prediction error is minimized by a hexagonal grid of sample points, under assumptions of stationarity and isotropy [261](see also Section 13.4). If boundary effects cannot be neglected, or if point data are available prior to sampling, then the grid that minimizes the prediction error will be irregular, and this can be found by simulated annealing [378] (see also Section 13.4).

Probability sampling, unlike the other modes, selects sample points at random locations. Therefore the probabilities of selecting the points are known, and these probabilities provide the basis for statistical analysis of the data. As explained in Section 13.3, there are many techniques for random selection of sampling points. Collectively, this approach to sampling is referred to as the design-based approach, as opposed the model-based approach, where the sample points are fixed instead of random and statistical analysis is based on a model of the spatial variation. The choice between these two approaches is an important statistical issue, which is dealt with in Section 13.2.

Sources of error

It is important to realize that the accuracy of the final result is not only determined by sampling error, i.e. the error due to the fact that sampling is limited to a finite number of points. Other sources of error are: sample treatment, measurement and 'non-response': a term used in the general statistical literature to indicate the situation where for some reason no data can be obtained from a sample element. In vegetation and soil sampling this occurs when a point in the field cannot be visited or when measurement is impossible for other reasons.

Although any reduction of the sampling error will lead to a smaller total error, there is little point in putting all effort in further reduction of the sampling error if

Chapter 13: Spatial sampling schemes 217

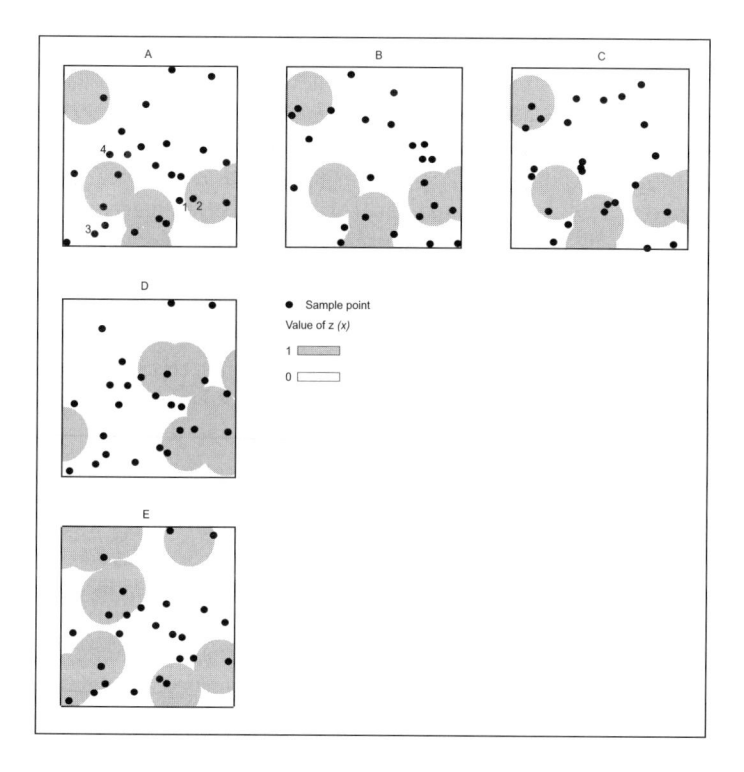

Figure 13.1: Repeated sampling in the design-based approach (A, B, C) and in the model-based approach (A, D, E). In the design-based approach the population is fixed and the sampling locations are random. In the model-based approach the sampling locations are fixed and the population is random. (From [52])

another source of error still has a higher order of magnitude. Therefore, in devising a sampling scheme, the relative importance of all error sources should be taken into consideration.

13.2 Design-based and model-based approach

There are two fundamentally different approaches to sampling: the design-based approach, followed in classical survey sampling, and the model-based approach, followed in geostatistics [316, 101]). The difference between the two approaches is illustrated in Fig. 13.1 with a simple example, taken from [52]: a square area is sampled at 25 points and a 0/1 variable measured to estimate the fraction of the area with value 1. Fig. 13.1A shows a spatial distribution of the 0/1 variable and a configuration of 25 sample points. Averaging the observed values at these points yields an estimate of the fraction.

Now both approaches quantify the uncertainty of such an estimate by considering what would happen if sampling were repeated many times in a hypothetical experiment. Obviously, if in this experiment neither the pattern of values nor the locations of the sample points were changed there would be no variation, so one or the other has to be varied. The two approaches differ in which of the two is varied. The design-based approach evaluates the uncertainty by repeated sampling with different sets of sample points, while considering the pattern of values in the area as unknown but fixed. The sets of sample locations are generated according to a chosen random sampling design. The row of figures (A, B and C) represents three possible outcomes. As opposed to this, the model-based approach evaluates the uncertainty by repeated sampling with a fixed set of sample points, while varying the pattern of values in the area according to a chosen random model of the spatial variation. For this approach the column of figures (A, D and E) represents three possible outcomes. (Note that the target quantity in this approach is no longer constant: it varies among realizations from the model. The standard statistical terminology therefore speaks of prediction instead of estimation.)

The experiment can remain truly hypothetical in most instances because probability calculus enables to determine what happens on average over all possible realizations. In more intricate situations this is infeasible, however, and repeated sampling has to be simulated numerically, either varying the sample points or the pattern of values, or both.

The fact that the two approaches use a different source of randomness has several important practical as well as theoretical consequences. The main consequence is that the statistical inference from the sample data is entirely different. In the design-based approach estimation, testing and prediction are based on the selection probabilities as determined by the random design. This means that in calculating weighted averages, the data are assigned weights determined by their selection probabilities, not by their geographical co-ordinates. In the model-based approach, inference is based on a stochastic model of the spatial variation. Here the weights of the data are determined by spatial correlations, which are given by the model as a function of geographical co-ordinates.

Before deciding on the details of a sampling design, a choice between the design-based and the model-based approach should be made. It goes beyond the scope of this book to discuss this issue in detail, only an outline is given. An extensive discussion is presented in [52]. The 'ideal' circumstances for application of the design-based approach are as follows.

i. The required result is an estimate of the frequency distribution in the area as a whole, or a parameter of this distribution, such as the mean and the standard deviation.
ii. At least 5 or 10 sample points can be afforded, depending on the spatial variation.
iii. It is practically feasible to locate these points at randomly selected places.
iv. It is important to obtain an unbiased estimate.
v. It is important to obtain an objective assessment of the uncertainty of the estimate.

Around this 'ideal' there is a range of circumstances in which the design-based approach is still preferable to the model-based approach.

The 'ideal' circumstances for application of the model-based approach are as follows.
i. The required result is prediction of values at individual points or the entire spatial distribution in the area.
ii. A large number of sample points can be afforded, depending on the spatial variation. The model usually implies stationarity assumptions and a variogram, which should be estimated from about 100 - 150 sample points [391].
iii. A reliable model of the spatial variation is available.
iv. High spatial correlations exist in the area.

As before, around this 'ideal' there is a range of circumstances in which the model-based approach is still preferable to the design-based approach. A typical intermediate situation is where averages are required for a number of sub-regions or 'blocks', in which only sparse sampling can be done. Brus and De Gruijter [52] explore this in a case study.

13.3 Design-based strategies

The purpose of this section is to give insight in how design-based sampling strategies work and how they can be applied in research projects. The text attempts to help understanding the basic principles at an intuitive level and is not meant as an exposé of sampling theory. A somewhat practically oriented textbook on design-based sampling strategies is [69], from which most of the material presented here is borrowed or derived. A comprehensive textbook on sampling theory is [316].

The general pattern in the development of sampling strategies is to take the simplest strategy (Simple Random Sampling, see below) as a starting point, with complete randomization of all sample points. Then restrictions on randomization are looked for, such that this would reduce the sampling variance or the cost of operation, or both. Different types of restrictions can be distinguished, each giving rise to a different type of sampling design.

Before discussing the basic designs, the statistical concept of 'sampling design' itself need to be defined more precisely. In the spatial context it is defined as a function that assigns a probability of selection to any set of points in the study area. For instance, the sampling design for Simple Random Sampling with sample size 25, assigns equal selection probabilities to every possible set of 25 points in the area and zero probability to any other set. (Note that a design assigns probabilities to sets of points, not to individual points.) A sampling strategy is defined as a combination (p, t) of a sampling design (p) and an estimator (t) for a given target parameter (T), such as the mean of the area. Statistical quality measures, like bias and variance, can only be defined and evaluated for these combinations, not for a design or an estimator on its own.

The following sections describe each basic strategy by discussing the type of randomization restriction, a technique for selecting samples according to the design, a simple example, the inference from sample data, the determination of sample sizes and advantages and disadvantages. We repeat from the previous section that design-based statistical inferences such as given below are valid, regardless of the structure of the spatial variation, because they do not make any assumption about this structure.

13.3.1 Scope of design-based strategies

A typical application of design-based strategies is to estimate the areal mean of a directly measured quantitative variable. However, the scope of these strategies is much wider than this. Extensions are possible in three directions: derived variables, other parameters, and smaller (sub-)areas.

Firstly, the target variable need neither be quantitative, nor directly measured. If the target variable is measured at a nominal or ordinal scale, then the sample data consist of class labels, and these can be analyzed statistically by first transforming them into 0/1 indicator variables. The presence or absence of a given class is thereby re-coded as 1 and 0, respectively. Of course, if there are k mutually exclusive classes, only $k-1$ indicator variables are needed. The mean of an indicator variable can be interpreted as the fraction of the area in which the class occurs.

Transformation into indicator variables can also be applied to quantitative variables in order to estimate the areal fraction in which the variable exceeds a given threshold. This technique can be extended to estimate the entire Spatial Cumulative Distribution Function (SCDF) of a quantitative variable. In that case areal fractions are estimated for a series of threshold values.

Apart from the simple 0/1 transformations, the target variable may be the output of a more or less complicated model for which the input data is collected at the sample points. Another important case of indirect determination is in validation studies, where the target variable represents an error, i.e. the difference between a measured value and a value predicted by a process model or a spatial distribution model, such as a map. A common example is the error resulting from a classification algorithm applied to remotely sensed images. The errors determined at the sample points can be used to estimate their spatial mean (which equals the bias), the mean absolute error, the mean squared error, or the entire SCDF of the errors.

Secondly, the target parameter need not be the spatial mean. For instance, it may also be a quantile, such as the median, the spatial variance, a tolerance interval, or a parameter of a model relating one or more predictor variables with a variable of interest. See e.g. [235] and [295] for design-based statistical inference on these and other target parameters.

Thirdly, the region for which estimation or testing of hypotheses is demanded need not be the entire area sampled; interest may also be in one or more sub-areas. There are two different methods of estimation and testing in sub-areas. The first is to sample the sub-areas independently from each other, in which case they act as 'strata' in a stratified sampling design (Section 13.3.3). In the second method the sampling design is independent from any division into sub-areas. Estimation in sub-areas is then only based on sorting the sample data afterwards according to the sub-areas in which the sample points happen to fall. In this case the sub-areas are referred to as 'domains of interest', or briefly 'domains'.

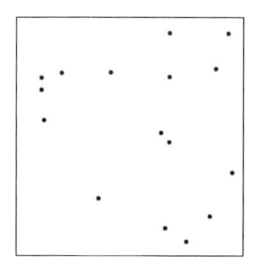

Figure 13.2: Notional example of Simple Random Sampling

13.3.2 Simple Random Sampling (SRS)
Restriction on randomization
No restrictions on randomization. All sample points are selected with equal probability and independently from each other.

Selection technique
An algorithm for SRS with sample size n, applicable to irregularly shaped areas, is as follows.
(a) Determine the minimum and maximum X and Y co-ordinates of the area: X_{min}, X_{max}, Y_{min} and Y_{max}.
(b) Generate independently from each other two (pseudo-)random co-ordinates, X_{ran} and Y_{ran}, from the uniform distribution on the interval (X_{min}, X_{max}) and $(Y_{min}$ and $Y_{max})$, respectively.
(c) Determine with a point-in-polygon routine whether the point (X_{ran}, Y_{ran}) falls in the area. Accept the point if it does; skip the point if it does not.
(d) Repeat step (b) and (c) until n points are selected.

Example: Fig. 13.1A, 13.1B and 13.1C show three realizations of SRS with 25 points; Fig. 13.2 is an example with $n = 16$. Notice the irregularity, the clustering and the empty spaces in the configurations.

Statistical inference
The spatial mean of the area, $\overline{Y}$, for a quantitative variable y is estimated by:

$$\overline{y} = \frac{1}{n}\sum_{i=1}^{n} y_i,$$

with n = sample size, and y_i = value of sample point i.

The strategy (SRS, $\overline{y}$) is 'p-unbiased'; this is a quality criterion defined as: $E_p[\overline{y}] = \overline{Y}$, where $E_p[.]$ denotes the statistical expectation over all possible sample realizations from a design p (in this case SRS). This means that if we would repeat sampling, measuring and calculating $\overline{y}$ in the same way again and again, we would find on average the true value $\overline{Y}$. (If measurement errors are present, then the unbiasedness still holds if the errors are purely random, i.e. zero on average.)

The variance of $\bar{y}$ is estimated by:

$$v(\bar{y}) = \frac{1}{n(n-1)} \sum_{i=1}^{n}(y_i - \bar{y})^2$$

and the standard deviation by

$$s(\bar{y}) = \sqrt{v(\bar{y})}.$$

If the data contain random measurement errors, then their contribution to the total estimation error is automatically included in the estimates $v(\bar{y})$ and $s(\bar{y})$.

The two-sided 100(1-a)% confidence interval for $\overline{Y}$ is given by:

$$\bar{y} \pm t_{1-\alpha/2} \cdot s(\bar{y}), \tag{13.1}$$

where $t_{1-\alpha/2}$ is the $(1-\frac{\alpha}{2})$ quantile of the Student distribution with $(n-1)$ degrees of freedom. This confidence interval is based on the assumption that y, and as a consequence $\bar{y}$, is normally distributed. If the distribution deviates clearly from normality, the data should be first transformed to normality, for instance by taking the logarithm. The interval boundaries thus found are then back-transformed to the original scale. Transformation is not necessary if n is large, because then $\bar{y}$ is approximately normally distributed according to the Central Limit Theorem.

The above formulas for estimating means can also be used for areal *fractions*. The fraction of the area where a qualitative variable q has a given value, for instance 'very suitable', can be estimated by first generating a 0/1 indicator variable from the sample data, with value 1 if $q=$ 'very suitable', and 0 otherwise. Then the above equations are simply applied to this indicator variable. The only exception is the calculation of confidence intervals because the indicator variable is clearly not normally distributed. The sample fraction has a Binomial distribution, and with small samples ($n < 20$) this distribution should be used to construct confidence intervals. With larger samples the distribution is close enough to normality and formula 13.1 will be accurate enough for most practical applications.

The above formulas can also be used for estimation in a domain (Section 13.3.1), if it contains sample points. A domain may or may not have a known geographical delineation. An example of the latter is where the mean biomass of a given vegetation type is to be estimated, and no map of the vegetation types at an appropriate scale is available. This mean biomass can be estimated if, in addition to the biomass, the vegetation type is recorded at the sample points.

The mean of a quantitative variable y in domain j, $\overline{Y}_j$, is simply estimated by averaging over the sample points that fall in this domain:

$$\widehat{\overline{Y}_j} = \frac{1}{n_j} \sum_{k=1}^{n_j} y_{jk},$$

where $n_j =$ number of sample points in domain j, and $y_{jk} =$ value at point k in domain j.

Variances, standard deviations and confidence intervals are calculated in the same way as for the area. The same applies to estimation of fractions and SCDF's in domains.

Chapter 13: Spatial sampling schemes

Sample size
The sample size needed to estimate a mean such that, with a chosen large probability $1-\alpha$, the relative error $\left|\frac{\bar{y}-\bar{Y}}{\bar{Y}}\right|$ is smaller than a chosen limit r, can be calculated by:

$$n = \left(\frac{u_{1-\alpha/2} \cdot S}{r\bar{Y}}\right)^2,$$

with: $u_{1-\alpha/2}$ = the $(1-\alpha/2)$ quantile of the standard normal distribution; S = standard deviation of y in the area.

In this formula $\frac{S}{\bar{Y}}$ is the coefficient of variation of y in the area. Of course, this parameter is not known exactly beforehand. Instead, a prior estimate is substituted, which can be obtained from a pilot or previous sampling in the same area, from sampling in a similar area, or from general knowledge of the spatial variation.

If instead of the relative error we wish the absolute error $|\bar{y}-\bar{Y}|$ to be smaller than a chosen limit d, we need sample size:

$$n = \frac{u_{1-\alpha/2}^2 \cdot S^2}{d^2}.$$

The sample size needed to estimate a fraction P such that, with a chosen large probability $1-\alpha$, the absolute error $|\bar{y}-\bar{Y}|$ is smaller than a chosen limit d, can be calculated by

$$n = \frac{u_{1-\alpha/2}^2 \cdot p(1-p)}{d^2},$$

where p is a prior estimate of P.

Advantage
The simplicity of this type of design enables relatively simple and straightforward statistical analyses of the sample data, also with non-standard estimation and testing problems.

Disadvantages
(i) The sampling variance is usually larger than with most other types of design at the same cost, and (ii) because large empty spaces can occur between the sampling points, estimation in domains may be impossible.

13.3.3 Stratified Sampling (StS)
Restriction on randomization
The area is divided in sub-areas, called 'strata', in each of which SRS is applied with sample sizes chosen beforehand.

Selection technique
The algorithm for SRS is applied to each stratum separately.

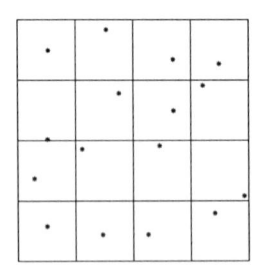

Figure 13.3: Notional example of Stratified Sampling

Example
Fig. 13.3 shows an example with 16 square strata and 1 point in each stratum. Notice the more even spreading compared with SRS in Fig. 13.2.

Statistical inference
Means, areal fractions and SCDF's (after 0/1 transformation) of the area are estimated by:

$$\bar{y}_{\text{St}} = \frac{1}{A} \sum_{h=1}^{L} A_h \cdot \bar{y}_h,$$

with L = number of strata; A_h = area of stratum h; A = total area, and $\bar{y}_h$ = sample mean of stratum h.

The strategy (StS, $\bar{y}_{\text{St}}$) is p-unbiased. Provided all sample sizes are > 1, the variance of $\bar{y}_{\text{St}}$ can be estimated by:

$$v(\bar{y}_{\text{St}}) = \frac{1}{A^2} \sum_{h=1}^{L} A_h^2 \cdot v(\bar{y}_h),$$

where $v(\bar{y}_h)$ is the estimated variance of $\bar{y}_h$:

$$v(\bar{y}_h) = \frac{1}{n_h(n_h - 1)} \sum_{i=1}^{n_h} (y_{hi} - \bar{y}_h)^2,$$

with n_h = sample size in hth stratum.

The standard deviation is estimated by $s(\bar{y}_h) = \sqrt{v(\bar{y}_h)}$. Confidence intervals are calculated in the same way as with SRS, see Eq. 13.1.

The method of estimating means, fractions or SCDF's (after 0/1 transformation) in a domain depends on whether the areas of the domain within the strata are known. If they are, then the mean of the jth domain, $\overline{Y}_j$, is estimated by

$$\widehat{\overline{Y}}_j = \frac{1}{A_j} \sum_h A_{hj} \cdot \bar{y}_{hj} \tag{13.2}$$

with A_{hj} = area of domain j within stratum h; A_j = total area of domain j, and $\bar{y}_{hj}$ = sample mean of domain j within stratum h. The variance of $\widehat{\bar{Y}}_j$ is estimated by

$$v(\widehat{\bar{Y}}_j) = \frac{1}{A_j^2} \sum_h A_{hj}^2 \cdot v(\bar{y}_{hj}),$$

where

$$v(\bar{y}_{hj}) = \frac{1}{n_{hj}(n_{hj}-1)} \sum_{i=1}^{n_{hj}} (y_{hij} - \bar{y}_{hj})^2,$$

with n_{hj} = number of sample points falling in domain j within stratum h.

If the areas of the domain within the strata are not known, they have to be estimated from the sample. Unbiased estimates to be substituted in 13.2 are:

$$\widehat{A}_{hj} = A_h \cdot \frac{n_{hj}}{n_h} \quad \text{and} \quad \widehat{A}_j = \sum_h \widehat{A}_{hj}.$$

The variance is now larger, because of the error in the estimated areas. It is estimated by

$$v(\widehat{\bar{Y}}_j) = \frac{1}{\widehat{A}_j^2} \sum_h \frac{A_h^2}{n_h(n_h-1)} \left[\sum_i (y_{hij} - \bar{y}_{hj})^2 + n_h \left(1 - \frac{n_{hj}}{n_h}\right) \left(\bar{y}_{hj} - \widehat{\bar{Y}}_j\right)^2 \right].$$

Sample sizes

The sample sizes in the strata may be chosen to minimize the variance $V(\bar{y}_{\mathrm{st}})$ for a given maximum allowable cost, or to minimize the cost for a given maximum allowable variance. A simple linear cost function is:

$$C = c_o + \sum c_h n_h,$$

with c_o = overhead cost, and c_h = cost per sample point in stratum h.

If we adopt this function, the optimal *ratios* of the sample sizes to the total sample size n are:

$$\frac{n_h}{n} = \frac{A_h S_h / \sqrt{c_h}}{\sum (A_h S_h / \sqrt{c_h})},$$

where the S_h are prior estimates of the standard deviations in the strata. This formula implies that a stratum gets a larger sample, if it is larger or more variable or cheaper to sample.

The total sample size affordable for a fixed cost C, given that optimal allocation to the strata is applied, is:

$$n = \frac{(C - c_o) \sum (A_h S_h / \sqrt{c_h})}{\sum A_h S_h \sqrt{c_h}}.$$

The total sample size needed to keep the variance below a maximum value V_m, again presuming that optimal allocation to the strata is applied, is:

$$n = \frac{1}{V_m} \cdot \sum (W_h S_h \sqrt{c_h}) \cdot \sum W_h S_h / \sqrt{c_h},$$

where $W_h = A_h/A$. If the cost per point is equal for the strata, this reduces to:

$$n = \frac{1}{V_m} \cdot \left(\sum W_h S_h \right)^2.$$

If, instead of V_m, an absolute error d has been specified with an allowed probability of exceeding α, then V_m can be derived from d and α, according to $V_m = d/u_{1-\alpha/2}$, where $u_{1-\alpha/2}$ is the $1-\alpha/2$ quantile of the standard normal distribution.

When estimating areal fractions rather than means of quantitative variables, the above formulas for sample sizes can still be applied if S_h is replaced by $\sqrt{P_h(1-P_h)}$, where P_h is a prior estimate of the fraction in stratum h.

Advantages
There are two possible reasons for stratification. The first is that the efficiency as compared with SRS may be increased, i.e. smaller sampling variance at the same cost, or lower cost with the same variance. In this case the stratification is chosen such that the expected gain in efficiency is maximized. In practice this can be achieved by forming strata that are as homogeneous as possible. Also, if the cost per sample point varies strongly within the area, for instance with distance from roads, it is efficient to stratify accordingly and to sample the 'cheap' strata more densely. Another reason for stratification may be that separate estimates for given sub-areas are needed. If the strata coincide with these sub-areas of interest then, as opposed to SRS, one has control over the accuracy of the estimates by allocating sufficient sample sizes to the strata.

Disadvantage
With inappropriate stratification or sub-optimal allocation of sample sizes, there could be loss rather than gain in efficiency. This can occur if the stratum means differ little or if the sample sizes are strongly disproportional to the surface areas of the strata. If, for instance, one has many small strata with unequal area and a small sample in each, then these sample sizes are bound to be strongly disproportional because they must be integer numbers.

13.3.4 Two-stage Sampling (TsS)
Restriction on randomization
As with StS, the area is divided in a number of sub-areas. Sampling is then restricted to a number of randomly selected sub-areas, in this case called primary units. Note the difference with StS where all sub-areas (strata) are sampled. In large scale surveys this principle is often generalized to multistage sampling. (Three-stage crop sampling, for instance, could use sub-areas from RS images as primary units, fields as secondary units, and sample plots as tertiary units.)

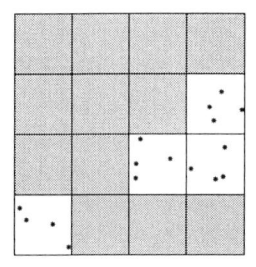

Figure 13.4: Notional example of Two-stage Sampling

Selection technique
A version is described by which the primary units (PU's) are selected with replacement and with probabilities proportional to their area. An algorithm to make n such selections from all N PU's in the area is as follows:
(a) Determine the areas of all PU's, $A_1, ..., A_N$, and their cumulative sums, $S_1, ..., S_N$, with $S_k = \sum_{i=1}^{k} A_i$.
(b) Generate a (pseudo-)random number x from the uniform distribution on the interval $(0, S_N)$.
(c) Select the PU of which the corresponding S_k is the first in the series that exceeds x.
(d) Repeat step (b) and (c) until n PU's are selected.

An alternative, sometimes more efficient algorithm works with a geographical representation of the area and its PU's:
(a) Select a random point in the area as in SRS.
(b) Determine with a point-in-polygon routine in which PU the point falls, and select this PU.
(c) Repeat step (b) and (c) until n selections have been made.

In the second stage, a pre-determined number of sample points, m_i, is selected within each of the PU's selected in the first stage. This is done in the same way as with SRS. If the geographical algorithm is applied, the random points used to select the PU's may also be used as sample points. If a PU has been selected more than once, an independent sample of points must be selected for each time the PU was selected.

Example
Fig. 13.4 shows four square PU's selected in the first stage, and four points in each in the second stage. Notice the stronger spatial clustering compared with SRS in Fig. 13.2. This is just a simple, notional example. It should be noted, however, that the PU's may be defined in any way that seems appropriate, and that the number of sample points may vary among units.

Statistical inference

Means, areal fractions and SCDF's (after 0/1 transformation) of the area are estimated by the remarkably simple estimator:

$$\bar{y}_{\mathrm{Ts}} = \frac{1}{n}\sum_{i=1}^{n}\bar{y}_i \qquad (13.3)$$

with n = number of PU selections, and $\bar{y}_i$ = sample mean of the PU from selection i.

The strategy (TsS, $\bar{y}_{\mathrm{Ts}}$) is p-unbiased. The variance is simply estimated by:

$$v(\bar{y}_{\mathrm{Ts}}) = \frac{1}{n(n-1)}\sum_{i=1}^{n}(\bar{y}_i - \bar{y}_{\mathrm{Ts}})^2.$$

Notice that neither the areas of the PU's, A_i, nor the secondary sample sizes m_i occur in these formulas. This simplicity is due to the fact that the PU's are selected with replacement and probabilities proportional to size. The effect of the secondary sample sizes on the variance is implicitly accounted for. (To understand this, consider that the larger m_i is, the less variable $\bar{y}_i$, and the smaller its contribution to the variance.)

The standard deviation is estimated by $s(\bar{y}_{\mathrm{Ts}}) = \sqrt{v(\bar{y}_{\mathrm{Ts}})}$. Confidence intervals are calculated in the same way as with SRS, see Eq. 13.1.

The method of estimating means, areal fractions and SCDF's in domains depends on whether the area of the domain, A_j, is known or not. If it is known, then the mean of the jth domain, $\overline{Y}_j$, is estimated by:

$$\widehat{\overline{Y}}_j = \frac{\widehat{Y}_j}{A_j} \qquad (13.4)$$

where $\widehat{Y}_j$ is an estimate of the total (spatial integral) of variable y over domain j. To estimate this total, we first define a new variable y', which equals y everywhere in the domain, but is zero elsewhere. The total of y over domain j equals the total of y' over the area, and this is estimated as A times the estimated mean of y', following Eq. greq3:

$$\widehat{Y}_j = A \cdot \bar{y}'_{\mathrm{Ts}} = \frac{A}{n}\sum_{i=1}^{n}\bar{y}'_i,$$

where $\bar{y}'_i$ is the sample mean of the transformed variable y' from PU selection i. The variance of the domain mean is estimated by:

$$v(\widehat{\overline{Y}}_j) = \left(\frac{A}{A_j}\right)^2 \cdot \frac{1}{n(n-1)}\sum_{i=1}^{n}(\bar{y}'_i - \bar{y}'_{\mathrm{Ts}})^2.$$

If the area of the domain is not known, it has to be estimated from the sample. An unbiased estimate to be substituted for A_j in Eq. 13.4 is:

$$\widehat{A}_j = \frac{A}{n}\sum_{i=1}^{n}\frac{m_{ij}}{m_i},$$

with m_{ij} = number of points in PU selection i and domain j. Hence, the ratio estimator:

$$\widehat{\overline{Y}}_{Rj} = \frac{\widehat{Y}_j}{\widehat{A}_j} = \frac{\sum_{i=1}^{n} \overline{y}'_i}{\sum_{i=1}^{n} m_{ij}/m_i},$$

with estimated variance:

$$v(\widehat{\overline{Y}}_{Rj}) = \left(\frac{A}{\widehat{A}_j}\right)^2 \cdot \frac{1}{n(n-1)} \sum_{i=1}^{n} \left(\overline{y}'_i - \widehat{\overline{Y}}_{Rj}\frac{m_{ij}}{m_i}\right)^2.$$

Sample sizes

The primary and secondary samples sizes n and m_i can be optimally determined via dynamic programming, given a budget or variance requirement, any cost function and prior estimates of the within- and between-unit variances; see [110].

A simple approximation is by taking the m_i constant, say $m_i = m$. This is reasonable if the PU's have roughly the same area and internal variability. The variance of the mean is now

$$v(\overline{y}_{Ts}) = \frac{1}{n}\left(S_B^2 + \frac{1}{m}S_W^2\right) \qquad (13.5)$$

where S_B^2 and S_W^2 are the between-unit and the pooled within-unit variance, respectively. Given the linear cost function $C = c_1 n + c_2 nm$, the sample sizes minimizing the variance under the constraint that the cost does not exceed a budget C_m, can be found using the Lagrange multiplier method:

$$n = \frac{S_B C_m}{S_W\sqrt{c_1 c_2} + S_B c_1}$$

and

$$m = \frac{S_W}{S_B}\sqrt{\frac{c_1}{c_2}}.$$

Conversely, minimizing the cost under the constraint that the variance does not exceed a maximum V_m:

$$n = \frac{1}{V_m}\left(S_W S_B \sqrt{\frac{c_2}{c_1}} + S_B^2\right)$$

and m as above.

If, instead of V_m, an absolute error d has been specified with an allowed probability of exceeding α, then V_m can be derived from d and α, according to $V_m = d/u_{1-\alpha/2}$, where $u_{1-\alpha/2}$ is the $1-\alpha/2$ quantile of the standard normal distribution.

When estimating areal fractions rather than means of quantitative variables, the above formulas for sample sizes can still be applied if S_B is interpreted as a prior estimate of the standard deviation between the fractions in the units P_i, and S_W is replaced by a prior estimate of the square root of the average of $P_i(1 - P_i)$ over the units.

Advantage
The spatial clustering of sample points created by TsS has the operational advantage of reducing the travel time between points in the field. Of course, the importance of this depends on the scale and the accessibility of the terrain. The advantage may be amplified by defining the PU's such that they reflect dominating accessibility features like roads and land ownerships.

Disadvantage
The spatial clustering generally leads to lower precision, given the sample size. However, the rationale is that due to the operational advantage a larger sample size can be afforded for the same budget, so that the initial loss of precision is outweighed.

13.3.5 Cluster Sampling (ClS)
Restriction on randomization
Pre-defined sets of points are selected, instead of individual points as in SRS, StS and TsS. These sets are referred to as 'clusters'.

Selection technique
In principle the number of clusters in the area is infinite, so it is impossible to create all clusters beforehand and to sample from this collection. However, only clusters which are selected need to be created, and selection of a cluster can take place via selection of one of its points. Hence the following algorithm:

(a) Select a random point in the area as in SRS; use this point as a 'starting point'.
(b) Find the other points of the cluster to which the starting point belongs, by applying predetermined geometric rules corresponding with the chosen cluster definition.
(c) Repeat step (a) and (b) until n clusters have been selected.

A condition for this algorithm to be valid is that the geometric rules are such that always the same cluster is created regardless of which of its points is used as starting point. A well-known technique satisfying this condition is random transect sampling with equidistant sample points on straight lines with a fixed direction. Given this direction, the random starting point determines the line of the transect. The other sample points are found by taking a pre-chosen distance in both directions from the starting point, until the line crosses the boundary of the area. Clusters thus formed will generally consist of a variable number of points, and the probability of selecting a cluster is proportional to the number of points in it. (This is taken into account in the statistical inference.)

Example
Fig. 13.5 shows four transects, each with four equidistant points. To limit the length of the transects, the area has first been dissected with internal boundaries perpendicular to the transects. Notice the spatial clustering and the regularity compared with SRS, StS and TsS (Fig. 13.2, 13.3 and 13.4). This is just a simple, notional

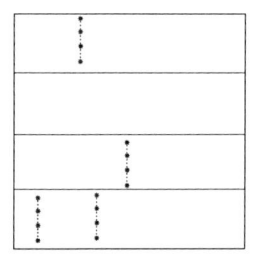

Figure 13.5: Notional example of Cluster Sampling

example. It should be noted, however, that the clusters may be defined in any way that seems appropriate.

Statistical inference
For this type of design the same formulas are used as for TsS, clusters taking the role of primary sampling units. For clarity the inference is presented again, together with the 'cluster interpretation' of the quantities.

Means, areal fractions and SCDF's (after 0/1 transformation) of the area are estimated by the estimator:

$$\bar{y}_{\text{Cl}} = \frac{1}{n} \sum_{i=1}^{n} \bar{y}_i \tag{13.6}$$

with n = number of clusters, and $\bar{y}_i$ = mean of cluster i.

The strategy (ClS, $\bar{y}_{\text{Cl}}$) is p-unbiased. The variance is estimated by:

$$v(\bar{y}_{\text{Cl}}) = \frac{1}{n(n-1)} \sum_{i=1}^{n} (\bar{y}_i - \bar{y}_{\text{Cl}})^2 .$$

Notice that the size of the clusters (number of points) don't occur in these formulas. This simplicity is due to the fact that the clusters are selected with probabilities proportional to size. The effect of the cluster size on the variance is implicitly accounted for. (To understand this, consider that the larger the clusters are, the smaller the variance among their means must be.)

The standard deviation is estimated by $s(\bar{y}_{\text{Cl}}) = \sqrt{v(\bar{y}_{\text{Cl}})}$. Confidence intervals are calculated in the same way as with SRS, see Eq. 13.1.

The method of estimating means, areal fractions and SCDF's in domains depends on whether the area of the domain, A_j, is known or not. If it is known, then the mean of the jth domain, $\overline{Y}_j$, is estimated by:

$$\widehat{\overline{Y}}_j = \frac{\widehat{Y}_j}{A_j} \tag{13.7}$$

where $\widehat{Y}_j$ is an estimate of the total (spatial integral) of variable y over domain j. To estimate this total, we first define a new variable y', which equals y everywhere in the domain, but is zero elsewhere. The total of y over domain j equals the total of y' over the area, and this is estimated as A times the estimated mean of y', following Eq. 13.6:

$$\widehat{Y}_j = A \cdot \overline{y}'_{\text{Cl}} = \frac{A}{n} \sum_{i=1}^{n} \overline{y}'_i,$$

where $\overline{y}'_i$ is the mean of the transformed variable y' in cluster i. The variance of the domain mean is estimated by:

$$v(\widehat{Y}_j) = \left(\frac{A}{A_j}\right)^2 \cdot \frac{1}{n(n-1)} \sum_{i=1}^{n} (\overline{y}'_i - \overline{y}'_{\text{Cl}})^2.$$

If the area of the domain is not known, it has to be estimated from the sample. An unbiased estimate to be substituted for A_j in Eq. 13.7 is:

$$\widehat{A}_j = \frac{A}{n} \sum_{i=1}^{n} \frac{m_{ij}}{m_i},$$

with m_{ij} = number of points in cluster i and domain j. Hence, the ratio estimator:

$$\widehat{\overline{Y}}_{Rj} = \frac{\widehat{Y}_j}{\widehat{A}_j} = \frac{\sum_{i=1}^{n} \overline{y}'_i}{\sum_{i=1}^{n} m_{ij}/m_i},$$

with estimated variance:

$$v(\widehat{\overline{Y}}_{Rj}) = \left(\frac{A}{\widehat{A}_j}\right)^2 \cdot \frac{1}{n(n-1)} \sum_{i=1}^{n} \left(\overline{y}'_i - \widehat{\overline{Y}}_{Rj} \frac{m_{ij}}{m_i}\right)^2.$$

Sample size

The number of clusters needed to keep the variance of the estimated mean below a given maximum V_m is given by $n = \frac{S_B^2}{V_m}$, where S_B^2 is a prior estimate of the variance between cluster means. Clearly, this variance depends on the number of points in the clusters and their spatial configuration. If prior information on the spatial variability is available in the form of a variogram, the method described in Section 13.3.8 can be used to estimate S_B^2 for a given cluster definition.

If, instead of V_m, an absolute error d has been specified with an allowed probability of exceeding α, then V_m can be derived from d and α, according to $V_m = d/u_{1-\alpha/2}$, where $u_{1-\alpha/2}$ is the $1-\alpha/2$ quantile of the standard normal distribution.

When estimating areal fractions rather than means of quantitative variables, the above formula for n can still be applied if S_B^2 is interpreted as a prior estimate of the variance between cluster fractions.

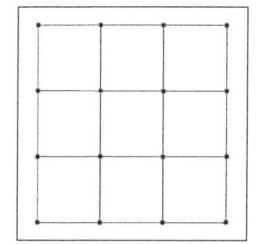

Figure 13.6: Notional example of Systematic Sampling

Advantages
Like in TsS, the spatial clustering of sample points has the operational advantage of reducing the travel time between points in the field. In addition, the regularity may reduce the time needed to locate consecutive points in the cluster. Of course, the importance of these advantages depend on the scale, the accessibility of the terrain and the navigation technique used.

Disadvantages
As with TsS, the spatial clustering generally leads to lower precision, given the sample size. Again, the rationale is that due to the operational advantages a larger sample size can be afforded for the same budget, so that the initial loss of precision is outweighed. If the spatial variation has a dominant direction, the precision can be optimized by taking transects in the direction of the greatest change.

Another disadvantage is that the sample size, i.e. the total number of points in the clusters which happen to be selected, is generally random. This may be undesirable for budgetary or logistic reasons. The variation in sample size can be reduced by defining clusters of roughly equal size.

13.3.6 Systematic Sampling (SyS)
Restriction on randomization
As with ClS, random selection is applied to pre-defined sets of points, instead of individual points as in SRS, StS and TsS. The difference with ClS is that only one cluster is selected. In this sense, SyS is a special case of ClS. (Note that the term 'cluster' as used here does not refer to geographical compactness, but to the fact that if one point of a cluster is included in the sample, all other points are included too.)

Selection technique
The selection algorithm for ClS is used with $n = 1$.

Example
Fig. 13.6 shows a random square grid. Notice the more even spatial spreading and the greater regularity compared with all other types of designs (Fig. 13.2 – 13.5).

Statistical inference
Means, areal fractions and SCDF's (after 0/1 transformation) of the area are simply estimated by the sample mean $\bar{y}$, as with SRS. The strategy (SyS, $\bar{y}$) is p-unbiased. This condition holds only if the grid is randomly selected, as is prescribed by the selection technique given above. With 'centered grid sampling', on the other hand, the grid is purposively placed around the center of the area, so that the boundary zones are avoided. This is a typical model-based strategy (see Section 13.4), which is p-biased.

Unfortunately, no unbiased variance estimators exist for this type of design. Many variance estimators have been proposed in the literature; all are based on assumptions about the spatial variation. A well-known procedure is Yates's method of balanced differences [403]. An overview of variance estimation is given by [69]. A simple, often applied procedure is to calculate the variance as if the sample was obtained by SRS. If there is no pseudo-cyclic variation this over-estimates the variance, so in that case the accuracy assessment will be on the safe side.

Means, areal fractions and SCDF's (after 0/1 transformation) in a domain are simply estimated by the sample mean in this domain:

$$\bar{y}_j = \frac{1}{m_j} \sum_{i=1}^{m_j} y_{ij},$$

where m_j is the number of grid points falling in domain j.

Sample size
As indicated above, the sample size is random in general. The average sample size is determined by the choice of the grid size. A rough approach to this choice is to determine the sample size in the same way as for SRS (Section 13.3.2) and to reduce this with a empirical factor (for instance 2) to account for better precision of SyS relative to SRS. The average required grid size for a square grid is then $\sqrt{A/m}$. However, if an estimated variogram is available, it is more accurate to apply the method described in Section 13.3.8.

Advantages
Because only one cluster is selected, the clusters should be pre-defined such that each of them covers the area as good as possible. This is achieved with clusters in the form of regular grids: square, triangular or hexagonal. The statistical precision can thus be maximized through the definition of the grid. In addition, SyS has the same operational advantage as ClS: the regularity of the grid may reduce the time needed to locate consecutive points in the field. Again, the importance of this depends on the scale, the accessibility of the terrain and the navigation technique used.

Disadvantages
Because this type of design does not produce any random repetition, no unbiased estimate of the sampling variance is available. If the spatial variation in the area

Chapter 13: Spatial sampling schemes

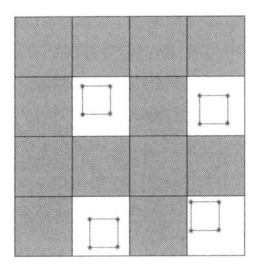

Figure 13.7: Notional example of Two-stage Sampling with Systematic Sampling in both stages

is pseudo-cyclic, the variance may be severely underestimated, thus making a false impression of accuracy. An operational disadvantage may be that the total travel distance between sample points is relatively long, due to the even spreading of the points. Finally, SyS has the same disadvantage as ClS: the sample size (number of grid points that happen to fall inside the area) is generally random, which may be undesirable for budgetary or logistic reasons. The possible variation in sample size will often be larger than with ClS, and it will be more difficult to reduce this variation.

13.3.7 Advanced design-based strategies
Apart from the basic strategies outlined in the previous sections, a large number of more advanced strategies have been developed. This section outlines some of the major possibilities.

Compound strategies
The basic strategies of the previous sections can be combined in many ways to form compound strategies. One example is given in Fig. 13.7, where TsS has been applied, however with SyS in both stages instead of SRS. In this case a square grid of 2 × 2 PU's was selected, and then a square grid of 2 × 2 points in each of the selected PU's. Notice that the total between-point distance is reduced as compared with SyS in Fig. 13.6, that the risk of interference with possible cyclic variation has practically vanished, and that the operational advantage of regularity in the configuration still largely exists.

Fig. 13.8 shows another example of a compound strategy: Stratified Cluster Sampling with four strata and two clusters in each stratum. The clusters are perpendicular transects, each with two points at a fixed distance. Notice that, due to the stratification, a more even spreading is obtained as compared with ClS in Fig. 13.5, while the operational advantage of regularity still exists. See [100] for an account of perpendicular random transect sampling and an application in quality assessment of soil maps.

As alluded in the examples above, the reason for combining two or more basic strategies is always an enhancement of advantages or mitigation of disadvantages of the basic strategies. As a final example, consider the situation in which the high

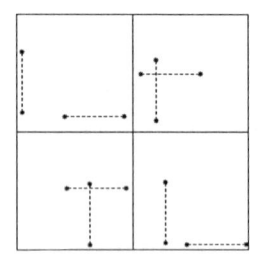

Figure 13.8: Notional example of Stratified Cluster Sampling

precision and the operational advantage of regularity in SyS is wanted, however, it is desirable that the precision can be quantified from the data, without recourse to assumptions about the spatial variability. A possible solution is to adapt the Two-stage/Systematic compound strategy of Fig. 13.7. In order to enable model-free variance estimation, the PU's could be selected at random instead of systematically, while maintaining grid sampling in the second stage. In that case, the variance can be estimated in the same way as with basic TsS.

In devising a compound strategy, very often there are good reasons to stratify the area first, and then to decide which designs will be applied in the strata. It is not necessary to have the same type of design in each stratum. As long as the stratum means and their variances are estimated without bias, these estimates can be combined into unbiased overall mean and variance estimates using the formulas given in Section 13.3.3.

If a variogram for the area is available, the variance of a compound strategy can be predicted prior to sampling, using the Monte-Carlo simulation technique presented in Section 13.3.8. In the case of stratification this technique can be applied to each stratum separately, using different variograms if necessary.

Spatial systematic strategies
Most strategies discussed so far are spatial in the sense that primary sampling units and clusters are defined on the basis of geographical co-ordinates. Also strata are usually defined that way. Given these definitions, however, the randomization restrictions do not refer to the co-ordinates of sample points. A category of more inherently spatial strategies exists of which the randomization restrictions make explicitly use of X and Y co-ordinates or distances in geographical space. Two examples are given.

Fig. 13.9 shows a 'systematic unaligned' sample. This techniques was proposed by [300]. The area is first divided into square strata and one point is selected in each stratum, however, not independently. A random X co-ordinate is generated for each row of strata, and a random Y co-ordinate for each column. The sample point in a stratum is then found by combining the co-ordinates of its row and column. Notice in Fig. 13.9 the irregular, but still fairly even spread of the points.

Fig. 13.10 shows a 'Markov chain' sample, a technique discussed by [48]. Again, notice the irregular but fairly even spread of the points. The underlying principle is

Chapter 13: Spatial sampling schemes

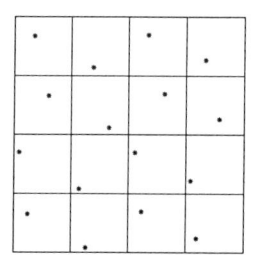

Figure 13.9: Notional example of Systematic Unaligned Sampling

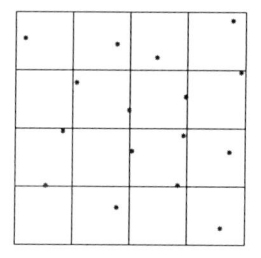

Figure 13.10: Notional example of Markov Chain Sampling

that the differences between the co-ordinates of consecutive points are not fixed, as with systematic unaligned samples, but stochastic. These differences have a variance which is determined through a parameter ϕ, chosen by the user. Thus Markov Chain designs form a class in which one-per-stratum StR and systematic unaligned designs are special cases, with $\phi = 0$ and $\phi = 1$, respectively. The example in Fig. 13.10 was generated with $\phi = 0.75$.

As illustrated by the examples, the purpose of this type of strategies is to allow enough randomness to avoid the risk of interference with periodic variations and linear artifacts like roads, ditches, cables and pipelines, while still maintaining as much as possible an even spread of the points over the area.

Regression estimators
Suppose that an ancillary variable x is available which is roughly linearly related to the target variable y and known everywhere in the area, for instance from remote sensing or a digital terrain model. Then this information can be exploited by using a 'regression estimator'. For a simple random sample this is

$$\bar{y}_{\text{Lr}} = \bar{y} + b(\overline{X} - \bar{x}),$$

where:
$\bar{y}$: sample mean of target variable;
$\bar{x}$: sample mean of ancillary variable, measured at the same points as y;
$\overline{X}$: areal mean of ancillary variable;

b : least squares estimate of the regression coefficient:

$$b = \frac{\sum_{i=1}^{n}(y_i - \overline{y})(x_i - \overline{x})}{\sum_{i=1}^{n}(x_i - \overline{x})^2}$$

For large samples (say $n > 50$) the variance can be estimated by [69]:

$$v(\overline{y}_{\text{Lr}}) = \frac{1}{n(n-2)} \sum_{i=1}^{n}[(y_i - \overline{y}) - b(x_i - \overline{x})]^2 .$$

If the ancillary variable is not known everywhere in the area, but can be measured cheaply in a large sample, then the relationship can be used by measuring y only on a random sub-sample, and again applying a regression estimator. This technique is known in the sampling literature as 'double sampling' or 'two-phase sampling'. Instead of the areal mean, $\overline{X}$, we now have the large sample mean $\overline{x}'$, so that

$$\overline{y}_{\text{Lr}} = \overline{y} + b(\overline{x}' - \overline{x}),$$

with estimated variance [69]:

$$v(\overline{y}_{\text{Lr}}) = s_{y.x}^2 \left[\frac{1}{n} + \frac{(\overline{x}' - \overline{x})^2}{\sum(x_i - \overline{x})^2} \right] + \frac{s_y^2 - s_{y.x}^2}{n'},$$

where $s_{y.x}^2$ is the estimated residual variance:

$$s_{y.x}^2 = \frac{1}{n(n-2)} \left[\sum_{i=1}^{n}(y_i - \overline{y})^2 - b^2(x_i - \overline{x})^2 \right] .$$

The regression estimators given above have been generalized to stratified sampling and to the case with more than one ancillary variable. They have a great potential for natural resource inventory, but their application in practice seems underdeveloped.

13.3.8 Model-based prediction of design-based sampling variances

If prior information on the spatial variability is available in the form of a variogram, the following method can be used to predict the sampling variance of any design-based strategy. The core of the method is the general equation for predicting the variance of a design-based estimated mean from a variogram [109]:

$$E_\xi[V_p(\widehat{\overline{Y}})] = \overline{\gamma}_A - E_p[\lambda' \cdot \Gamma_s \cdot \lambda], \qquad (13.8)$$

where:

$E_\xi[.]$: statistical expectation over realizations from the model ξ underlying the chosen variogram;

$E_p[.]$: statistical expectation over realizations from the design p;
$V_p(.)$: variance over realizations from the design p (the usual sampling variance in the design-based approach);
$\overline{\gamma}_A$: mean semi-variance between two random points in the area;
λ : the vector of design-based weights of the points of a sample selected according to design p (For instance, if one cluster of 3 points and one of 2 points were selected, the weights in calculating the mean would be (cf. Eq. 13.6): 1/6, 1/6, 1/6, 1/4, 1/4.);
Γ_s : matrix of semi-variances between the points of a sample selected according to design p.

The first term, $\overline{\gamma}_A$, is calculated by numerical integration or by Monte-Carlo simulation, repeatedly selecting a pair of random points, calculating its semivariance, and averaging. The second term can also be evaluated by Monte-Carlo simulation, repeatedly selecting a sample according to design p, calculating its mean semi-variance $\lambda' \cdot \Gamma_s \cdot \lambda$, and averaging. This generic procedure is computationally demanding but it is the only option for compound and spatial systematic strategies (Section 13.3.7). For the basic strategies, however, much more efficient algorithms are possible, making use of the structure of the design types. The following special prediction equations can namely be derived from the general Equation 13.8.

Simple Random Sampling

In the case of SRS, Equation 13.8 simplifies to:

$$E_\xi[V_p(\widehat{\overline{Y}})] = \frac{1}{n}\overline{\gamma}_A.$$

Stratified Sampling

For StS, Equation 13.8 becomes:

$$E_\xi[V_p(\widehat{\overline{Y}})] = \sum_{h=1}^{L} \frac{1}{n_h}\overline{\gamma}_{Ah},$$

where $\overline{\gamma}_{Ah}$ is the mean semi-variance between two random points in stratum h. Different variograms can be used for the strata.

Two-stage Sampling

For TsS and m_i constant the sampling variance is given by Eq. 13.5. The variance components in this equation are the between-unit and the pooled within-unit variance, S_B^2 and S_W^2. These components can be predicted from the two terms in Eq. 13.8. The first term predicts the total variance, $S_T^2 = S_B^2 + S_W^2$, while the second term predicts $S_W^2/2$ if we take $n = 1$ and $m = 2$. In other words, the second term is calculated by repeatedly selecting one unit and two random points in it. The result is the mean semi-variance between pairs of random points within units, denoted by $\overline{\gamma}_U$. The sampling variance is then predicted by:

$$E_\xi[V_p(\widehat{\overline{Y}})] = \frac{1}{n}\left(\overline{\gamma}_A - \frac{2(m-1)}{m}\overline{\gamma}_U\right).$$

Cluster Sampling

The sampling variance with ClS equals the between-cluster variance, S_B^2, divided by the number of clusters, n. To predict S_B^2 for a given cluster definition, we apply Eq. 13.8 to ClS with $n = 1$. In other words, the second term is calculated by repeatedly selecting only one cluster. Within each cluster the points have equal weight $(1/m_i)$, so that $\lambda' \cdot \Gamma_s \cdot \lambda$ simplifies to the unweighted mean:

$$\lambda' \cdot \Gamma_s \cdot \lambda = \frac{1}{m_i^2} \sum_{k=1}^{m_i} \sum_{l=1}^{m_i} \gamma_{kl} = \frac{2}{m_i^2} \sum_{k=1}^{m_i-1} \sum_{l=k+1}^{m_i} \gamma_{kl},$$

because Γ_s is symmetric with zero diagonal. The result is the mean semi-variance between pairs of points within clusters, denoted by $\overline{\gamma}_C$. The sampling variance is then predicted by:

$$E_\xi[V_p(\widehat{\overline{Y}})] = \frac{1}{n}\left(\overline{\gamma}_A - \overline{\gamma}_C\right).$$

Of course, in the special case that all clusters have the same size and shape, $\lambda' \cdot \Gamma_s \cdot \lambda$ needs to be calculated only once.

Systematic Sampling

As SyS is ClS with $n = 1$, the sampling variance can be predicted by:

$$E_\xi[V_p(\widehat{\overline{Y}})] = \overline{\gamma}_A - \overline{\gamma}_C.$$

Again, in the special case that all clusters have the same size and shape, $\lambda' \cdot \Gamma_s \cdot \lambda$ needs to be calculated only once.

13.4 Model-based strategies

In the model-based approach the emphasis is on identifying suitable stochastic models of the spatial variation, which are then primarily used for prediction, given the sample data. This subject is treated in Chapters 5 to 8 of this volume. The models can also be used to find efficient sampling designs, but the main focus is on model building and inference, not on sampling design. This is natural, because the approach was developed to cope with prediction problems in the mining industry, where the data had already been collected via convenience or purposive sampling (Section 13.1.5). Nevertheless, stochastic models of the spatial variation have been successfully used in optimizing spatial sampling configurations for model-based strategies. Three different forms can be distinguished.

Firstly, if no prior point data from the area are available, the model can be used to determine the optimal sampling grid for point kriging or block kriging, given an accuracy requirement. It has been shown [254] that if the spatial variation is second order stationary and isotropic, then equilateral triangular grids usually render the most accurate predictions, closely followed by square grids. In case of anisotropy the grid should be stretched in the direction with the smallest variability.

Chapter 13: Spatial sampling schemes   241

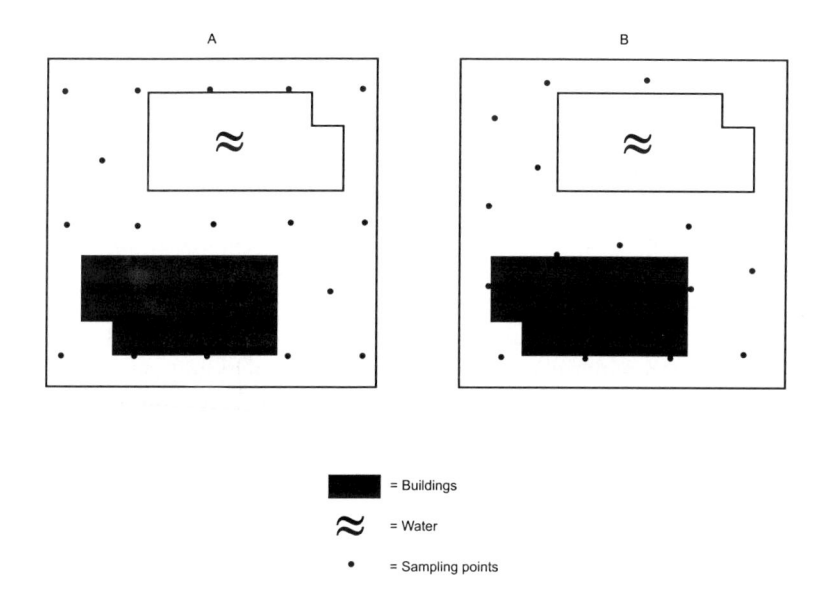

Figure 13.11: Examples of two model-based strategies: (A) centered equilateral triangular grid, and (B) configuration optimized by spatial simulated annealing with the Minimization of the Mean of Shortest Distances criterion. (From [378]).

McBratney et al. [261] presented a method to determine the optimal grid spacing for point kriging, given a variogram; a program and examples can be found in [260]. A similar method to determine the optimal grid spacing for block kriging is given by [262]. These methods are intended for large areas with a compact shape, so that boundary effects can be disregarded.

Secondly, if point data from the area pre-exist, the model can be used to find good locations for additional sampling. To that end a contour map of the kriging variance is made; additional sampling is then projected preferably in regions with high variance as this provides the largest reduction of uncertainty. This technique is practical and has found wide-spread application. It is only approximative, however, in the sense that it does not lead to an exact optimal configuration of sampling points.

Thirdly, if the area is small or irregularly shaped, then boundary effects cannot be disregarded and computationally more intensive methods are needed. Van Groeningen and Stein [378] present such a method, based on spatial simulated annealing. Fig. 13.11 shows an example of a point configuration optimized by their method.

The area contains two kinds of inclusions which cannot be sampled: a building in the South and water in the North. In this example, the soil under the building is part of the research area, say for soil sanitation, while the water is not. The optimized configuration shows that sample points are attracted by the 'research inclusion', but repelled by the 'non-research inclusion'. For comparison an equilateral triangular grid is shown, with the points removed that cannot be sampled. Using this method it is very easy to account for pre-existing data points; at the start they are simply added to the new points and their locations are kept fixed during the optimization process. The method then renders an optimized configuration, as opposed to the approximative method described above. Another advantage of this method is that it is versatile, because different quality criteria for optimization can be build in easily.

The scope of model-based strategies is wider than that of design-based strategies. Firstly, the data requirements are more relaxed. Data from convenience, purposive as well as probability sampling can be used for model-based inference, while design-based inference requires probability sampling. Secondly, model-based inference can be directed towards a wider class of target quantities, including local functions and functions defined by geographic neighborhood operations. An example of the latter is the total surface area of land patches consisting of a minimum number of adjacent pixels classified as suitable for a given land use. A local function which can only be predicted by a model-based strategy is, for instance, the spatial mean of a small domain (or 'block') with no sample points in it.

The price paid by the model-based approach for its larger versatility is full dependency on a stochastic model of which the validity is more or less arguable. If the alternative of the design-based approach is not applicable, this dependency just has to be accepted. However, where the scope of the two approaches overlap (Section 13.3.1), one has a choice as discussed in Section 13.2.

Chapter 14

Remote sensing and decision support systems

Ali Sharifi

Remote sensing is the process of extracting information from a data stream that involves the earth's surface, the atmosphere that lies in between it and the spacecraft, and the image forming sensors on board of spacecraft. Information extraction proceeds by the derivation and use of models that account for the transformations that take place at each of the three steps and at the processing stage [354]. The driving force behind this development has been the generation of information about the Earth's resources. In this process Geographic Information Systems have played a crucial role, as it has facilitated the organization and process of information extraction as well as providing capacities to use these information in combination with many others, coming from variety of sources for solving problems in the earth surface. Remote sensing therefore, can not be seen as an end in itself, and should be more correctly regarded as a part of GIS. Its data can form parts of dynamic data sets of the earth surface in contrast to the ancillary data sets, such as elevation, and soils, which are static. Its required processing capacities which attempts to reconstruct the scene model through model inversion of image, and remote sensing models could be considered as a part of the analytical capabilities of GIS. Remote sensing and GIS are therefore closely interrelated, and if we broadly define GIS, it can include remote sensing.

Currently Remote Sensing and GIS technology have offered a great potential to capture data through variety of Earth Observation Platforms, and integrate/relate them through their common spatial denominator. They also offer appropriate technology for data collection from Earth-surface, information extraction, data management, routine manipulation and visualization, but they lack well developed, required necessary analytical capabilities to support decision making processes. For improved decision-making, all these tools, techniques, models and decision-making procedure have to become integrated in an information processing system called "Decision Support System". This involves preparing and running the models, capturing, preparing, and organizing its inputs (over and over), developing and evaluating scenarios, interpreting and communicating its results, and making them fit in the existing decision environment. This chapter is an attempt to briefly introduce the concept of DSS, its principles, components and applications in the context of geo-information (remote sensing and GIS) technology.

14.1 Decision Making Process (DMP)

In the domain of decision-making, a decision-problem (choice problem) is said to exist where an individual or group perceives a difference between a present state and a desired state, and where:
1. The individual or group has alternative courses of action available
2. choice of action can have a significant effect on this perceived difference
3. The individual or group is uncertain a priori as to which alternative should be selected

This is an action-oriented definition of problem [1], meaning that if no alternative options are available, an undesired state is not defined as decision-problem. In this context, following [271], we define a decision process as a set of actions and dynamic factors that can begins with identification of stimulus for action and ends with specific commitments of action. The Decision-Makers "DM" are individuals or groups of individuals who directly or indirectly provide value judgments or opinions on the decision process necessary to define and choose between alternative courses of action [62]. The immediate vision of DM is that he behaves in an objectively rational way, in which he tries to find optimal course of action and has all the relevant information for his decision ready available. In this vision the DM is pictured as homo economicus. Many authors have criticized this vision. Bosman [47] summarizes the critique. The homo economicus vision was an implicit assumption of any theory on improving decision-making until Herbert Simon wrote his famous book on "The New Science of Management Decisions" [328]. Simon distinguishes the following two paradigms for decision-making.

1. Objective rationality assumes a rational DM who knows exactly the goal of decision making and how its outcome can be measured. More over, he has access to all information that he requires (all alternative options and their outcomes) and therefore can find the decision, which leads to the best possible solution " maximizing goal achievement level".

2. Procedural rationality or bounded rationality assumes that DM is looking for a course of action which is good enough. This is because in real life people usually make decisions that are based on attempt to attain only satisfactory level of goal attainment. According to this principle, DMs follows "satisfying" (good enough) decision-making behavior, based on search activity, to meet certain aspiration levels rather than optimizing behavior aimed at finding the best possible decision. This has been the most valuable contribution of Simon to the study of decision-making.

Each of these paradigms is supported by its own type of models and techniques, e.g., the rational model is supported by normative models and optimization techniques, whereas the bounded rationality is supported by descriptive models and simulation techniques.

Many model of decision making process have been proposed in the literature. We here examine two models, representing the two schools of thought. The oldest and simplest model, which represents the classical economic model of DM (rational model), is the one developed by Dewey [107]. This regards decision-making process as answering the three consecutive questions: what is the problem? What are the existing solutions? And which alternative is the best? The other, is the well-known model of Simon [328], which represents bounded or procedural rationality distinguishes the following phases.

1. The intelligence or problem formulation phase
   This involves scanning of the environment for situations (problem or opportunities) demanding a decision. Here data are obtained, processed, and examined for clues that may identify problems or opportunities.
2. The design phase
   This involves inventing, developing analyzing possible courses of action, which includes process of understanding the problem, generating solutions and test solutions for feasibility
3. The choice phase
   Involving evaluation and selection of an alternative or course of action from those available.

Much confusion exists between the terms " decision making" and "problem solving". Strictly speaking, decision making is not the same as problem-solving. Since DM stops when recommendation for action is made. In fact, problem solving puts the decision into effect and monitors the consequences, which includes other phases such as, preparation of plan, implementation, monitoring and control.

Decision problems can be categorized as "well-structured", "ill-structured", and "unstructured". We define problem well-structured (programmable), when all phases of the decision making- process can be formalized, and it is possible to prepare a decision rule or decision procedure such as a set of steps, a flowchart, a formula, or a procedure to collect and apply data to derive the best solution (at least a good enough solution). Since structured, programmable decisions can be pre-specified, many of these decisions can be delegated to lower-level staff, or even can be automated. If none of the phases of the decision making process could be formalized, we define the problem as unstructured or non-programmable. Such problems have no pre-established procedures, either because the decision is too infrequent to qualify for preparing a decision procedure, or because the decision procedure is not understood well enough, or is too changeable to allow a stable pre-established procedure. If the situation is in intermediate, meaning that only some of the phases of the decision making process can be formalized (have structure), then we call the problem, ill-structured or semi-programmable. Totally structured problems usually do not require much DMs analysis. In fact they are usually well served by standard operating procedures for straightforward problems, and Expert Systems for complex problem environment.

It should be realized that a problem is not structured or ill-structured per se: it all depends on the person and on the point on time, this is a subjective phenomenon. A problem may be ill-structured to some one at one point in time, and may be structured to him or some one else at later point in time. The challenge here is to find structure in problems that they seem to have no structure. Structuring a problem means breaking down the problem into a number of sub-problems that better fulfill the conditions set for well structured problems [47].

14.2 Spatial Decision Making

Spatial decision making is based on spatial analysis, which, focuses attention upon locations and distribution of phenomena; interactions of people, goods, and services between places and regions; spatial structure, arrangements, and organizations and spatial processes. The problems related to all these phenomena have a spatial dimension. Cowlard [81] defines geographical problems, as those issues and questions, which are the result of the interrelationships between people and their environment (earth surface). This can be anywhere within the range of decisions to maintain, improve or restore people-environment relationships. A typical geographical problems may involve ecological issues (relationships between people and physical environments), or locational issues (relationships between people and spatial environments), and require the DM to work with:

1. A wide range of physical and human factors and complex relationships between issues, between options and between evidences. There is usually a variation in interests, requiring value-judgments
2. Uncertainty in space and time: about future developments and rapidly changing situations and imperfect or incomplete evidences. Spatial issues require to address a range of scales, from local to global
3. Different cognitive styles and individual differences in human behavior, leading to conflicting objectives and views and a potentially wide impact of decisions
4. Short and long term views (more production is a good strategy in short term, however, that may not be a good one if one puts it in a long term perspective)

Geographical decision making therefore is an attempt to solve the complicated problems, which can arise from people-environment relationships. This can be facilitated through a systematic approach that is introduced in the following sub-section.

14.2.1 A systematic approach for solving spatial problems

The management science approach adopts the view that mangers can follow a fairly systematic process of solving problems. Therefore it is possible to use scientific approach to managerial decision making. This approach includes identification of problem or opportunity, gathering important data, building a model, experimenting with the model, analyzing results and making a sound decision. On the basis of management science approach, the Simon's model of decision making and [81] a systematic and logical route for an effective spatial decision making is set up (Fig. 14.1). This is mainly based on answering the following major questions:

1. What is the problem and what must be achieved?
2. What is the evidence and what does it show?

Chapter 14: Decision support systems

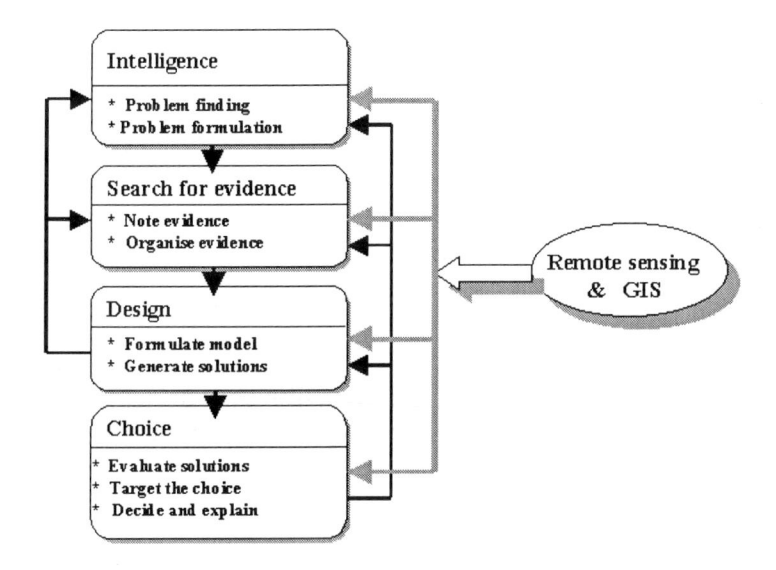

Figure 14.1: A systematic approach to support spatial decision making

3. What are the alternatives and what are their outcomes?
4. What decision ought to be taken?

As it can be seen from the Fig. 14.1, at each phase there may be a return to previous phases. This means that decision-making process may be a continuous process. This is certainly the case when we are engaged in research and development problem, in which normally a prototype product is first designed, developed and improved through iterations. The result of each phase has to be transferred to the DM in a manageable, communicable and quick form to control and verify the process. This can be supported by visualization and graphic presentation, which provide facility to communicate large amounts of information quickly and effectively [325].

Each of the above questions is answered through a sequence of actions, identifying a distinct stage in the systematic approach. In the following, each activity with the role that remote sensing and GIS can play in their accomplishment is briefly described.

Phase 1 Intelligence
During this phase the DM scan the environment to detect and clearly define the problem/opportunity and its related objectives. This gives a solid base and clear direction for the next phase, helps to concentrate ideas on the important aspects and may begin to suggest the evidence and techniques to be used at the next phase. This is carried out in two stages:

Stage 1 Problem finding: Problem finding conceptually defined as finding the difference between some existing situation and some desired state. In another word there is some model of what is desired, this is compared to the reality, differences are identified, evaluated to see whether they constitute a problem. The following models are suggested to generate expectations against which reality is measured.

1. Historical models in which the expectation is based on an extrapolation of past experience
2. Planning models in which the plan is the expectation
3. Scientific models simulating the desired situations
4. Other people's model in which expectation is based on expert opinions
5. Extra-organizational models in which the expectation is derived from competitors and outcome of similar environment

Furthermore the nature of the problem, issues or questions that are to be resolved should be understood, clearly defined and described. Remote sensing and GIS can strongly support this stage. They support assessments of the current and the desired situations and their comparisons to find a problem or an opportunity, e.g., assessment of the current and planned or ideal land use and their comparisons to explore presence of any problem.

Stage 2 Problem formulation: The purpose of problem formulation is to clarify problem, so that the design and choice phases operate on the right problem. Frequently, the process of clearly stating the problem is sufficient, in other cases some reduction of complexity is needed. According to Weber [387] problem decomposition tools are intrinsic component of problem solving. A complex problem needs to be decomposed into manageable solutions. Davis and Olson [96] suggest the following strategies for reducing complexity of problems:

1. Determining the boundaries (i.e., clearly identifying what is included in the problem)
2. Examining changes that may have caused the problem
3. Problem decomposition and factoring the problem into smaller problems
4. Focusing on the controllable elements
5. Furthermore, as results of this stage the goals, aims or objectives must be clarified.

Phase 2 Search for evidence

"Evidence" is defined as the total set of data/information that DM has at his disposal, including the skills, which are necessary to use them. It is therefore the key resource at all stages of decision making. The quality of evidence is very important aspect. Ideally DM hopes to have good quality evidence in abundant supply. Frequently, the evidence will be lacking, and DM has to enhance its quality before it is used in the analysis. Decision problem may include evidence that is difficult to measure or predict. Evidence may be in several forms. "Certain" where the evidence is the truth. These types of evidence are directly used in the analysis. "Risk evidence" is when the evidence do not represent the truth but with an estimate of the truth with a known probability. The expected value of these evidences can be used in the analysis procedure. "Uncertain evidence" is where the quality of evidence is not known. Normally, before use their risks are estimated and then treated as risk evidence in the analysis procedure.

Evidence, may have different types. "Facts" are objective evidence specific to the problem. These are measurable often-precise truth or an estimate of truth that rely on no assumptions. They are objective evidences and often used to test feasibility and answer the question, "is the decision practical"? "Values" are subjective evidence specific to the problem. These are opinions, views, attitudes, prejudices, assumptions and interpretations that are difficult to measure but have important influences on the DMP. They are often used to test desirability and answer the question " is the decision right"

"Knowledge" represents the DM's familiarity with the problem domain, existing theories, models, relevant concepts, experiences and any information relevant to the problem. "Experience" this is the DM's expertise in the DMP, including his ability to use the related tools and techniques and experiences in DM. The evidence may be in different forms and format, such as numerical, alphabetical, graphical, map, in sound (spoken form), aerial photographs, satellite images, etc.

Stage 3 Note the evidence: All related available evidences such as facts, values, knowledge or experiences are collected, evaluated for inaccuracies, biases and other characteristics which must be considered during the analysis. Remote sensing technology has proved to be an instrumental technique for collecting facts about different earth resource problems. Considerable bodies of knowledge regarding the application of remote sensing technology in support of discovering and resolving earth resource problems have been developed and presented as theory, model, tools, and techniques.

Stage 4 Organize the information: At this stage the collected information is organized and translated into more usable forms. This involves, selecting, tabulating, classifying, summarizing, scaling and reducing information into more manageable formats. At this stage GIS is particularly instrumental tool.

Phase 3 Design solutions
An important part of decision making is generation of alternatives to be considered at the choice phase. This creative process can be enhanced by alternative generation procedures and support mechanisms. It involves generating, developing, and analyzing possible courses of action. Commonly, this phase results in a model.

Stage 5 Analysis and model formulation: Inclusion of models in DMP allows to better understand the behavior of the real system. Modeling includes a detailed analysis of the problem situation to understand its general nature and its extent but also to establish causes, effects and trends. An important step is resource analysis, which helps to understand characteristics of resources, and the processes through which they are allocated and utilized. Thus, a conceptual model of the problem as perceived by the DM is made, and later improved to correspond as closely as possible to the real system. After verification and validation followed by correspondence and consistency checks the model is complete to a relevant level [343]. Depending on the type of problem various types of models can be used, e.g., simulation, optimization and heuristics models. As a result, the decision-maker has a clear picture of the evidence and a model that can be used to generate alternatives. Remote sensing and GIS allow a representation of complex spatial relations and structures that are common in spatial problems. They play a role in testing and validating models by spatial and non-spatial data. As such they support understanding of a range of earth resource processes like soil degradation, deforestation, and land use changes. They can also be used to assess impacts of projects by visualizing their effects and their spatial distribution, and support searching for feasible alternatives like feasible sites, routes, land use and land allocation.

Stage 6 Generate alternatives: With the developed models a range of alternative solutions is generated and tested for their feasibility. The number of alternatives depends on whether the decision-maker is an optimizer, is looking for the best solution or, is just looking for a good enough solution. Remote sensing and GIS can provide the required data for the model and visualization of the results for supporting their feasibilities.

Phase 4 Choice of solution
This is the final phase of decision making, possibly without a clear point in time, as many decisions may be taken gradually and over time. It includes three main stages:

Stage 7 Evaluation of options: Based on the identified criteria, each option is examined on its advantages, disadvantages and consequences, and on its effectiveness and efficiencies in meeting the objectives. All options are then compared and vaulted using Multi-criteria Evaluation "MCE" techniques. Remote sensing and GIS can support this stage by assessing impacts from different points of views like attainability, desirability, and veto criteria [384]. Integration of MCE and GIS can provide a very powerful tool in support of this stage.

Stage 8 Targeting the best choice: As a result of stage 7, the attractive options are selected. Since errors may have been introduced to the process at different stages, a sensitivity analysis evaluates the stability and robustness of these options with respect to different assumptions and variables. Visualization of the results is important to assess the acceptability and desirability of the choice.

Stage 9 Make the decision and explain the result: The effective/attractive solutions are selected and explained. If the systematic approach has been followed, the result must be suitable to compile a well-argued decision report. Cowlard [81] considers this the most important stage where visualization is the key element, as it can communicate large amounts of information effectively and quickly.

14.2.2 Methods and techniques to support spatial decisions

Libraries of tools and techniques have been developed to support various phases of decision-making. We distinguish analytical functions, tools and techniques that can be applied both for a thematic and for a spatial data analysis. Examples are tabulation, scaling, matrix manipulation, classification, networking, charting, mapping, decision tree, pay-off matrices, environmental impact assessment, cost-benefit analysis, different modeling techniques, and multiple criteria evaluation techniques [81, 96, 287, 207, 369, 384, 281, 117, 306]. The supporting techniques at the problem formulation phase are those that help searching the environment for problems and opportunities. This may include a number of disciplinary models, which are used to derive the benchmark for comparisons. The design phase mostly includes disciplinary models tailored to a specific problem, which are used to generate alternative solutions. The disciplinary models, which are problem specific, can not be covered under the general tools and techniques, therefore not discussed in this chapter. The support for choice phase, however is generic, and may be the same for all problems. Normally, no matter what the spatial problem is (site selection, or land use allocation), the DM need to consider the impacts of choice alternatives along multiple dimensions in order to choose the best alternative. This process which involves policy priorities, trade-offs, and uncertainties is mainly aided by Multiple Criteria Decision-Making "MCDM" methods. Therefore, MCDM methods are briefly described here.

Colson and De Bruyn [73] define MCDM as a world of concepts, approaches, models and methods to help decision-makers to describe, evaluate, sort, rank, select or reject objects on the basis of evaluation according to several criteria. They distinguish two main classes of methods:

1. 1 Multiobjective decision making methods (MODM), sometimes viewed as a natural extension to mathematical programming, where several objective functions are considered simultaneously;
2. Multi-attribute decision making methods (MADM, Fig. 14.2), where the set of options is finite and discrete.

A common goal to both is an attempt to help the DMs (DM) make good (efficient) decisions that makes them satisfied. In the "multi-objective" case, the DM is faced with several conflicting objectives and a number of continuous decision variables bounded by mathematical constraints. Here the solution is found in two stages. In the first stage the efficient set is selected from the inefficient one. That is to separate the Pareto-optimal feasible solutions from the none Pareto-optimal ones. The second stage consists of searching for an optimum compromise solution acceptable to DM among the efficient solutions. To undertake the second stage it is necessary to incorporate in one way or another the preference of the DM. A number of methods following these principles are developed and presented in literature, e.g. [306], [71], and [173].

The "multi-attribute" includes several attributes in the DMP. An attribute is a characteristic of an option/object, which can be evaluated objectively or subjectively by one or several persons according to a measurement scale. On the basis of the value of these attributes also called "criteria", and the priorities that the

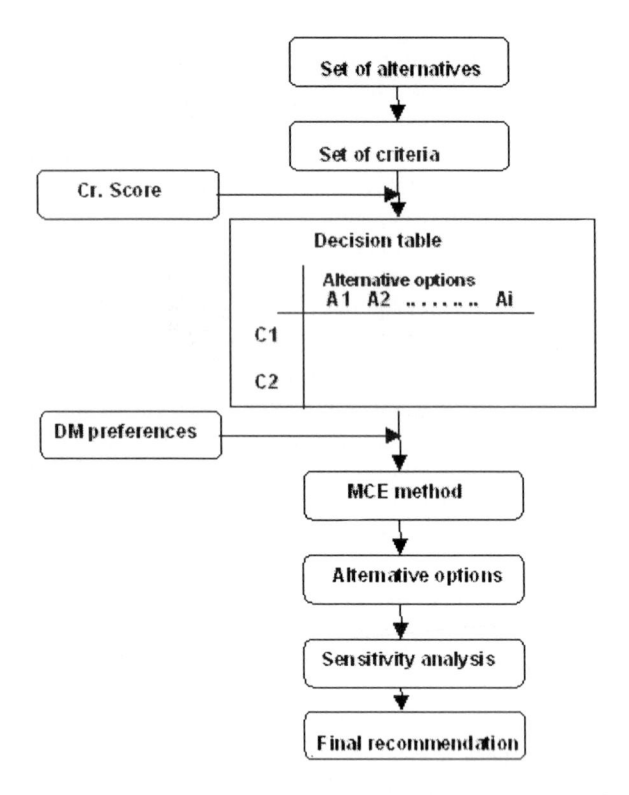

Figure 14.2: General model of MADM (modified from [205])

DM assigns to each one, also mostly referred as weight, the alternative options are evaluated and one of the following types of result is generated:
1. Identification of the best alternative(one recommended alternative)
2. Rejection of "bad" ones or identification of good ones (reduce decision space to several good alternatives)
3. Complete ordering of alternatives (ranking them from best to worst).

Most of the method builds an aggregation of the evaluations, or preferences associated with each attribute, so that a unique preference structure can be derived on the whole set.

According to the aggregation method and to elucidation of the preferences, three approaches can be distinguished:
1. Aggregation of attributes to form unique meta-criteria, the so-called compensatory approach. Weak performance of an option on one attribute can be compensated by good performance in other criteria. This approach is cognitively demanding as it requires the stakeholder to express criterion priority as weights or priority functions. Examples are: weighed summation, and multi-attribute utility methods [205, 207]. This can be easily implemented through overlay processing provided by RS and GIS.

2. Out-ranking, based on pairwise comparisons of all alternative options and their out-ranking relations, the so-called partially compensatory approach. It is based on the assumption that some compensations are admissible, while others not. The ELECTRE methods is a typical example of this approach [313].
3. Non-compensatory approaches without compensation between criteria. Common in these approaches is a stepwise reduction of the set of alternatives without deficiency trade off. An example is the dominance method, where elimination is based entirely on option scores. An option is dominated and subsequently eliminated, if another option is better on one or more criteria and is equal on the remaining criteria [205]. This is easily implemented in RS and GIS using logical operators.

Different MADM methods often represent different approaches to decision making, and choosing among methods may depend on the type of problem, the stakeholder values and other factors. It has been recognized that selection of the right method is a MADM on itself (for more information see the Advisor algorithm developed by [291], [282] and [205].

14.2.3 Uncertainty in decision making process

Uncertainty arises from lack of information or knowledge of some aspects of a decision problem. The main factors of uncertainty are the role of chance, the ambiguity and insufficient understanding of a problem and in making an abstraction of the actual problem. The ambiguity may arise from misperception or misconception or from inadequate descriptions and expression of the reality. The ambiguity may also appear from imprecision in formulation of the preference, or their assessment when comparing two options. These points can be accounted for in out-ranking methods via creation of the extended types of criteria (Quasi-criteria; pseudo criteria) and via acceptance of some incompatibilities [73].

MADM methods are composed of several components: the criteria, criteria-scores, criteria priorities and multi-criteria evaluation techniques "MCE". Because of ambiguity in one or all components, MCE results are subject to uncertainty, consisting of uncertainty in each component. Following [384] we distinguish criteria uncertainty, assessment uncertainty, priority uncertainty and method uncertainty.

Criteria uncertainty is concerned with the choice of evaluation criteria. Voogd [384] suggests three classes of criteria. Attainability criteria guarantee that a selected option is realistic, or is an option that can be realized. Veto criteria guarantee that the selected option fulfills the minimum requirements. Desirability criteria guarantee that the selected options is acceptable by all stakeholders. In these connections the following problems can be noted:
1. Are all the relevant criteria taken into account.
2. Is the definition of criteria in agreement with its intention, this is especially true when the main criteria is translated to more sub-criteria (measurable criteria so called indicators) may result in such a distorted criterion that hardly correspond to the main intention of the original criteria. For example, in remote sensing "criterion" agricultural production sometimes is broken

down into measurable indicators such as NDVI, agricultural area, water use, evapotranspiration etc.
3. Are the criteria equally distributed among the various dimensions of choice possibilities, e.g., not all environmental or all economic or all social

Assessment uncertainty refers to ambiguity in each criterion score, being the degree to which an option meets a criterion. Assessment uncertainty can be dealt with in four ways [384]:

1. Use of probability functions for each criterion a probability distribution estimates the expected degree of accuracy, reflecting its possible variation. Hence, the probability distribution of each criterion score has to be known, and the various probability distributions have to be independent. In actual studies, both conditions are hardly realized.
2. Sensitivity analysis models how a change in one criterion-score influences the evaluation result. As such it assesses sensitivity of critical scores that lead to uncertainty in the final evaluation. A sensitivity analysis assumes that various scores are mutually independent. Janssen and Herwijnen [206] developed and implemented a practical approach for sensitivity analysis. The DM specifies the maximum permissible deviation between the actual criterion value and the estimated value. The results for 100 values are listed in a probability table showing how many times option ranks in different positions. Another type of sensitivity analyzes, so-called certainty interval, compares two options with respect to a critical criteria. For example if option A better than option B, then the question is how much the critical criterion should change for B to become better than A.
3. Re-scaling to a lower measurement level transforms quantitative scores to values at an ordinal scale. By reducing the measurement accuracy, its reliability increases.
4. Feedback to research: If there is much debate on the values of some criterion-scores, then more research is needed to remove or decrease the uncertainty.

The priority uncertainty: priorities or weights, reflecting the relative importance of criteria and criterion-scores, can have a major impact on the final results. This specifies a high uncertainty, in particular when weights are quantitatively assessed. To deal with this, the same approaches as discussed for assessment uncertainty can be employed.

The method uncertainty: to increase reliability, every decision problem is evaluated using different evaluation methods. Each method is based on its own assumptions. The method will have a twofold impact on the evaluation. On the one hand it influences the structure of evaluation, like a concordance analysis versus weighted summation, whereas on the other hand it influences the content of the structure, e.g., the way in which dominance functions are specified in a concordance analysis. Voogd [384] claims that method uncertainty increases as the number of options and/or the number of criteria increases. Janssen and Herwijnen [206] suggest that if more than 70% of the methods supports one option then that is a good measure for dominance of that option.

14.3 Decision Support Systems

Decision Support Systems "DSS" are a sub-class of management information systems which support analysts, planners, and managers in the decision making process. They can reflect different concept of decision making and different decision situations. DSS are especially useful for semi-structured or unstructured problems where problem solving is enhanced by an interactive dialogue between the system and the user. Their primary feature is harnessing computer power to aid the DM to explore the problem, and increase the level of understanding about decision environment through access to data and models appropriate to the decision. They are aimed at generating and evaluating alternative solutions in order to gain insight into the problems, trade-offs between various objectives and support decision making process.

The primary intention of DSS is to assist specific DMs, individually or in groups, rather than the entire organization. This allow custom design of the system, in which DMs can use the system interactively to build and more importantly, to change analytic models of the decision problem. Interactive use allows immediate changes in assumed parameters with rapid feedback, encouraging the learning process that is impossible when the DM has to wait extended period of time for output. Therefore, interaction and support for transferring the results of analysis to the DMs in a communicable, manageable, easy understandable and quick way is a central feature of any effective man-machine-system. A real time dialogue allows the user to define and explore a problem incrementally in response to immediate answers from the system. Fast and powerful systems with modern processor technology can offer the possibility to simulate the dynamic processes with animated output and provide high degree of responsiveness that is essential to maintain a successful dialogue and direct control over the software.

DSS functions ranges from information retrieval and display, filtering and pattern recognition, extrapolation, multi-attribute utility theory, optimization techniques, inference and logical comparison to complex modeling. Decision support paradigms may include predictive models, which gives unique answers but with limited accuracy and validity. Scenario analysis relaxes the initial assumptions by making them more conditional, but at the same time more dubious. Prescriptive analysis of decisions emphasizes the development, evaluation and application of techniques to facilitate decision-making. These studies rely upon logic of mathematics and statistics and utilize the concepts of utility and probability to analyze decision problems. The concept of utility relates to the expression of preferences among alternative options, while probability serves to evaluate the likelihood of these preferences being utilized. The techniques adopted in these approaches incorporate explicit statements of preferences of DMs. Such preferences are represented by weighting scheme, constraints, goal, utilities, and other parameters. They analyze and support decision through formal analysis of alternative options, their attributes vis-à-vis evaluation criteria, goals or objectives, and constraints.

Group decision support systems focus on expediting the exchange of ideas among participants (brainstorming), stimulating quieter members to participate, and organizing collective thought into a workable consensus. In this context a set of par-

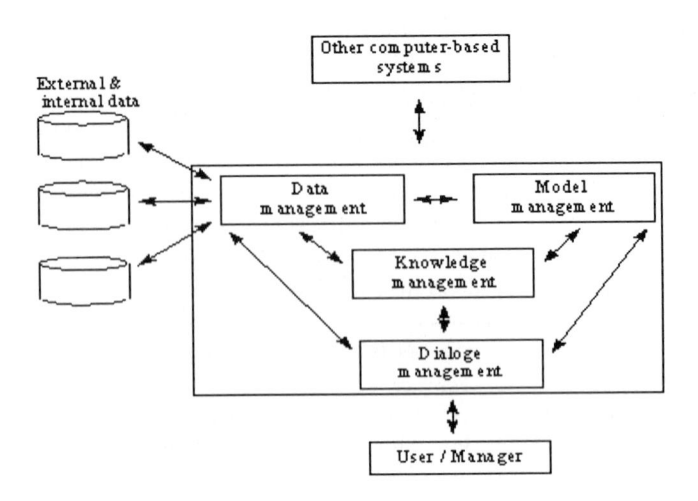

Figure 14.3: Components of a DSS (after [369])

ticipatory multiple criteria and multiple objective evaluation techniques are needed that aim to place the GIS analyst as a mediator between the computer technology package and the DM. GIS here can be seen as a possible vehicle for problem solving and decision making while accommodating the multiplicity of stakeholders in the decision making process. Research and development in this line has led to the development of Collaborative Spatial Decision Support Systems (CSDSS). Such types of system enables group to work together and participate in decision making concerning spatial issues by providing a set of generic tools, e.g., exchange of numerical, textual, graphical information, generation of solutions, group evaluation, consensus building and voting. This types of systems enable the stake-holders to collaborate in decision making process with no limitation of space and time, they can contribute to the group decision at any time and from any location. Examples of possible application may be environmental restorations, conservation planning, multiple resource use, and land use planning.

Many widely accepted definitions of DSS have database, model-base and interface components. Turban [369] further elaborates these components and designs a DSS architect as shown in Fig. 14.3. In this design DSS is composed of four sub-systems: "Data management subsystem" which contains relevant data for the situation and is managed by the database management systems. It may includes three type of databases, mainly, internal data as generated by the organization, the data from external sources, and private data which are specific to DM. "Model base management subsystem" which is the most important components of the system that forms the analytical capability of DSS and includes all statistical, financial, analytical, quantitative and qualitative models (library of models). It is composed of model base, model base management, modeling language, model directory, model execution, and integration facilities. "Dialog management" facilitates the commu-

Chapter 14: Decision support systems   257

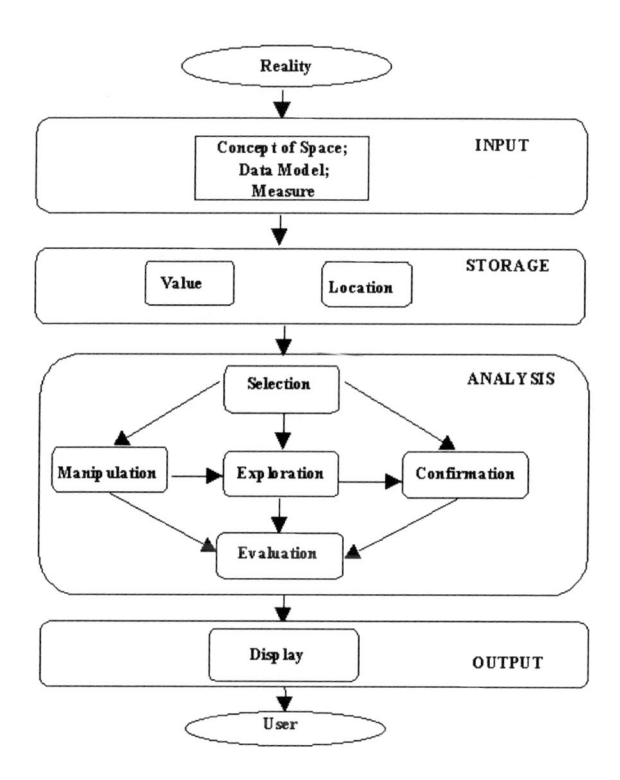

Figure 14.4: Functions of GIS (after [13])

nication between the user and system. "Knowledge management" Which plays a central role and maintains information about models, rules for model selection and retrieval, rules for constructing and use of models, procedure for experimentation, the existing data and the information requirements of each model.

14.3.1 GIS and Decision Support Systems

Anselin and Getis [13] identify four analytical function groups in their conceptual GIS model: selection, manipulation, exploration and confirmation (Fig. 14.4). Selection involves the query or extraction of data from the thematic or spatial databases. Manipulation entails transformation, partitioning, generalization, aggregation, overlay and interpolation procedures. Selection and manipulation in combination with visualization can be a powerful analysis tools. Data exploration encompasses those methods which try to obtain insight into trends, spatial outliers, patterns and associations in data without having a preconceived theoretical notion about which relations to be expected [368, 13]. The data driven sometimes called data mining approach is considered as very promising, due to the fact that theory in general in many disciplines are poor and moreover, spatial data is becoming increasingly available (rapid move from the data poor environment to a data rich environment). According to [317], during the early days of DSS, the challenge was to provide DMs access to enough information to allow them to make a better decision. Now, the challenge is not to provide enough information for DMs; rather it is how to screen the information, select those that are useful/relevant to the decision, without overwhelming or misleading the DM. Hence the challenge of DSS today, is to provide facilities to select the necessary and useful information for the decision problem. On this basis the concept of "data warehouse" has emerged and used in DSS (for more information see [317]). Confirmative analysis, however is based on a priori hypothesis of spatial relations which are expected and formulated in theories, models and statistical relations (technique driven). Confirmative spatial methods and techniques originate from different disciplines like operation research, social geography, economic models and the environmental sciences. The four analytical functions can be considered as a logical sequence of spatial analysis. The further integration of the maps/results from spatial analysis is an important next step to support decision making, which is called evaluation [13]. The lack of enough functionality especially in explorative and confirmative analysis and evaluation in GIS packages has been the topic of many debates in the scientific communities and as a result techniques to support this steps are gaining attention nowadays. In this context several studies has demonstrated the usefulness of integrating multi-objective decision techniques with GIS, and few vendors have incorporated some analytical techniques in their GIS packages (IDRISI, ESRI).

To alleviate the above problems GIS should be upgraded by DSS functionality in a user friendly and easy to use environment. However, there is a trade-off between the efficiency and ease of use, and the flexibility of the system. The more options are predetermined and available from the menu of choices, the more defaults are provided, the easier it becomes to use a system for an increasingly small sets of tasks. There is also trade-off between the ease of understanding and the precision of the results. Providing a visual or symbolic presentation changes the quality of the information in the course of transformation from quantitative to qualitative data sets. Finally, the easier the system the harder is to make and maintain.

It is the integration of GIS and DSS, in combination with simulation or optimization models, related databases, and expert system tools, that form a powerful, attractive and user-friendly decision support tools for a large spectrum of planning and management problems. The research community especially has sought ways to enhance the analytical capability of GIS, through the integration of spatial data handling, modeling and decision support functionality to support management functions. Various logical ways of coupling GIS and disciplinary models are discussed in the literature [127]. Vassilios and Despotakis [380] argued that the integration might be achieved in one of the following way:

1. Enhance GIS to include DSS functionality
2. Enhance DSS to include GIS functionality. These two possibilities are also referred as tight coupling

Chapter 14: Decision support systems

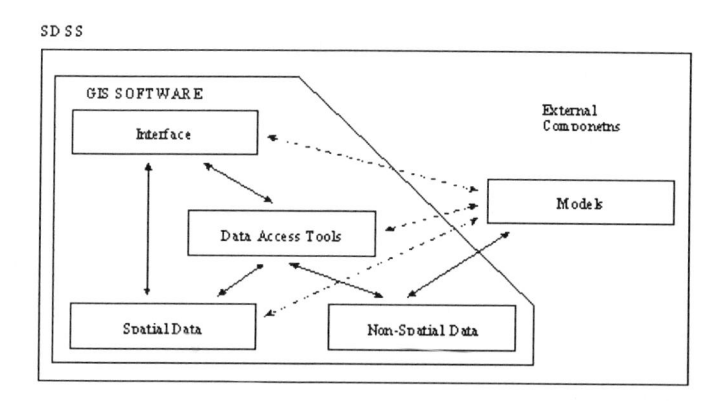

```
SDSS
┌─────────────────────────────────────────────────┐
│  ┌─────────────────────┐                        │
│  │ GIS SOFTWARE        │         External       │
│  │  ┌──────────┐       │         Componetns     │
│  │  │Interface │◄·· ·· │                        │
│  │  └──────────┘       │                        │
│  │    ▲   ↘            │      ┌──────────┐     │
│  │    │     ↘          │····► │  Models  │     │
│  │    ▼       ↘        │      └──────────┘     │
│  │  ┌─────────────┐ ◄··┤         ↗ ↗           │
│  │  │Data Access  │    │        ↗ ↗            │
│  │  │   Tools     │────┼───────                │
│  │  └─────────────┘    │                        │
│  │    ↙   ↘            │                        │
│  │   ↙     ↘           │                        │
│  │  ┌────────┐  ┌──────────────┐               │
│  │  │Spatial │◄·│Non-Spatial   │               │
│  │  │ Data   │  │    Data      │               │
│  │  └────────┘  └──────────────┘               │
│  └─────────────────────────────────────────────┐│
└─────────────────────────────────────────────────┘
```

Figure 14.5: SDSS-integrating models with GIS (after [229])

3. Combine both GIS and DSS by connecting the two systems externally. This can be achieved by: 1) Loose coupling, GIS and the model communicate through exchange of files. It is normally time consuming and prone to error. 2) Close coupling, GIS and models are connected through a common user interface which is taking care of the data sharing between the two systems

Sprague [344] identified three levels of software technology: specific DSS, DSS generators, and DSS tools. A specific DSS is a tailor-made system to support specific DM dealing with specific set of problems. A DSS generator is a package of hardware and software that provides capabilities to easily and quickly build a specific DSS. A DSS tools are the elements which facilitate the development of specific DSS generator as well as a specific DSS. In this architect, specific DSS can be build either directly from DSS tools or from the DSS generator. The difference would be in the flexibility that is provided by the DSS generators. This architect is very convenient to be related to GIS. Based on earlier description of GIS, we realize that it can easily be considered as a DSS generator. Such type of architect is designed by [229] and presented in Fig. 14.5. Here SDSS is built through integration of models with GIS. Simpler integration of models is achieved by only allowing the modeling routines access to non-spatial data (full-line in Fig. 14.5). Full integration however requires those models to be able to make use of all the features of the GIS (dashed and full line). However, it should be remembered that, the needs are best handled not by integrating all forms of geographic analysis in one package but by providing appropriate linkages and hooks to allow software components to act in a federation [156]. The advantage of such approach is that different integration of functionality group can take place serving different user needs. Moreover the integration can be guided more by functional than technological considerations, as it takes place irrespective of hard/software used. To achieve this, greater degree of openness of spatial data formats is required (Open Geo-data Inter-operability Specification). In this connection one of the tasks is the identification of the generic GIS functionality/services [274]. Such approach becomes closer to the conceptual design of DSS given by [344]: specific DSS, the DSS generator and DSS tools.

Developments in the context of OGIS will complement the DSS toolbox with generic spatial data handling functionality and thus facilitate the development of SDSS.

Bibliography

[1] Ackoff, R. L. (1981). The art and science of mess management. *TIMS Interfaces* **11**, 20–26.
[2] Adams, J. B. (1974). Visible and near-infrared diffuse reflectance: Spectra of pyroxenes as applied to remote sensing of solid objects in the solar system. *Journal of Geophysical Research* **79**, 4,829–4,836.
[3] Adams, J. B. (1975). Interpretation of visible and near-infrared diffuse reflectance spectra of pyroxenes and other rock forming minerals. *In* Karr, C. (ed). em Infrared and Raman Spectroscopy of Lunar and Terrestrial Materials. Academic Press, New York, pp. 91–116.
[4] Adams, J. B., Smith, M. O. & Johnston, P. E. (1985). Spectral mixture modelling: a new analysis of rock and soil types at the Viking Lander 1 site. *Journal of Geophysical Research* **91**, 8,098–8,112.
[5] Addink, E. & Stein, A. (1999). A comparison of conventional and geostatistical methods to replace clouded pixels in NOAA-AVHRR images. *International Journal of Remote Sensing* **20**, 961-977.
[6] Adomeit, E. M., Jupp, D.L. B., Margules, C. I. & Mayo, K. K. (1981). The separation of traditionally mapped land cover classes by Landsat data. *In* Gillison, A. N. & Anderson, D. J. (eds). *Vegetation classification in Australia*, ANU Press, Canberra, pp. 150–165.
[7] Ahern, F. J., Brown, R. J., Goodenough, D. G. & Thomson, K.P. B. (1980). A simulation of Thematic Mapper performance in an agricultural application. *In: Proceedings of the 6th Canadian Symposium on Remote Sensing*. Halifax, Nova Scotia, pp. 585–596.
[8] Ahmed, S. & De Marsily, G. (1987). Comparison of geostatistical methods for estimating transmissivity using data on transmissivity and specific capacity. *Water Resources Research* **23**, 1717–1737.
[9] Alabert, F. G. (1987). The practice of fast conditional simulations through the LU decomposition of the covariance matrix. *Mathematical Geology* **19**, 369–386.
[10] Aldred, B. K. (1972). *Point-In-Polygon Algorithms.* IBM, Peterlee UK.
[11] Almeida, A.S. & Journel, A. G. (1994). Joint simulation of multiple variables with a Markov-type coregionalization model. *Mathematical Geology* **26**, 565–588.
[12] Anderson, J. R., Hardy, E. E., Roach, J. T. & Witmer, R. E. (1976). *A Land Use And Land Cover Classification System For Use With Remote Sensor Data.* U.S.G. S., Washington.

[13] Anselin, L. & Getis (1992). Spatial statistical analysis and geographic information systems. In Fisher, M. M., Nijkamp, P. (eds). *Geographic Information Systems, Spatial Modelling And Policy Evaluation*, Springer Verlag, Berlin.

[14] Aplin, P., Atkinson, P. M. & Curran, P. J. (1999). Fine spatial resolution satellite sensor imagery for land cover mapping in the UK. *Remote Sensing of Environment (in press)*.

[15] Arbia, G. (1989). *Spatial Data Configuration in Statistical Analysis of Regional Systems*. Kluwer, Dordrecht.

[16] Ardo, J. & Pilesjo, P. (1992). On the accuracy of the global positioning system - a test using a hand held receiver. *International Journal of Remote Sensing* **13**, 329–323.

[17] Atkinson, P. M. (1991). Optimal ground-based sampling for remote sensing investigations: estimating the regional mean. *International Journal of Remote Sensing* **12**, 559–567.

[18] Atkinson, P. M. (1993). The effect of spatial resolution on the experimental variogram of airborne MSS imagery. *International Journal of Remote Sensing* **14**, 1,005–1,011.

[19] Atkinson, P. M. (1997). Scale and spatial dependence. In Van Gardingen, P. R., Foody, G. M. & Curran, P. J. (eds). *Scaling-up: From Cell to Landscape*. Cambridge University Press, Cambridge, pp. 35–60.

[20] Atkinson, P. M. & Curran, P. J. (1995). Defining an optimal size of support for remote sensing investigations. *IEEE Transactions on Geoscience & Remote Sensing* **33**, 768–776.

[21] Atkinson, P. M. & Curran, P. J. (1997). Choosing an appropriate spatial resolution for remote sensing. *Photogrammetric Engineering & Remote Sensing* **63**, 1,345–1,351.

[22] Atkinson, P. M., Curran, P. J. & Webster, R. (1990). Sampling remotely sensed imagery for storage, retrieval and reconstruction. *Professional Geography* **37**, 345–353.

[23] Atkinson, P. M., Webster, R. & Curran, P. J. (1992). Cokriging with ground-based radiometry *Remote Sensing of Environment* **41**, 45–60.

[24] Atkinson, P. M., Webster, R. & Curran, P. J. (1994). Cokriging with airborne mss imagery. *Remote Sensing of Environment* **50**, 335–345.

[25] Atkinson, P. M., Cutler, M.E. J. & Lewis, H. (1997). Mapping sub-pixel proportional land cover with AVHRR imagery. *International Journal of Remote Sensing* **18**, 917–935.

[26] Atteia, O., Dubois, J. P & Webster, R. (1994). Geostatistical analysis of soil contamination in the Swiss Jura. *Environmental Pollution* **86**, 315–327.

[27] August, P., Michaud, J., LaBash, C. & Smith, C. (1994). GPS for environmental applications: accuracy and precision of locational data. *Photogrammetric Engineering & Remote Sensing* **60**, 41–45.

[28] Bailey T. C. & Gatrell A. C. (1995). *Interactive Spatial Data Analysis* Longman, Harlow.

[29] Baltsavias, E. P. (1996). Digital ortho-images - a powerful tool for the ex-

traction of spatial- and geo-information. *ISPRS Journal of Photogrammetry & Remote Sensing* bf 51, 63–77.

[30] Band, L. E. (1993). Effect of land-surface representation on forest water and carbon budgets. *Journal of Hydrology* **150**, 749–772.

[31] Band, L. E. & Moore, I. D. (1995). Scale - Landscape attributes and Geographical Information Systems. *Hydrological Processes* **9**, 401–422.

[32] Bastin, L. (1997). Comparison of fuzzy c-means classification, linear mixture modelling and MLC probabilities as tools for unmixing coarse pixels. *International Journal of Remote Sensing* bf 18, 3,629–3,648.

[33] Baumgardner, M. F., Stoner, E. R., Silva, L. F. & Biehl, L. L. (1985). Reflectance properties of soils. *In* Brady, N. (ed). *Advances of Agronomy*. Academic Press, New York, pp 1–44.

[34] Bell, A. (1986). Satellite mapping of the Great Barrier Reef. *Ecos*, 12–15.

[35] Belward, A. S. & Lambein, E. (1990). Limitations to the identification of spatial structures from AVHRR data. *International Journal of Remote Sensing* **11**, 921–927.

[36] Bentley, J. (1975). Multidimensional binary search trees used for associative searching. *Communications of the ACM* **18**, 509–517.

[37] Berry, B.J. L. & Baker, A. M. (1968). Geographic sampling *In* Berry, B.J. L. & Marble, D. F., eds. *Spatial Analysis: A Reader In Statistical Geography*. Prentice-Hall, New Jersey, pp. 91–109.

[38] Berry, J. K. (1993). Cartographic modeling: The analytical capabilities of GIS. *In* Goodchild, M. F., Parks, B. O. & Steyaert, L. T. (eds). *Integrating GIS and Environmental Modeling*. Oxford University Press, London, pp. 58-74.

[39] Bezdek, J. C. (1981). *Pattern Recognition with Fuzzy Objective Function Algorithms*. Plenum Press, New York.

[40] Bezdek, J. C., Ehrlich, R. & Full, W. (1984). FCM: The fuzzy c-means clustering algorithm. *Computers & Geosciences* **10**, 191–203.

[41] Bhatti, A. U., Mulla, D. J. & Frazier, B. E. (1991). Estimation of soil properties and wheat yields on complex eroded hills using geostatistics and thematic mapper images. *Remote Sensing of Environment* **37**, 181–191.

[42] Bian, L. & Walsh, S. J. (1993). Scale dependencies of vegetation and topography in a Mountainous environment in Montana. *The Professional Geographer* **45**, 1–11.

[43] Bishop, Y. M., Fienburg, S. E. & Holland, P. W. (1975). *Discrete Multivariate Analysis: Theory and Practice*. MIT Press, Cambridge Massachusetts.

[44] Boardman, J. W. (1989). Inversion of imaging spectrometry data using singular value decomposition. *In: Proceedings 12th Canadian Symposium on Remote Sensing IGARSS'89, Vol. 4.* pp. 2,069–2,072.

[45] Boardman, J. W., Kruse, F. A. & Green, R. O. (1995). Mapping target signatures via partial unmixing of AVIRIS data. *In: Proceedings of the Fifth JPL Airborne Earth Science Workshop*, JPL Publication 95-1, Vol.1. pp. 23–26.

[46] Bolstad, P. V., Gessler, P., Lillesand, T.M. (1990). Positional uncertainty in manually digitized map data. *International Journal of Geographical Information Systems* **4**, 399–412.

[47] Bosman A. (1986). Relations between specific decision support sytems. *In* E. R. McLean, H. G. Sol (eds). *Decision support systems: A decade in perspective*, Elsevier Amsterdam, pp. 69–85.

[48] Breidt, F. J. (1995). Markov chain designs for one-per-stratum spatial sampling. *Survey Methodology* **21**, 63–70.

[49] Brooker, P. (1985). Two dimensional simulation by turning bands. *Mathematical Geology* **17**, 81–90.

[50] Brown, L. (ed). (1993). *The New Shorter Oxford English Dictionary*. Clarendon Press, Oxford.

[51] Bruniquel-Pinel, V. & Gastellu-Etchegorry, J. P. (1998). Sensitivity of texture of high resolution images of forest to biophysical and acquisition parameters. *Remote Sensing of Environment* **65**, 61–85.

[52] Brus, D. J. & De Gruijter, J. J. (1997). Random sampling or geostatistical modelling? Choosing between design-based and model-based sampling strategies for soil (with Discussion). *Geoderma* **80**, 1–59.

[53] Buiten, H. J. (1993). Geometrical and mapping aspects of remote sensing. *In* Buiten, H. J., Clevers, J.G.P. W. (eds). *Land Observation by Remote Sensing: Theory and Applications*. Gordon and Breach Science Publishers, Yverdon, pp. 297–321.

[54] Burgess, T. M. & Webster, R. (1980). Optimal interpolation and isarithmic mapping of soil properties I. The semi-variogram and punctual kriging. *Journal of Soil Science* **31**, 315–331.

[55] Burgess, T. M., Webster, R. & McBratney, A. B. (1981). Optimal interpolation and isarithmic mapping of soil properties IV. Sampling strategy. *Journal of Soil Science* **32**, 643–659.

[56] Burns, R. G. (1970). *Mineralogical Application to Crystal Field Theory*. Cambridge University Press, Cambridge.

[57] Burrough, P. A. & Frank, A. U. (1995). Concepts and paradigms in spatial information: are current geographical information systems truly generic? *International Journal of Geographical Information Systems* **9**, 101–116.

[58] Burrough, P. A. (1996). Natural objects with indeterminate boundaries. *In* Burrough, P. A., Frank, A. U. (eds). *Geographic Objects with Indeterminate Boundaries*. Taylor & Francis, London, pp. 3–28.

[59] Burrough, P. A., Van Gaans, P.F. M. & Hootsmans, R. (1997). Continuous classification in soil survey: spatial correlation, confusion and boundaries. *Geoderma* **77**, 115–135.

[60] Burrough, P. A. & McDonnell, R. A. (1998). *Principles of Geographical Information Systems*. Oxford University Press, Oxford.

[61] Campbell, J. B. (1996). *Introduction to Remote Sensing*. Taylor & Francis, London.

[62] Chankong V. & Haimes, Y. Y. (1983). *Multiobjective Decision Making: Theory and Methodology*. North Holand, Amsterdam.

[63] Chrisman, N. R. (1984). The role of quality information in the long-term functioning of a geographic information system. *Cartographica* **21**, 79–87.

[64] Christakos, G. (1992). *Random Field Modeling in Earth Sciences*. Academic Press, New York.
[65] Clark, C. D. (1990). Remote sensing scales related to the frequency of natural variation: an example from paleo-ice-flow in Canada. *IEEE Transactions on Geoscience & Remote Sensing* **28**, 503–515.
[66] Clark, I. (1977). Regularization of a semi-variogram. *Computers & Geosciences* **3**, 341–346.
[67] Clark, R. N., King, T.V. V., Kleijwa, M., Swayze, G. A. & Vergo, N. (1990). High spectral resolution reflectance spectroscopy of minerals. *Journal of Geophysical Research* **95**, 12,653–12,680.
[68] Clark, W.A. V. & Avery, K. L. (1976). The effects of data aggregation in statistical analysis. *Geographical Analysis* **8**, 428–438.
[69] Cochran, W. G. (1977). *Sampling Techniques* Wiley, New York.
[70] Cohen, J. (1960). A coefficient of agreement for nominal scales. *Educational & Psychological Measurement* **20**, 37–46.
[71] Cohon, J. L. (1978). *Multiobjective Programming and Planning*. Academic Press, New York.
[72] Colby, J. D. (1991). Topographic normalization in rugged terrain. *Photogrammetric Engineering & Remote Sensing* **57**, 531–537.
[73] Colson G. & De Bruyn C. (1989). Models and methods in multiple objectives decision making. *Mathematical Computer Modelling* **12**, 1,201–1,211.
[74] Colwell, R. N. (1983). *Manual of Remote Sensing*. ASPRS, Falls Church Virginia.
[75] Condit, H. R. (1970). The spectral reflectance of American soils. *Photogrammetric Engineering & Remote Sensing* **36**, 955–966.
[76] Congalton, R. G. & Mead, R. A. (1983). A quantitative method to test for consistency and correctness in photointerpretation. *Photogrammetric Engineering and Remote Sensing* **49**, 69–74.
[77] Congalton, R. G., Oderwald, R. G. & Mead, R.A. (1983). Assessing Landsat classification accuracy using discrete multivariate analysis statistical techniques. *Photogrammetric Engineering and Remote Sensing* **49**, 1671–1678.
[78] Congalton, R. G. (1988). A comparison of sampling schemes used in generating error matrices for assessing the accuracy of maps generated from remotely sensed data. *Photogrammetric Engineering & Remote Sensing* **54**, 593–600.
[79] Congalton, R. G. (1991). A review of assessing the accuracy of classifications of remotely sensed data. *Remote Sensing of Environment* **37**, 35–46.
[80] Coughlan, J. C. & Dungan, J. L. (1996). Combining remote sensing and forest ecosystem modeling: An example using the Regional HydroEcological Simulation System (RHESSys). In Gholz, H. L., Nakane, K. & Shinoda, H. (eds). *The Use of Remote Sensing in the Modeling of Forest Productivity at Scales from the Stand to the Globe*. Kluwer Academic Publishers, Dordrecht, pp. 139-158.
[81] Cowlard K. A. (1998). *Decision-Making in Geography: a Manual of Method and Practice*. Hodder & Stoughton, London.
[82] Cressie, N.A. C. (1985). Fitting variogram models by weighted least squares. *Mathematical Geology* **17**, 563–586.

[83] Cressie, N. A. C. (1991). *Statistics for Spatial Data*. Wiley, New York.
[84] Crowley, J. K. (1991). Visible and near-infrared (0.4-2.5 μm) Reflectance spectra of playa evaporite minerals. *Journal of Geophysical Research* **96**, 16,231–16,240.
[85] Csillag, F. (1987). A cartographer's approach to quantitative mapping of spatial variability from ground sampling to remote sensing of soils. *In: Proceedings, AUTO-CARTO 8*. ACSM/ASPRS, Falls Church, Virginia, pp. 155–164.
[86] Curran, P. J. & Williamson, H. D. (1986). Sample size for ground and remotely sensed data. *Remote Sensing of Environment* **20**, 31–41.
[87] Curran, P. J., Williamson, H. D. (1987). Airborne MSS data to estimate GLAI. *International Journal of Remote Sensing* **8**, 57–74.
[88] Curran, P. J. (1988). The semi-variogram in remote sensing: an introduction. *Remote Sensing of Environment* **3**, 493–507.
[89] Curran, P. J. & J. L. Dungan (1989). Estimation of signal to noise: a new procedure applied to AVIRIS data. *IEEE Transactions on Geoscience & Remote Sensing* **27**, 620–628.
[90] Curran, P. J. & Guyot, G. (1997). Applications/optical domain. *In* Guyot, G., Phulpin, T. (eds). *Physical Measurements and Signatures in Remote Sensing*. Balkema, Rotterdam, pp. 893–894.
[91] Curran, P. J., Foody, G. M. & Van Gardingen, P. R. (1997). Scaling-up. *In* Van Gardingen, P. R., Foody, G. M. & Curran, P. J. (eds) *Scaling-up: From Cell to Landscape*. Cambridge University Press, Cambridge, pp. 1–5.
[92] Curran, P. J. & Atkinson, P. M. (1998). Geostatistics and remote sensing. *Progress in Physical Geography* **22**, 61–78.
[93] Curran, P. J., Milton, E. J., Atkinson, P. M. & Foody, G. M. (1998). Remote sensing: From data to understanding. *In* Longley, P. A., Brooks, S. M., McDonnell, R. & Macmillan, B. (eds). *Geocomputation: A Primer*. Wiley & Sons, London, pp. 33–59.
[94] Curry, L. (1966). A note on spatial association. *The Professional Geographer* **18**, 97–99.
[95] Dang, Q. L., Margolis, H. A., Sy, M., Coyea, M. R., Collatz, G. J. & Walthall, C. L. (1997). Profiles of photosynthetically active radiation, nitrogen and photosynthetic capacity in the boreal forest: Implications for scaling from leaf to canopy. *Journal of Geophysical Research* **102**, 28,845–28,859.
[96] Davis, G. B., Olson, M. H. (1985). *Management Information Systems, Conceptual Foundations, Structures and Development*, 2nd edition. McGraw-hill, New York.
[97] Davis, M. M. (1987). Production of conditional simulations via the LU triangular decomposition of the covariance matrix. *Mathematical Geology* **19**, 91–98.
[98] De Bruin, S. & Gorte, B.G. H., submitted. Probabilistic image classification using geological map delineations applied to land cover change detection. *Geoderma*.
[99] De Bruin, S. & Stein, A. (1998). Soil-landscape modelling using fuzzy c-means clustering of attribute data derived from a Digital Elevation Model (DEM). *Geoderma* **83**, 17–33.

[100] De Gruijter, J. J. & Marsman, B. A. (1985). Transect sampling for reliable information on mapping units. In Nielsen, D. R., Bouma, J. (eds). *Proceedings of the SSSA-ISSS Workshop on Spatial Variability, Las Vegas, U.S.A., Nov. 30 - Dec. 1, (1984)*. Pudoc, Wageningen, pp. 150-165.

[101] De Gruijter, J. J. & Ter Braak, C.J. F. (1990). Model-free estimation from spatial samples: a reappraisal of clas-sical sampling theory. *Mathematical Geology* **22**, 407–415.

[102] De Jong, S. M. & Burrough, P. A. (1995). A fractal approach to the classification of Mediterranean vegetation types in remotely sensed data. *Photogrammetric Engineering & Remote Sensing* **61**, 1,041–1,053.

[103] Desjardins, R. L., MacPherson, J. I., Mahrt, L., Schwepp, P., Pattey, E., Neumann, H., Baldocchi, D., Wofsey, S., Fitzjarrald, D., McCaughey, H. & Joiner, D. W. (1997). Scaling up flux measurements for the boreal forest using aircraft-tower combinations. *Journal of Geophysical Research* **102**, 29,125–29,133.

[104] De Vries, P. G. (1986). *Sampling theory for forest inventory*. Springer, Berlin.

[105] Deutsch, C. V. & Cockerham, P. W. (1994). Practical considerations in the application of simulated annealing to stochastic simulation. *Mathematical Geology* **26**, 67–82.

[106] Deutsch, C. V. & Journel, A. G. (1998). *GSLIB Geostatistical Software Library and User's Guide*. Oxford University Press, New York.

[107] Dewey, J. (1910). *How we think*. D.C. Heath, New York.

[108] Diggle, P. J., Tawn, J. A. and Moyeed, R. A. (1998). Model-based geostatistics. *Applied Statistics* **47**, 299–350.

[109] Domburg, P., De Gruijter, J. J. & Brus, D. J. (1994). A struc-tured approach to designing soil survey schemes with predic-tion of sampling error from variograms. *Geoderma* **62**, 151–164.

[110] Domburg, P., De Gruijter, J. J. & Van Beek, P. (1997). Designing efficient soil survey schemes with a knowledge-based system using dynamic programming. *Geoderma* **75**, 183–201.

[111] Douven, W.J.A. M. (1997). *Improving the accessibility of spatial information for environmental management*. PhD thesis, Free University, Amsterdam.

[112] Dowd, P. A. (1992). A review of recent developments in geostatistics. *Computers and Geosciences* **17**, 1481–1500.

[113] Duda, R. O. & Hart, P. E. (1973). *Pattern Classification and Scene Analysis*. John Wiley & Sons, New York.

[114] Dungan, J. L., Peterson, D. L. & Curran, P. J. (1994). Alternative approaches for mapping vegetation quantities using ground and image data. In W. K. Michener, J. W. Brunt & S. G. Staffort (eds). *Environmental Information Management and Analysis: Ecosystems to Global Scales*. Taylor & Francis, London, pp..

[115] Dungan, J. L. (1998). Spatial prediction of vegetation quantities using ground and image data. *International Journal of Remote Sensing* **19**, 267–285.

[116] Dungan, J. L. & Coughlan, J. C. (1999). Quantifying uncertainty in regional soil maps using ecological theory and remotely sensed vegetation. *The Professional Geographer (submitted)*.

[117] Eastman R, Kyem, A. Toledano & Jin Weigen (1993). *GIS and Decision Making, Exploration in Geographic Information Technology, Volume 4*. United Nations Institute for Training and Research, New York.

[118] Edwards, A.W. F. (1972). *Likelihood : an account of the statistical concept of likelihood and its application to scientific inference.* Cambridge University Press, Cambridge.

[119] Efron, B. & Tibshirani, R. (1993). An Introduction to the Bootstrap. Chapman and Hall, New York.

[120] Ehlers, M. & Welch, R. (1987). Stereocorrelation of Landsat TM images. *Photogrammetric Engineering & Remote Sensing* **53**, 1,231 –1,237.

[121] Ehlers, M. (1997). Rectification and registration. *In* Star, J. L., Estes, J. E. & McGwire, K. C. (eds). *Integration of Geographic Information Systems & Remote Sensing.* Cambridge University Press, Cambridge, pp.13–36.

[122] Elvidge, C. D. (1990). Visible and near infrared reflectance characteristics of dry plant material. *International Journal of Remote Sensing* **11**, 1,775–1,795.

[123] Englund, E. J., Weber, D. D. & Leviant, N. (1992). The effects of sampling design parameters on block selection. *Mathematical Geology* **24**, 329–343.

[124] Epema, G. F. (1992). Atmospheric conditions and its influence on reflectance of bare soil surfaces in southern Tunesia. *International Journal of Remote Sensing* **13**, 853–868.

[125] ESRI (1994). *Cell-based Modeling with GRID.* Environmental Systems Research Institute, Redlands, CA.

[126] Famiglietti, J. S., Wood, E. F., Sivaplan. M. & Thongs, D. J. (1992). A catchment scale water balance model for FIFE. *Journal of Geophysical Research* **97**, 18,997–19,007.

[127] Fedra, K. (1993). GIS and environmental modelling. *In* Goodchild, M. F., Park B. O. & Steyaert, L. T. (eds). *Environmental Modelling with GIS.* Oxford University Press, New York, pp. 35–50.

[128] Fisher, P. F. (1994). Visualization of the reliability in classified remotely sensed images. *Photogrammetric Engineering and Remote Sensing* **60**, 905–910.

[129] Fisher, P. F. (1997). The pixel: a snare and a delusion. *International Journal of Remote Sensing* **18**, 679–685.

[130] Fix, E. & Hodges, J. (1951). *Nonparametric Discrimination: Consistency Properties.* Technical Report Report Number 4, USAF School of Aviation Medicine, Randolph Field, Texas.

[131] Foody, G. M. (1992). A fuzzy sets approach to the representation of vegetation continua from remotely sensed data: an example from lowland heath. *Photogrammetric Engineering & Remote Sensing* **58**, 221–225.

[132] Foody, G. M., Campbell, N. A., Trodd, N. M. & Wood, T. F. (1992). Derivation and applications of probabilistic measures of class membership from the maximum-likelihood classification. *Photogrammetric Engineering & Remote Sensing* **58**, 1,335–1,341.

[133] Foody, G. M. & Curran, P. J. (1994). Scale and environmental remote sensing. *In* Foody, G. M. & Curran, P. J. (eds) *Environmental Remote Sensing from Regional to Global Scales.* Wiley, Chichester, pp. 223–232.

[134] Foody, G. M. & Arora, M. K. (1996). Incorporating mixed pixels in the training, allocation and testing of supervised classifications. *Pattern Recognition Letters* **17**, 1,389–1,398.
[135] Foody, G. M., Lucas, R. M., Curran, P. J. & Honzak, M. (1997). Non-linear mixture modelling without end-members using an artificial neural network. *International Journal of Remote Sensing* **18**, 937–953.
[136] Foody, G. M., Lucas, R. M., Curran, P. J. & Honzák, M. (1996). Estimation of the areal extent of land cover classes that only occur at a sub-pixel level. *Canadian Journal of Remote Sensing* **22**, 428–432.
[137] Ford, G. E. & Zanelli, C. I. (1985). Analysis and quantification of errors in the geometric correction of satellite images. *Photogrammetric Engineering & Remote Sensing* **51**, 1,725–1,734.
[138] Fotheringham, A. S. (1989). Scale-independent spatial analysis. *In* Goodchild, M. F. & Gopal, S. (eds). Accuracy of Spatial Databases, Taylor and Francis, London, pp. 221–228.
[139] Fotheringham, A. S. & Wong, D. (1991). The modifiable areal unit problem in multivariate statistical analysis. *Environment & Planning A* **23**, 1,025–1,044.
[140] Fotheringham, A. S. (1992). Exploratory spatial data analysis and GIS. *Environment & Planning A* **24**, 1675–1678.
[141] Fotheringham, A. S., Charlton, M. E. & Brunsdon, C. (1995). The zone definition problem and location-allocation modelling. *Geographical Analysis* **27**, 60–77.
[142] Fotheringham, A. S. (1998). Trends in quantitative methods II: Stressing the computational. *Progress in Human Geography* **22**, 283–292.
[143] Friedman, J., Baskett, F. & Shustuk, L. (1975). An algorithm for finding nearest neighbors. *IEEE Transactions on Computers*, C-24: 1,000–1,006.
[144] Friedman, J., Bentley, J. & Finkel, R. (1977). An algorithm for finding best matches in logarithmic expected time. *ACM Transactions on Mathematical Software*, 3:209–226.
[145] Froidevaux, R. (1993). Probability field simulation. *In* Soares, A. (ed). Geostatistics Troia '92. Kluwer Academic Publishers, Dordrecht, pp. 73–83.
[146] Gahegan, M. N. (1994). A consistent user-model for a GIS incorporating remotely sensed data. *In* Worboys, M. F. (ed). *Innovations in GIS, Selected Papers from the First National Conference on GIS Research UK*. Taylor & Francis, London, pp. 65–74.
[147] Gates, D. M., Keegan, H. J., Schleter, J. C. & Weidner, V. R. (1965). Spectral properties of plants. *Applied Optics* **4**, 11–20.
[148] Gatrel, A. C. (1985). Any space for spatial analysis? *In* Johnston, R. J. (ed). The Future of Geography. Methuen, London, pp. 190–208.
[149] Gehlke, C. E. & Biehl, K. (1934). Certain effects of grouping upon the size of the correlation coefficient in census tract material. *Journal of the American Statistical Association Supplement* **29**, 169–170.
[150] Ginevan, M. E. (1979). Testing land-use map accuracy - another look: *Photogrammetric Engineering & Remote Sensing* **45**, 1,371–1,377.
[151] Gleick, J. (1987). *Chaos: Making a New Science*. Viking Penguin, New York.
[152] Gobron, N., Pinty, B., Verstratete, M. & Govaerts, Y. (1999). The MERIS

Global Vegetation Index (MGVI): Description and preliminary application. *International Journal of Remote Sensing (in press)*.
[153] Goel, N. S. (1988). Models of vegetation canopy reflectance and their use in estimation of biophysical parameters from reflectance data. *Remote Sensing Reviews* **4**, 1–222.
[154] Goetz, A.F. H. (1991). Imaging Spectrometry for studying earth, air, fire and water. *EARSeL Advances in Remote Sensing* **1**, 3–15.
[155] Gohin, F. & Langlois, G. (1993). Using geostatistics to merge in situ measurements and remotely sensed observations of sea surface temperature. *International Journal of Remote Sensing* **14**, 9–19.
[156] Goodchild, M. F. (1993). The state of GIS for environmental problem-solving. *In* Goodchild, M. F., Park B. O. & Steyaert, L. T. (eds). *Environmental modelling with GIS*. Oxford University Press, New York, pp. 8–15.
[157] Goodchild, M. F., Chih-Chang, L. & Leang, Y. (1994). Visualizing fuzzy maps. *In* Hearnshaw, H. M. & Unwin, D. J. (eds). *Visualisation in geographical information systems*. Wiley, Chichester, pp. 158–167.
[158] Goovaerts, P. (1997). *Geostatistics for Natural Resources Evaluation*. Oxford University Press, New York.
[159] Goovaerts, P. (1998a). Accounting for estimation optimality criteria in simulated annealing. *Mathematical Geology* **30**, 511–534.
[160] Goovaerts, P. (1998b). *Impact of the simulation algorithm, magnitude of ergodic fluctuations and number of realizations on the spaces of uncertainty of flow predictions*. Unpublished annual report No. 11, Stanford Center for Reservoir Forecasting, Stanford University.
[161] Gorte, B.G. H. (1998). *Probabilistic Segmentation of Remote Sensing Images*. PhD thesis, Wageningen Agricultural University, Wageningen.
[162] Gorte, B.G. H. & Stein, A. (1998). Bayesian classification and class area estimation of satellite images using stratification. *IEEE Transactions on Geoscience & Remote Sensing*, 36, 803–812.
[163] Govaerts, Y. M. (1996). *A Model of Light Scattering in Three Dimensional Plant Canopies: A Monte Carlo Ray Tracing Approach*. Joint Research Centre, Ispra. ECSC-EC-EAEC, Brussels.
[164] Grace, J., Van Gardingen, P. R. & Luan, J. (1997). Tackling large-scale problems of scaling-up. *In* Van Gardingen, P. R., Foody, G. M. & Curran, P. J. (eds). *Scaling-up: From Cells to Landscape*. Cambridge University Press, Cambridge, pp. 7–16.
[165] Green, M. & Flowerdew, R. (1996). New evidence on the modifiable areal unit problem. *In* Longley, P. & Batty, M. (eds) *Spatial Analysis: Modelling in a GIS Environment*. Geoinformation International, Cambridge, pp. 41–54.
[166] Grove, C. I., Hook, S. J. & Paylor II, E. D. (1992). *Laboratory Reflectance Spectra of 160 Minerals, 0.4 to 2.5 Micrometers*. NASA-JPL Publication 92-2, Pasadena.
[167] Hall, D. & Ball, G. (1965). *ISODATA: a Novel Method of Data Analysis and Pattern classification*. Technical report, Stanford Research Institute, Menlo Park CA.

[168] Hall, D. & Khanna, D. (1977). The ISODATA method of computation for relative perception of similarities and differences in complex and real computers. In Enslein, K., Ralston, A., & Wilf, H., (eds). *Statistical methods for Digital Computer, vol. 3*. Wiley, New York, pp. 340–373.

[169] Handcock, M. S. & Stein, M. L. (1993). A Bayesian analysis of kriging. *Technometrics* **35**, 403–410.

[170] Hapke, B. (1981). Bidirectional reflectance spectroscopy, 1, theory. *Journal of Geophysical Research* **86**, 3,039–3,054.

[171] Haslett, J., Bradley, R., Craig, P. & Unwin, A. (1991). Dynamic graphics for exploring spatial data with application to locating global and local anomalies. *American Statistician* **45**, 234–242.

[172] Hay, A. M. (1979). Sampling design to test land use accuracy. *Photogrammetric Engineering & Remote Sensing* **45**, 529–533.

[173] Hazell, P. B. R. & Norton, R. (1986). *Mathematicla Programming for Economi Analysis in Agriculture*. Macmillan, New York.

[174] Heipke, C. (1995). State-of-the-art of digital photogrammetric workstations for topographic applications. *Photogrammetric Engineering & Remote Sensing* **61**, 49–56.

[175] Helterbrand, J. D. & Cressie, N. (1994). Universal cokriging under intrinsic coregionalization. *Mathematical Geology* **26**, 205–226.

[176] Hendricks-Franssen, H.J.W. M., Van Eijnsbergen, A. C. & Stein, A. (1997). Use of spatial prediction techniques and fuzzy classification for mapping of soil pollutants. *Geoderma* **77**, 243–262.

[177] Heuvelink, G.B. M. (1993). *Error Propagation in Quantitative Spatial Modelling: Applications in Geographical Information Systems*. Netherlands Geographical Studies No. 163. Koninklijk Nederlands Aardrijkskundig Genootschap, Utrecht.

[178] Heuvelink, G.B. M. (1998). *Error Propagation in Environmental Modelling with GIS*. Taylor & Francis, London.

[179] Heuvelink, G.B. M. & Burrough, P. A. (1993). Error propagation in cartographic modeling using Boolean logic and continuous classification. *International Journal of Geographical Information Systems* **7**, 231–246.

[180] Hinton, J. C. (1996). GIS and remote sensing integration for environmental applications. *International Journal of Geographical Information Systems* **10**, 877–890.

[181] Hoffer, R. M. & staff (1975). *Computer-aided analysis of SKYLAB multispectral scanner data in mountainous terrain for land use, forestry, water resource, and geological applications*. Laboratory for Applications of Remote Sensing, Purdue University, West Lafayette, Ind.

[182] Hogg, R. V. & Tanis, E. A. (1983). *Probability and Statistical Inference*. Macmillan Publishing, New York.

[183] Holt, D., Steel, D. G. & Tranmer, M. (1996). Area homogeneity and the modifiable areal unit problem. *Geographical Systems* **3**, 181–200.

[184] Hudson, W. D. (1987). Evaluation of several classification schemes for map-

ping forest cover types in Michigan. *International Journal of Remote Sensing* **8**, 1785-1796.

[185] Hunt, G. R. & Salisbury, J. W. (1970). Visible and near-infrared spectra of minerals and rocks, I. Silicate Minerals. *Modern Geology* **1**, 283-300.

[186] Hunt, G. R. & Salisbury, J. W. (1971). Visible and near-infrared spectra of minerals and rocks, II. Carbonates. *Modern Geology* **2**, 23-30.

[187] Hunt, G. R. & Salisbury, J. W. (1976a). Visible and near-infrared spectra of minerals and rocks: XI. Sedimentary rocks. *Modern Geology* **5**, 211-217.

[188] Hunt, G. R. & Salisbury, J. W. (1976b). Visible and near-infrared spectra of minerals and rocks: XII. Metamorphic rocks. *Modern Geology* **5**, 219-228.

[189] Hunt, G. R. (1977). Spectral signatures of particulate minerals in the visible and near-infrared. *Geophysics* **42**, 501-513.

[190] Hunt, G. R., Salisbury, J. W. & Lenhoff, C. J. (1971a). Visible and near-infrared spectra of minerals and rocks, III. Oxides and hydroxides. *Modern Geology* **2**, 195-205.

[191] Hunt, G. R., Salisbury, J. W. & Lenhoff, C. J. (1971b). Visible and near-infrared spectra of minerals and rocks, IV. Sulphides and sulphates. *Modern Geology* **3**, 1-14.

[192] Hunt, G. R., Salisbury, J. W. & Lenhoff, C. J. (1972). Visible and near-infrared spectra of minerals and rocks, V. Halides, arsenates, vanadates, and borates. *Modern Geology* **3**, 121-132.

[193] Hunt, G. R., Salisbury, J. W. & Lenhoff, C. J. (1973a). Visible and near-infrared spectra of minerals and rocks, VI. Additional Silicates. *Modern Geology* **4**, 85-106.

[194] Hunt, G. R., Salisbury, J. W. & Lenhoff, C. J. (1973b). Visible and near-infrared spectra of minerals and rocks, VII. Acidic igneous Rocks. *Modern Geology* **4**, 217-224.

[195] Hunt, G. R., Salisbury, J. W. & Lenhoff, C. J. (1973c). Visible and near-infrared spectra of minerals and rocks, VIII. Intermediate igneous Rocks. *Modern Geology* **4**, 237-244.

[196] Hunt, G. R., Salisbury, J. W. & Lenhoff, C. J. (1974). Visible and near-infrared spectra of minerals and rocks, IX. Basic and Ultrabasic igneous rocks. *Modern Geology* **5**, 15-22.

[197] Hunt, G. R., Salisbury, J. W. & Lenhoff, C. J. (1975). Visible and near-infrared spectra of minerals and rocks, X. Stoney Meteorites. *Modern Geology* **5**, 115-126.

[198] Hurbich, C. M. & Tsai, C. (1989). Regression and time series model selection in small samples. *Biometrika* **76**, 97-107.

[199] Hutchinson, C. F. (1982). Techniques for combining Landsat and ancillary data for digital classification improvement. *Photogrammetric Engineering & Remote Sensing* **48**, 123-130.

[200] Ichoku, C. & Karnieli, A. (1996). A review of mixture modelling techniques for sub-pixel land cover estimation. *Remote Sensing Reviews* **13**, 161-186.

[201] Irons, J. R., Weismiller, R. A. & Petersen, G. W. (1989). Soil reflectance. *In*

G. Asrar (ed). *Theory & Applications of Optical Remote Sensing*. Wiley, New York, pp 66–106.
[202] Isaaks, E. H. & Srivastava, R. M. (1989). *Applied Geostatistics*. Oxford University Press, Oxford.
[203] Isaaks, E. H. (1990). *The application of Monte Carlo methods to the analysis of spatially correlated data*. PhD Thesis, Stanford University, Stanford, CA.
[204] Iverson, L. R., Cook, E. A. & Graham, R. L. (1994). Regional forest cover estimation via remote sensing - the calibration center concept. *Landscape Ecology* **9**, 159–174.
[205] Jankowski, P. (1995). Integrating Geographic Information Systems and multiple-criteria decision-making methods. *International Journal of Geographic Information Systems* **9**, 251–273.
[206] Janssen, R. & Van Herwijnen, H. (1994). *DEFINTE: a system to support decisions on a finite set of alternatives, user manual*. Kluwer Academic Publishers, Dordrecht.
[207] Janssen, J. (1993). *Multiobjective Decision Support for Environmental Management*. Kluwer Academic Publisher, Dordrecht.
[208] Janssen, L.L. F. & Middelkoop, H. (1992). Knowledge-based crop classification of a Landsat Thematic Mapper image. *International Journal of Remote Sensing* **13**, 2,827–2,837.
[209] Janssen, L.L. F., Jaarsma, M. N. & Van der Linden, E.T. M. (1990). Integrating topographic data with remote sensing for land-cover classification. *Photogrammetric Engineering & Remote Sensing* **56**, 1,503–1,506.
[210] Janssen, L.L. F., Schoenmakers, R.P.H. M. & Verwaal, R. G. (1992). Integrated segmentation and classification of high resolution satellite images. *In* Molenaar, M., Jansen, L. & Van Leeuwen H. (eds). *Multisource Data Integration in Remote Sensing for Land Inventory Applications: Proceedings of the International Workshop IAPR TC7, September 7 - 9, (1992), Delft, The Netherlands*. Wageningen Agricultural University, Wageningen, pp. 65–84.
[211] Jensen, J. R., Cowen, D., Narumalani, S. & Halls, J. (1997). Principles of change detection using digital remote sensor data. *In* Star, J. L., Estes, J. E., McGwire, K. C. (eds). *Integration of Geographic Information Systems & Remote Sensing*. Cambridge University Press, Cambridge, pp. 37–54.
[212] Johnson, P. E., Smith, M. O., Taylor-George, S. & Adams, J. B. (1983). A semiempirical method for analysis of the reflectance spectra of binary mineral mixtures. *Journal of Geophysical Research* **88**, 3,557–3,561.
[213] Jolliffe, I. (1986). *Principal component analysis*. Springer, New York.
[214] Jones, A. R., Settle, J. J. & Wyatt, B. K. (1988). Use of digital terrain data in the interpretation of SPOT-1 HRV multispectral imagery. *International Journal of Remote Sensing* **9**, 669–682.
[215] Journel, A. G. (1974). Geostatistics for conditional simulation of ore bodies. *Economic Geology* **69**, 527–545.
[216] Journel, A. G. & Huijbregts, C. J. (1978). *Mining Geostatistics*. Academic Press, London.
[217] Journel, A. G. & Isaaks, E. H. (1984). Conditional indicator simulation:

Application to a Saskatchewan uranium deposit. *Mathematical Geology* **16**, 685–718.

[218] Journel, A. G. & Alabert, F. (1988). Focussing on spatial connectivity of extreme valued attributes: Stochastic indicator models of reservoir heterogeneities. SPE paper 18324.

[219] Journel, A. G. & Alabert, F. (1989a). Non-Gaussian data expansion in the earth sciences. *Terra Nova* **1**, 123–134.

[220] Journel, A. G. & Rossi, M. E. (1989b). When do we need a trend model in kriging? *Mathematical Geology* **21**, 715–738.

[221] Journel, A. G. (1990). *Fundamentals of Geostatistics in Five Lessons.* American Geophysical Union, Washington, D.C.

[222] Journel, A. G. & Deutsch, C. V. (1993). Entropy and spatial disorder. *Mathematical Geology* **25**, 329–355.

[223] Journel, A. G. (1996a). Modelling uncertainty and spatial dependence: Stochastic imaging. *International Journal of Geographical Information Systems* **10**, 517–522.

[224] Journel, A. G. (1996b). Conditional simulation of geologically averaged block permeabilities. *Journal of Hydrology* **183**, 23–25.

[225] Jupp, D.L. B., Strahler, A. H. & Woodcock, C. E. (1988). Autocorrelation and regularization in digital images I. Basic theory. *IEEE Transactions on Geoscience & Remote Sensing* **26**, 463–473.

[226] Jupp, D.L. B., Strahler, A. H. & Woodcock, C. E. (1989). Autocorrelation and regularization in digital images II. Simple image models. *IEEE Transactions on Geoscience & Remote Sensing* **27**, 247–258.

[227] Kalensky, Z. D. & Scherk, L. R. (1975). Accuracy of forest mapping from Landsat CCTs. In: *Proceedings of the 10th International Symposium on Remote Sensing of Environment, Volume 2*. ERIM, Ann Arbor, Mich., pp. 1,159–1,163.

[228] Kaufman, L. & Rousseeuw, P. (1990). *Finding Groups in Data, an Introduction to Cluster Analysis.* John Wiley & Sons, New York.

[229] Keenan, P. (1995). *Using a GIS as a DSS generator.* Working Paper MIS 95-9, Department of Management Information Systems, Graduate School of Business, University College Dublin.

[230] Van Kemenade, C., Poutré, J. L., & Mokken, R. (1998). *Density-based Unsupervised Classification for Remote Sensing.* Technical Report SEN-R9810, CWI, Amsterdam. Also availabe via http://www.cwi.nl/~hlp/PAPERS/RS/dense98.ps.

[231] Van Kemenade, C., Poutré, J. L., & Mokken, R. (1999). Density-based unsupervised classification for remote sensing (extended abstract). In Kanellopoulos, Wilkinson, & Moons, editors, *Proceedings of MAchine Vision In Remotely sensed Image Comprehension (MAVIRIC)*. Springer. Extended abstract of CWI report SEN-R9810.

[232] Kitanidis, P. K. (1983). Statistical estimation of polynomial generalized covariance functions and hydrologic applications. *Water Resources Research* **19**, 909–921.

[233] Kitanidis, P. K. (1997). *Introduction to Geostatistics: Applications to Hydrogeology.* Cambridge University Press, Cambridge.

[234] Klinkenberg, B. & Goodchild, M. F. (1992). The fractal properties of topography: a comparison of methods. *Earth Surface Processes and Landforms* **17**, 217–234.

[235] Krishnaiah, P. R. & Rao, C. R. (eds). (1988). *Sampling. Handbook of Statistics, Vol. 6.* North-Holland, Amsterdam.

[236] Künsch, H., Papritz, A. & Bassi, F. (1997). Generalized cross-covariances and their estimation. *Mathematical Geology* **29**, 779–799.

[237] Kullback, S. (1954). *Information and Statistics.* Wiley, New York.

[238] Kupfersberger, H. & Bloschl, G. (1995). Estimating aquifer transmissivities – on the value of auxiliary data. *Journal of Hydrology* **165**, 95–99.

[239] Lam, N. S. (1983). Spatial interpolation methods: a review. *The American Cartographer* **10**, 129–149.

[240] Laslett, G. M., McBratney, A. B., Pahl, P. J. & Hutchinson, M. F. (1987). Comparison of several spatial prediction methods for soil pH. *Journal of Soil Science* **38**, 325–341.

[241] Lathrop, R. G. & Lillesand, T. M. (1986). Use of Thematic Mapper data to assess water quality in Green Bay and Central Lake Michigan. *Photogrammetric Engineering & Remote Sensing* **52**, 671–680.

[242] Lathrop, R. G., Aber, J. D., Bognar, J. A., Ollinger, S. V., Casset, S. & Ellis, J. M. (1994). GIS development to support regional simulation modeling of north-eastern (USA) forest ecosystems *In* Michener, W., Brunt, J. & Stafford, S. (eds). *Environmental Information Management and Analysis: Ecosystem to Global Scales.* Taylor & Francis, London, pp. 431–453.

[243] Laurini, R. & D. Thomas (1992). *Fundamentals of Spatial Information Systems.* London, Academic Press.

[244] Lillesand, T. M. & R. W. Kiefer (1994). *Remote Sensing and Image Interpretation, third edition.* John Wiley & Sons, Chichester.

[245] Lira, J., Marzolf, G.R., Marrocchi, A. & Naugle, B. (1992). A probabilistic model to study spatial variations of primary productivity in river impoundments. *Ecological Applications* **2**, 86–94.

[246] Lovejoy, S. & Schertzer, C. (1995). How bright is the coast of Brittany? *In* Wilkinson, G. G., Kanellopoulos, I. & Mégier, J. (eds). *Fractals in Geoscience & Remote Sensing.* Joint Research Centre, ECSC-EC-ESAC, Brussels, pp. 102–151.

[247] MacEachren, A. M. (1985). Accuracy of thematic maps - implications of choropleth symbolization. *Cartographica* **22**, 38–58.

[248] Maling, D.H. (1989). *Measurement from Maps: Principles and Methods of Cartometry.* Pergamon Press, New York.

[249] Mandelbrot, B. (1967). How long is the coastline of Britain? Statistical self-similarity and fractal dimension. *Science* **155**, 636–638.

[250] Mantoglou, A. (1987). Digital simulation of multivariate two and three dimensional stochastic processes with the spectral turning bands method. *Mathematical Geology* **19**, 129–150.

[251] Markham, B. L. & Barker, J. L. (1986). Landsat MSS and TM post-calibration

ranges, exoatmospheric reflectances and at-satellite temperatures. *EOSAT Landsat Technical Notes* **1**, 3–8.
[252] Markham, B. L. & Barker, J. L. (1987). Thematic mapper bandpass solar exoatmospheric irradiances. *International Journal of Remote Sensing* **8**, 517–523.
[253] Martin, D. (1998). Optimising census geography: The separation of collection and output geographies. *International Journal of Geographical Information Science* **12**, 673–685.
[254] Matérn, B. (1986). *Spatial Variation (2nd Edition)*. Springer-Verlag, New York.
[255] Mather, P. M. (1987). *Computer Processing of Remotely Sensed Images; An Introduction*. Wiley, Chichester.
[256] Matheron, G. (1965). *Les variables Régionalisées et leur Estimation*. Masson, Paris.
[257] Matheron, G. (1971). *The Theory of Regionalized Variables and Its Applications*. Les Cahiers du Centre de Morphologie Mathématique, No. 5, Centre de Géostatistique, Fontainebleau.
[258] Matheron, G. (1973). The intrinsic random functions and their applications. *Advances in Applied Probability* **5**, 439–468.
[259] Matheron, G. (1981). Splines and kriging: their formal equivalence. In Merriam, D. F. (ed). *Down-to-Earth Statistics: Solutions looking for Geological Problems*. Syracuse University Geological Contributions, Syracuse, pp. 77-95
[260] McBratney, A. B. & Webster, R. (1981). The design of optimal sampling schemes for local estimation and mapping of regionalized variables: II. Program and examples. *Computers & Geosciences* **7**, 335–365.
[261] McBratney, A. B., Webster, R. & Burgess, T. M. (1981). The design of optimal sampling schemes for local estimation and mapping of regionalized variables-I Theory and method. *Computers & Geosciences* **7**, 331–334.
[262] McBratney, A. B. & Webster, R. (1983a). How many observations are needed for regional estimation of soil properties? *Soil Science* **135**, 177–183.
[263] McBratney, A. B. & R. Webster (1983b). Optimal interpolation and isarithmic mapping of soil properties V. Co-regionalization and multiple sampling strategy. *Journal of Soil Science* **34**, 137–162.
[264] McBratney, A. B. & Moore, A. W. (1985). Applications of fuzzy sets to climatic classification. *Agricultural and Forest Meteorology* **35**, 165–185.
[265] McBratney, A. B. & Webster, R. (1986). Choosing functions for semivariograms of soil properties and fitting them to sampling estimates. *Journal of Soil Science* **37**, 617–639.
[266] McBratney, A. B. & De Gruijter, J. J. (1992). A continuum approach to soil classification by modified fuzzy k-means with extragrades. *Journal of Soil Science* **43**, 159–175.
[267] McGwire, K., Friedl, M. & Estes, J. E. (1993). Spatial structure, sampling design and scale in remotely sensed imagery of a California savanna woodland. *International Journal of Remote Sensing* **14**, 2,137–2,164.
[268] Meyer, P., Itten, K. I., Kellenberger, T., Sandmeier, S. & Sandmeier, R. (1993). Radiometric corrections of topgraphically induced effects on Landsat TM

data in an Alpine environment. *ISPRS Journal of Photogrammetry & Remote Sensing* **48**, 17–28.
[269] Middelkoop, H. & Janssen, L.L. F. (1991). Implementation of temporal relationships in knowledge based classification or satellite images. *Photogrammetric Engineering & Remote Sensing* **57**, 937–945.
[270] Milton, E. J., Rollin, E. M. & Emery, D. R. (1995). Advances in field spectroscopy. In Danson, R. M. & Plummer, S. E. (eds) *Advances in Environmental Remote Sensing*. Wiley & Sons, Chichester, pp. 9–32.
[271] Mintzberg, H., Raisinghani D. & Theoret A. (1976). The structure of unstructured decision processes. *Administrative Science Quarterly* **21**, 246–275.
[272] Molenaar, M. (1996). A syntactic approach for handling the semantics of fuzzy spatial objects. In Burrough, P. A. & Frank, A. U. (eds). *Geographic Objects with Indeterminate Boundaries*. Taylor & Francis, London, pp. 207–224.
[273] Molenaar, M. (1998). *An Introduction to the Theory of Spatial Object Modelling*. Taylor & Francis, London.
[274] Mularz, D. E., Smith, H. G. & Wobray, T. J. (1995). Realizing the NSDI-vision through community architecture. *Geo-information systems* **5(1)**, 52–53.
[275] Mustard, J. F. & Pieters, C. M. (1987). Quantitative abundance estimates from bidirectional reflectance measurements. *Journal of Geophysical Research* **92**, 617–626.
[276] Myers, D. E. (1982). Matrix formulation of co-kriging. *Mathematical Geology* **14**, 249–257.
[277] Myers, D. E. (1983). Estimation of linear combinations and cokriging. *Mathematical Geology* **15**, 633–637.
[278] Myers, D. E. (1989). To be or not to be stationary? that is the question. *Mathematical Geology* **21**, 347–362.
[279] Neprash, J. A. (1934). Some problems in the correlation of spatially distributed variables. *Journal of the American Statistical Association Supplement* **29**, 167–168.
[280] Newcomer, J. A. & Szajgin, J. (1984). Accumulation of thematic map errors in digital overlay analysis. *The American Cartographer* **11**, 58–62.
[281] Nijkamp, P., Rietveld, P. & Voogd, H. (1990). *Multicriteria Evaluation in Physical Planning*. Elsevier Science Publishers, Amsterdam.
[282] Nijkamp, P. (1990). Multicriteria analysis: a decision support system for sustainable environmental management. In Achibugi, F., & Nijkamp, P. (ed). *Economy & Ecology: Toward sustainable development*. Kluwer, Dordrecht, pp. 203–220.
[283] Novo, E. M. & Shimabukuro, Y. E. (1997). Identification and mapping of the Amazon habitats using a mixing model. *International Journal of Remote Sensing* **18**, 663–670.
[284] Ogunjemiyo, S., Schuepp, P. H., MacPherson, J. I. & Desjardins, R. L. (1997). Analysis of flux maps versus surface characteristics from Twin Otter grid flights in BOREAS 1994. *Journal of Geophysical Research* **102**, 29,135–29,145.
[285] Olea, R. A. & Pawlowsky, V. (1996). Compensating for estimation smoothing in kriging. *Mathematical Geology* **28**, 407–417.

[286] Oliver, M. A., Webster, R. & Slocum, K. (in press). Image filtering by kriging Analysis. In Gómez-Hernández, J., Soares, A. & Froidevaux, R. (eds), *geoENV II: Geostatistics for Environmental Applications*. Kluwer, Dordrecht, pp.
[287] Olson D. L. & Courtney J. F. (1992). *Decision Support Models and Expert Systems*. Macmilan, New York.
[288] Omar, Duraid A.-N. (1998). *Remote Sensing Tutorial*. School of Geography, University of Nottingham. http://www.geog.nottingham.ac.uk/ dee/rs.html.
[289] Openshaw, S. (1984). *Modifiable Areal Unit Problem*. CATMOG No. 38, GeoAbstracts, Norwich.
[290] Openshaw, S. & Taylor, P. J. (1979). A million or so coefficients: Three experiments on the modifiable areal unit problem. In Wrigley, N. (ed). *Statistical Applications in the Spatial Sciences*. Pion, London, pp. 127–144.
[291] Ozernoy, V. M. (1992). Choosing the best multicriteria decision-making method. *INFO* **30**, 159–171.
[292] Papritz, A., Künsch, H. R. & Webster, R. (1993). On the pseudo cross-variogram. *Mathematical Geology* **25**, 1015–1026.
[293] Papritz, A. & Moyeed, R. A. (1999). Linear and non-linear kriging methods: Tools for monitoring soil pollution. In Barnett, V., Turkman, K. F. & Stein, A. (eds) *Statistics for the Environment 4: Statistical Aspects of Health and the Environment*. Wiley, Chichester, pp. 303–336.
[294] Parratt, L. G. (1961). *Probability and Experimental Errors in Science*. Wiley, New York.
[295] Patil, G. P. & Rao, C. R. (eds). (1994). *Environmental Statistics, Handbook of Statistics, Vol. 12*. North-Holland, Amsterdam.
[296] Pedley, M. I. & Curran, P. J. (1991). Per-field classification: an example using SPOT HRV imagery. *International Journal of Remote Sensing* **12**, 2,181–2,192.
[297] Peuquet, D. J. (1984). A conceptual framework and comparison of spatial data models. In Marble, D. F., Calkins, H. W. & Peuquet, D. J. (eds). *Basic Readings in Geographic Information Systems*. SPAD Systems, Williamsville, NY, pp. 3.35–3.44.
[298] Peuquet, D. J. & Boyle, A. R. (1984). *Raster Scanning, Processing and Plotting of Cartographic Documents*. Spad Systems, Williamsville, NY.
[299] Pinar, A. (1994). *The Use of the Remotely Sensed Red Edge to Estimate Wheat Yield*. PhD thesis, University of Wales Swansea, Swansea.
[300] Quenouille, M. H. (1949). Problems in plane sampling. *Annals of Mathematical Statistics* **20**, 355–375.
[301] Rees, W. G. (1996). *Physical Principles of Remote Sensing*. Cambridge University Press, Cambridge.
[302] Richards, J. (1993). *Remote Sensing Digital Image Analysis, an Introduction*, 2nd edition. Springer-Verlag, Berlin.
[303] Ripley, B. D. (1996). *Pattern Recognition and Neural Networks*. Cambridge University Press, Cambridge.
[304] Ritchie, J. C. & Cooper, C. M. (1991). An algorithm for estimating sur-

face suspended sediment concentrations with Landsat MSS digital data. *Water Resources Bulletin* **27**, 373–379.
[305] Robinson, A. H. (1956). The necessity of weighting values in correlation of areal data. *Annals, Association of American Geographers* **46**, 233–236.
[306] Romero, C. & Rehman, T. (1989). *Multi Criteria Analysis for Agricultural Decisions*. Elsevier, Amsterdam.
[307] Rosenfield, G. H. & Melley, M. L. (1980). Applications of statistics to thematic mapping. *Photogrammetric Engineering & Remote Sensing* **46**, 1287–1294.
[308] Rosenfield, G. H., Fitzpatrick-Lins, K. & Ling, H. S. (1982). Sampling for thematic map accuracy testing. *Photogrammetric Engineering & Remote Sensing* **48**, 131–137.
[309] Rosenfield, G. H. & Fitzpatrick-Lins, K. (1986). A coefficient of agreement as a measure of thematic classification accuracy. *Photogrammetric Engineering & Remote Sensing* **52**, 223–227.
[310] Ross, G.J. S. (1990). *Maximum Likelihood Program*. Numerical Algorithms Group: Oxford.
[311] Rossi, E. R., Dungan, J. L. & Beck, L. R. (1994). Kriging in the shadows: Geostatistical interpolation for remote sensing. *Remote Sensing of Environment* **49**, 32–40.
[312] Rossi, R. E. (1998). How the support effect affects summary statistics. *Geostatistics (submitted)*.
[313] Roy B. (1978). ELECTER II, un alghortithme de classement fonde sur une representation floue des preferences en presence de criteres multiples. *Cah, CERO*, **20**, 32–43.
[314] Rubin, Y. & Gomez-Hernandez, J. (1990). A stochastic approach to the problem of upscaling of transmissivity in disordered media: Theory and unconditional simulations. *Water Resources Research* **26**, 691–701.
[315] Salton, G. & McGill, M. J. (1983). *Introduction to modern information retrieval*. McGraw-Hill, New York.
[316] Särndal, C. E., Swensson, B. & Wretman, J. (1992). *Model Assisted Survey Sampling*. Springer, New York.
[317] Sauter, V. (1997). *Decision Support System*. Wiley, New York.
[318] Sèze, G. & Rossow, W. B. (1991). Effects of satellite data resolution on measuring the space/time variations of surfaces and clouds. *International Journal of Remote Sensing* **12**, 921–952.
[319] Schanda, E. (1986). *Physical Fundamentals of Remote Sensing*. Springer, Berlin.
[320] Schetselaar, E. M. & Rencz, A. N. (1997). Reducing the effects of vegetation cover on airborne radiometric data using Landsat TM data. *International Journal of Remote Sensing* **18**, 1,503–1,515.
[321] Schowengerdt, R. A. (1996). On the estimation of spatial-spectral mixing with classifier likelihood functions. *Pattern Recognition Letters* **17**, 1,379–1,387.
[322] Scott, D. (1993). *Multivariate Density Estimation*. Wiley, New York.
[323] Settle, J. J. & Drake, N. A. (1993). Linear mixing and the estimation of ground cover proportions. *International Journal of Remote Sensing* **14**, 1,159–1,177.

[324] Shannon (1948). The mathematical theory of communication. *Bell Systems Technical Journal* **27**, 379–423.

[325] Sharifi, M. A. (1992). *Development of an Appropriate Resource Information System to Support Agricultural Management at Farm Enterprise Level.* PhD thesis, Wageningen Agricultural University, Wageningen.

[326] Shimabukuro, Y. E., Carvalho, V. C. & Rudorff, B.F. T. (1997). NOAA-AVHRR data processing for the mapping of vegetation cover. *International Journal of Remote Sensing* **18**, 671–677.

[327] Silverman, B. (1986). *Density Estimation for Statistics and Data Analysis.* Chapman and Hall, London.

[328] Simon, H. A. (1960). *The New Science of Management Decision.* Harper & Brothers, New York.

[329] Singer, R. B. & McCord, T. B. (1979). Mars: Large scale mixing of bright and dark surface materials and implications for analysis of spectral reflectance. In: *Proceedings of the 10th Lunar and Planetary Science Conference, Houston, U.S.A., 19-23 March (1979).* pp. 1,835–1,848.

[330] Singh, A. (1989). Digital change detection techniques using remotely-sensed data. *International Journal of Remote Sensing* **10**, 989–1,003.

[331] Singh, R. P. & Sirohim A. (1994). Spectral reflectance properties of different types of soil surfaces. *ISPRS Journal of Photogrammetry & Remote Sensing* **49(4)**, 34–30.

[332] Skidmore, A. K. (1989a). An expert system classifies eucalypt forest types using Landsat Thematic Mapper data and a digital terrain model. *Photogrammetric Engineering & Remote Sensing* **55**, 1,449–1,464.

[333] Skidmore, A. K. (1989b). , Unsupervised training area selection in forests using a nonparametric distance measure and spatial information. *International Journal of Remote Sensing* **10**, 133–146.

[334] Skidmore, A. K. & Turner, B. J. (1989). Assessing the accuracy of resource-inventory maps. In G. Lund (ed). *Proceedings of Global Natural Resource Monitoring and Assessments, Venice, Italy, volume 2.* ASPRS, Baltimore, pp. 524–535.

[335] Skidmore, A. K. & Turner, B. J. (1992). Assessing map accuracy using line intersect sampling. *Photogrammteric Engineering & Remote Sensing* **58**, 1,453–1,457.

[336] Skidmore, A. K., Gauld, A. & Walker, P. W. (1996). A comparison of GIS predictive models for mapping kangaroo habitat. *International Journal of Geographical Information Systems* **10**, 441–454.

[337] Skidmore, A. K., Varekamp, C., Wilson, L., Knowles, E. & Delaney, J. (1997). Remote sensing of soils in a eucalypt forest environment. *International Journal of Remote Sensing* **18**, 39–56.

[338] Skole, D. & Tucker, C. (1993). Tropical deforestation and habitat fragmentation in the Amazon – satellite data from 1978 to 1988. *Science* **260**, 1,905–1,910.

[339] Smith, M. O., Johnston, P. E. & Adams, J. B. (1985). Quantitative determination of mineral types and abundances from reflectance spectra using principal component analysis. *Journal of Geophysical Research* **90**, 797–804.

[340] Smith, R.C. G., Prathapar, S.A., Barrs, H. D. & Slavich, P. (1989). Use of a

thermal scanner image of a water stressed crop to study soil spatial variability. *Remote Sensing of Environment* **29**, 111–120.

[341] Smith, E. A., Hsu, A. Y., Crosson, W. L., Field, R. T., Fritschen, L. J., Gurney, R. J., Kanemasu, E. T., Kustas, W. P., Nie, D., Shuttleworth, W. J., Verma, S. B., Weaver, H. L. & Wesely, M. L. (1992). Area-averages surface fluxes and their time-space variability of the FIFE experimental domain. *Journal of Geophysical Research* **97**, 18,599–18,622.

[342] Sohn, Y. & McCoy, R. M. (1997). Mapping desert shrub rangeland using spectral unmixing and modeling spectral mixtures with TM data. *Photogrammetric Engineering & Remote Sensing* **63**, 707–716.

[343] Sol, H. G. (1982). *Simulation in Information System Development*. PhD thesis, University of Groningen, Groningen.

[344] Sprague, R. H. (1980). A framework for development of decision support systems. *Management information science, quarterly* **4**, 1–26.

[345] Srivastava, R.M. (1992). Reservoir characterization with probability field simulation. In: *Annual Technical Conference of the Society of Petroleum Engineers*. SPE 24753, Washington, D.C., pp. 927-938.

[346] Srivastava, R.M. (1994). The visualization of spatial uncertainty. *In* Yarus. J. M. & Chambers, R. L. (eds). *Stochastic Modeling and Geostatistics: Principles, Methods, and Case Studies*. American Association of Petroleum Geologists, Tulsa, Oklahoma, pp. 339–345.

[347] Steel, D. G. & Holt, D. (1996). Rules for random aggregation. *Environment & Planning A* **28**, 957–978.

[348] Stein, A. & Corsten, L. C. A. (1991). Universal kriging and cokriging as a regression procedure. *Biometrics* **47**, 575–587.

[349] Stein, A., Varekamp, C., Van Egmond, C. & Van Zoest, R. (1995). Zinc concentrations in groundwater at different scales. *Journal of Environmental Quality* **24**, 1205–1214.

[350] Stein, A., Bastiaanssen, W.G. M., De Bruin, S., Cracknell, A. P., Curran, P. J., Fabbri, A. G., Gorte, B.G. H., Van Groenigen, J. W., Van der Meer, F. D. & Saldoña, A. (1998). Integrating spatial statistics and remote sensing. *International Journal of Remote Sensing* **19**, 1,793–1,814.

[351] Stoner, E. R. & Baumgardner, M. F. (1981). Characteristic variations in reflectance of surface soils. *Soil Science Society of America Journal* **45**, 1,161–1,165.

[352] Story, M. & Congalton, R. G. (1986). Accuracy assessment: a user's perspective. *Photogrammetric Engineering & Remote Sensing* **52**, 397–399.

[353] Strahler, A. H. (1980). The use of prior probabilities in maximum likelihood classification of remotely sensed data. *Remote Sensing of Environment* **10**, 135–163.

[354] Strahler, A. H., Woodcock, C. E. & Smith, J. A. (1986). On the nature of models in remote sensing. *Remote Sensing of Environment* **20**, 121–139.

[355] Switzer, P. (1993). The spatial variability of prediction errors. *In* Soares, A. (ed). *Geostatistics Troia '92*. Kluwer Academic Publishers, Dordrecht, pp. 261–272.

[356] Teillet, P. M. (1986). Image correction for radiometric effects in remote sensing. *International Journal of Remote Sensing* **7**, 1,637–1,651.

[357] Teillet, P. M., Gaulthier, R. P., Staenz, K. & Fournier, R. A. (1997). BRDF equifinality studies in the context of forest canopies. In Guyot, G. & Phulpin, T. (eds). *Physical Measurements and Signatures in Remote Sensing*. Balkema, Rotterdam, pp. 163–170.

[358] Thapa, K. & Bossler, J. (1992). Accuracy of spatial data used in geographic information systems. *Photogrammetric Engineering & Remote Sensing* **58**, 835–841.

[359] Thomas, I. L. & Allcock, G. M. (1984). Determining the confidence interval for a classification. *Photogrammetric Engineering & Remote Sensing* **50**, 1491–1496.

[360] Thomas, E. N. & Anderson, D. L. (1965). Additional comments on weighting values in correlation analysis of areal data. *Annals, Association of American Geographers* **55**, 492–505.

[361] Thompson, S. K. (1992). *Sampling*. Wiley, New York.

[362] Thompson, S. K. & Seber, G.A. F. (1996). *Adaptive Sampling*. Wiley, New York.

[363] Todd, W. J., Gehring, D. G. & Haman, J. F. (1980). Landsat wildland mapping accuracy. *Photogrammetric Engineering & Remote Sensing* **46**, 509–520.

[364] Toll, D. L. (1984). An evaluation of simulated Thematic Mapper data and Landsat MSS data for discriminating suburban and regional land use and land cover. *Photogrammetric Engineering & Remote Sensing* **50**, 1,713–1,724.

[365] Townshend, J.R. G. (1981). The spatial revolving power of Earth resources satellites. *Progress in Physical Geography* **5**, 32–55.

[366] Townshend, J.R. G. & Justice, C. O. (1988). Selecting the spatial resolution of satellite sensors required for global monitoring of land transformations. *International Journal of Remote Sensing* **9**, 187–236.

[367] Tran, T. T. (1996). The missing scale and direct simulation of block effective properties. *Journal of Hydrology* **183**, 37–56.

[368] Tukey, J. W. (1977). *Exploratory Data Analysis*. Adison Wesley, Boston.

[369] Turban, E. (1995). *Decision Support and Expert Systems: Management Support Systems*. Macmillan, New York.

[370] Van der Meer, F. (1994a). Extraction of mineral absorption features from high-spectral resolution data using non-parametric geostatistical techniques. *International Journal of Remote Sensing* **15**, 2,193–2,214.

[371] Van der Meer, F. (1994b). Sequential indicator conditional simulation and indicator kriging applied to discrimination of dolomitization in GER 63-channel imaging spectrometer data. *Nonrenewable Resources* **3**, 146–164.

[372] Van der Meer, F. (1995). Estimating and simulating the degree of serpentinization of peridotites using hyperspectral remotely sensed imagery. *Non-renewable resources* **4**, 84–98.

[373] Van der Meer, F. (1997). Mineral mapping and Landsat Thematic Mapper classification using spectral unmixing. *Geocarto International* **12(3)**, 27–40.

[374] Van der Meer, F. (1998). Mapping dolomitization through a co-regionalization

of simulated field and image-derived reflectance spectra: A proof-of-concept study. *International Journal of Remote Sensing* **19**, 1615–1620.

[375] Van der Wel, F.J. M., Van der Gaag, L. C. & Gorte, B.G. H. (1998). Visual exploration of uncertainty in remote-sensing classification. *Computers and Geosciences* **24**, 335–343.

[376] Van Genderen, J. L. & Lock, B. F. (1977). Testing land-use map accuracy. *Photogrammetric Engineering & Remote Sensing* **43**, 1,135–1,137.

[377] Van Genderen, J. L., Lock, B. F. & Vass, P. A. (1978). Remote sensing: statistical testing of thematic map accuracy. *Remote Sensing of Environment* **7**, 3–14.

[378] Van Groenigen, J. W. & Stein, A. (1998). Spatial Simulated Annealing for constrained optimization of spatial sampling schemes. *Journal of Environmental Quality* **27**, 1,078–1,086.

[379] Van Meirvenne, M., Pannier, J., Hofman, G. & Louwagie, G. (1996). Regional characterization of the long-term change in soil organic carbon under intensive agriculture. *Soil Use and Management* **12**, 86–94.

[380] Vassilios, K. & Despotakis (1991). Sustainable Development Planning using Geographic Information Systems. PhD thesis, National Technical University of Athens, Athens, Greece.

[381] Vckovski, A. (ed). (1998). Interoperability in GIS. *International Journal of Geographical Information Science* **12**, 297–425.

[382] Ver Hoef, J. M. & Cressie, N. (1993). Multivariable spatial prediction *Mathematical Geology* **25**, 219–240.

[383] Verstraete, M. M., Pinty, B. & Curran, P. J. (1999). MERIS potential for land applications. *International Journal of Remote Sensing (in press)*.

[384] Voogd, H. (1983). *Multicriteria Evaluation for Urban and Regional Planning*. Pion Limited, London.

[385] Wackernagel, H. (1995). *Multivariate Geostatistics: An Introduction with Applications*. Springer Verlag, Berlin.

[386] Wahl, F. M. (1987). *Digital Image Signal Processing*. Artech House, Boston.

[387] Weber, B. J. (1986). Systems to think with: A response to vision for decision support systems. *Journal of Management Information Systems* **II(4)**, 85–97.

[388] Webster, R. & McBratney, A. B. (1989). On the Akaike Information Criterion for choosing models for variograms of soil properties. *Journal of Soil Science* **40**, 493–496.

[389] Webster, R., Curran, P. J. & Munden, J. W. (1989). Spatial correlation in reflected radiation from the ground and its implications for sampling and mapping by ground-based radiometry. *Remote Sensing of Environment* **29**, 67–78.

[390] Webster, R. & Oliver, M. A. (1990). *Statistical Methods for Soil and Land Resources Survey*. Oxford University Press, Oxford.

[391] Webster, R. & Oliver, M. A. (1992). Sample adequately to estimate variograms of soil properties. *Journal of Soil Science* **43**, 177–192.

[392] Welch, R. (1982). Spatial resolution requirements for urban studies. *International Journal of Remote Sensing* **3**, 139–146.

[393] Wen, R. & Sinding-Larsen, R. (1997). Image filtering by factorial-kriging—sensitivity analysis and application to Gloria side-scan sonar images. *Mathematical Geology* **29**, 433–468.

[394] Wessman, C. A., Aber, J. D., Peterson, D. L. & Melillo, J. M. (1988). Foliar analysis using near infrared reflectance spectroscopy. *Canadian Journal of Forest Research* **18**, 6–11.

[395] Wood, T. F. & Foody, G. M. (1989). Analysis and representation of vegetation continua from Landsat Thematic Mapper data for lowland heaths. *International Journal of Remote Sensing* **10**, 181–191.

[396] Woodcock, C. E. & Strahler, A. H. (1987). The factor of scale in remote sensing. *Remote Sensing of Environment* **21**, 311–322.

[397] Woodcock, C. E., Strahler, A. H. & Jupp, D.L. B. (1988a). The use of variograms in remote sensing I: Scene models and simulated images. *Remote Sensing of Environment* **25**, 323–348.

[398] Woodcock, C. E., Strahler, A. H. & Jupp, D.L. B. (1988b). The use of variograms in remote sensing II: Real digital images. *Remote Sensing of Environment* **25**, 349–379.

[399] Woodsford, P. A. (1996). Spatial database update - A key to effective automation. *In* Kraus, K. & Waldhäusl, P. (ed). *International Society for Photogrammetry and Remote Sensing 18th Congress. International Archives of Photogrammetry and Remote Sensing, 31(B4)*. RICS Books, Coventry, pp. 955–961.

[400] Worboys, M. F. (1994). Introduction. *In* Worboys, M. F. (ed). *Innovations in GIS, Selected Papers from the First National Conference on GIS Research UK*. Taylor & Francis, London, pp. 1–3.

[401] Wright, J. K. (1942). Map makers are human: comments on the subjective in maps. *The Geographical Review* **32**, 527–544.

[402] Wrigley, N. (1995). Revisiting the modifiable areal unit problem and the ecological fallacy. *In* Cliff, A. D., Gould, P. R., Hoare, A. G. & Thrift, N. J. (eds) *Diffusing Geography*. Blackwell, Oxford, pp. 49–71.

[403] Yates, F. (1981). *Sampling Methods for Censuses and Surveys, 4th edition*. Griffin, London.

[404] Yavada, U. L. (1986). A rapid and non-destructive method to determine chlorophyll in intact leaves. *Horticultural Science* **21**, 1,449–1,450.

[405] Yule, G. U. & Kendall, M. G. (1950). *An Introduction to the Theory of Statistics*. Charles Griffin, London.

[406] Zadeh, L. A. (1965). Fuzzy Sets. *Information & Control* **8**, 338–353.

[407] Zhang, R., Warrick, A. W. & Myers, D. E. (1990). Variance as a function of sample support size. *Mathematical Geology* **22**, 107–122.

[408] Zimmerman, D. L. & Zimmerman, M. B. (1991). A comparison of spatial semivariogram estimators and corresponding ordinary kriging predictors. *Technometrics* **33**, 77–91.

[409] Zonneveld, I. S. (1974). Aerial photography, remote sensing and ecology. *ITC Journal* **4**, 553–560.